Biotechnology in the Public Sphere

A European Sourcebook

edited by John Durant, Martin W Bauer
and George Gaskell

This book has been produced as part of a
Concerted Action funded by the European
Community (Contract BI04-CT95-0043)

Science Museum

British Library Cataloguing-in-Publication Data
A catalogue record for this publication is available from the British Library

Set from Pagemaker in Postscript Monotype Plantin Light by Jerry Fowler
Printed in England by the Cromwell Press
Cover design: Ian Youngs

© 1998 The Board of Trustees of the Science Museum

ISBN 1 900747 09 X

Science Museum, Exhibition Road, London SW7 2DD

Contents

Abbreviations used in the text   **iv**

List of contributors   **v**

Acknowledgements   **x**

Part I
Introduction

The representation of biotechnology:
policy, media and public perception   **3**

Part II
National profiles

Austria   **15**

Denmark   **29**

Finland   **43**

France   **51**

Germany   **63**

Greece   **77**

Italy   **89**

The Netherlands   **103**

Poland   **118**

Sweden   **130**

Switzerland   **144**

United Kingdom   **162**

Europe   **177**

Part III
Public perceptions

Public perceptions of biotechnology in 1996:
Eurobarometer 46.1   **189**

Part IV
Conclusion

Biotechnology in the public sphere:
a comparative review   **217**

Appendices

1. Eurobarometer standard 46.1. Technical specifications   **231**

2. Eurobarometer 1996 survey results   **239**

3. Guidelines for the time-series press analysis, 1973–96   **276**

4. Sampling procedures   **278**

5. Coding frame for media analysis   **283**

6. Basic frequencies of corpus variables of biotechnology
in the European opinion-leading press, 1973–96   **290**

7. Reliability of content analysis   **297**

Index   **301**

Abbreviations used in the text

BSE	Bovine spongiform encephalopathy
BST	Bovine somatotropin
DG	Directorate General
DNA	Deoxyribonucleic acid
EC	European Community
EEC	European Economic Community
EFB	European Federation of Biotechnology
EU	European Union
FDA	Food and Drug Administration (US)
GM	Genetically modified or genetic modification
GMO	Genetically modified organism
GDP	Gross Domestic Product
GNP	Gross National Product
INRA	International Research Associates
IPR	Intellectual property rights
IT	Information technology
IVF	*In-vitro* fertilisation
MP	Member of Parliament
NGO	Non-governmental organisation
OECD	Organization for Economic Cooperation and Development
R&D	Research and development
rDNA	Recombinant DNA
TA	Technology assessment
UN	United Nations
UNESCO	United Nations Educational, Scientific, and Cultural Organization
WHO	World Health Organization

List of contributors

Each contributor's national research team in the EC Concerted Action Programme is indicated in brackets.

Agnes Allansdottir
(Italy)

Social psychologist at the University of Siena. Her main fields of research are public understanding of science, media science and social representations.

Sebastiano Bagnara
(Italy)

Professor of Psychology and director of the course in science communication at the University of Siena.

Martin W Bauer
(United Kingdom)

Lecturer in Social Psychology and Research Methodology at the London School of Economics and a Research Fellow at the Science Museum, London. Recent research projects include a re-analysis of Eurobarometer 1989, public perceptions of the Human Genome Project in the UK, the construction of indicators for public understanding of science in Europe and the UK Science and Technology Media Monitor, 1945–85. He is editor of the book *Resistance to New Technology – Nuclear Power, Information Technology and Biotechnology* (1995).

Marie-Louise von Bergmann-Winberg
(Sweden)

Professor of Political Science and Dean of the Faculty of Interdisciplinary Research at Mid Sweden University. Her doctoral degree from the Swedish School of Economics in Helsinki, Finland, is in political science, especially comparative politics and welfare theory. She is also President of the EU European Institute for Federalism Research.

Anne Berthomier
(France)

Doctoral student at the Laboratoire Communication et Politique, Paris. Her thesis concerns ways in which the media present biotechnology.

Heinz Bonfadelli
(Switzerland)

Professor and Head of the Department of Communication, University of Zürich. He is researching the use and effects of mass media, e.g, the knowledge-gap hypothesis.

Daniel Boy
(France)

Political scientist, specialising in the study of public attitudes towards science and of Green movements in Europe. He is at the Centre d'Études de la Vie Politique Française (Cevipof) in Paris.

Eleanor Bridgman
(United Kingdom)

Studied human sciences at the University of Sussex and completed a Master's course in science and technology policy at the Science Policy Research Unit. Her Master's thesis is on the impact of labelling of genetically modified food on R&D in the food biotechnology industry. She is Project Manager of the European Public Concerted Action 'Biotechnology and the Public'.

Suzanne de Cheveigné
(France)

Sociologist, specialising in the study of the relationships between science, media and society. She is at the Centre National de la Recherche Scientifique (CNRS), Laboratoire Communication et Politique in Paris.

John Durant
(United Kingdom)

Director of Science Communication at the Science Museum, London and Professor of Public Understanding of Science at Imperial College of Science, Technology and Medicine. He chairs the EFB Task Group on public perceptions of biotechnology. Recent research projects include a UK survey of public understanding of science, a study of public perceptions of the Human Genome Project, and the UK Science and Technology Media Monitor, 1945–85

Björn Fjæstad (Sweden)	Adjunct Professor of Science Communication at Mid Sweden University and leader of the Swedish national project. His doctoral degree, from the Stockholm School of Economics in 1973, is in economic psychology. He is also the editor and publisher of the Swedish popular science journal *Forskning & Framsteg*.
Helle Frederiksen (Denmark)	Junior researcher in the Department of Communication, Education and Computer Science, Roskilde University. She holds Master's degrees in linguistics and computer science. Her main theory areas include argumentation analysis and discourse analysis, and she has carried out studies in media coverage of technology and risk problems.
George Gaskell (United Kingdom)	Reader in Social Psychology and Director of the Methodology Institute, the London School of Economics. His research interests include cognitive aspects of survey methodology, biotechnology and the British public, and television and young people in the 1990s.
Jean-Christophe Galloux (France)	Professor of Law at the University of Versailles. He heads the Research Centre on Business Law and High Technologies (DANTE). He is an expert on biotechnology law to the EC Commission and is also a member of the Advisory Group on Ethics (AGE) of EUROPABIO.
Hélène Gaumont Prat (France)	Lecturer at the University of Versailles and researcher at the Research Centre on Business Law and High Technologies (DANTE). She specialises in biotechnology law.
Alexander Goerke (Germany)	Studied communication science at the University of Muenster. His dissertation (1997) was entitled 'Risk Journalism and Risk Society'. Since 1998 he has been at the University of Jena where his main research areas are risk communication, theory of journalism and international communication.
Petra Grabner (Austria)	Studied political sciences, history and sociology at the University of Salzburg. Her Master's thesis was on Austrian genetic engineering policy. She is Assistant Professor at the Department of Political Science, University of Salzburg, and a collaborator in the projects 'Biotechnology in Austrian Public Discourse' and 'Fusion: Democracy and Autonomy in a Complex, Centralizing Technology'.
Jan Gutteling (Netherlands)	Social psychologist working at the Department of Communication Studies at the University of Twente. His research focus is on the planning, design and evaluation of risk communication, on which he has published various articles and a monograph.
Jürgen Hampel (Germany)	Studied sociology at the University of Mannheim and received his PhD at the Free University of Berlin. He has held scientific positions at the University of Mannheim, the Wissenschaftszentrum Berlin für Sozialforschung and the Technical University of Chemnitz-Zwickau. Since 1994 he has been at the Center of Technology Assessment in Baden-Württemberg, where his main research focus is the perception and sociology of technology.
Anneke Hamstra (Netherlands)	Worked at the SWOKA Institute for Strategic Consumer Research between 1987 and 1997, where she was project leader on Constructive Technology Assessment. Her main scientific activities concentrated on consumer acceptance of biotechnology. In 1998, Carla Smink took over her position when she left SWOKA to head the Consumer Test Centre at Philips Domestic Appliances & Personal Care in Drachten.

Petra Hieber (Switzerland)	Graduated in History and English from the University of Basle, where she concentrated on social and gender history. She now works for a privately-owned office for public relations, counselling and planning in the fields of health care, biotechnology and ecology, focusing on research into the history of biotechnology in Switzerland.
Erling Jelsøe (Denmark)	Associate Professor in the Department of Environment, Technology and Social Studies, Roskilde University. His fields of interest include developments within the food sector, technological change and consumer issues, as well as social impacts and regulation of new biotechnologies.
Matthias Kohring (Germany)	Studied communication science at the University of Muenster, where his dissertation (1996) was on the function of science journalism. From 1996 to 1997 he was at the Technical University of Ilmenau, and since 1998 he has been Scientific Assistant at the University of Jena. His main research areas are theory of journalism and mass media, crisis public relations, and trust in media.
Jesper Lassen (Denmark)	Junior researcher in the Department of Environment, Technology and Social Studies, Roskilde University. He has a PhD in consumer influence on food quality and has participated in a research programme concerning the interaction between technology and society since 1988, including a project on the impact of modern biotechnology on consumers.
Sigrid Lehner (Austria)	Studied psychology at the University of Vienna. Her Master's thesis was on national stereotypes and prejudice. She is a collaborator in the project 'Biotechnology in Austrian Public Discourse' and is located at the University of Linz.
Martina Leonarz (Switzerland)	MA student, finishing her thesis on film theories and communications. Works at a privately-owned office for applied social research in science journalism and risk communication.
Miltos Liakopoulos (United Kingdom)	Doctoral student at the London School of Economics. His research focus is on British public attitudes to biotechnology and British media coverage of biotechnology. Research interests include social debates on science and the introduction of new technology in society, and public participation structures in technological decision-making. Recent projects include the survey analysis of public understanding of science and technology in Europe.
Athena Marouda- *Chatjoulis* (Greece)	Social psychologist with training in group psychotherapy. Director of Operations and Scientific Development, Centre for Counselling and Development (a Greek consulting company), part-time lecturer at the University of Thessaly and research consultant at the Institute of Applied Psychology, Department of Communication and Mass Media, University of Athens. She participates in several European projects. Her research areas include decision-making, human resources and organisational and group behaviour.
Werner A Meier (Switzerland)	Senior Researcher at the Department of Communication, University of Zürich. His special interests are risk communication, media economy and media policy.
Cees Midden (Netherlands)	Professor of Psychology at the Department of Technology Management at Eindhoven University of Technology. His research focus is on man-technology interactions as these become apparent in the development of new products and systems, in societal and market introductions and in the consumption and use of products. He has published various books and many articles on environmental consumer behaviour, the perception and communication of technological risks, and the social diffusion of innovations.

Anna Olofsson (Sweden)	Doctoral student of sociology at Mid Sweden University. Her thesis is on the framing of genetic technology by the Swedish mass media over the past 25 years.
Susanna Olsson (Sweden)	Doctoral student in sociology at Mid Sweden University. Her thesis is on public perception of technology and the environment.
Fabio Pammolli (Italy)	Lecturer in economics at the Istituto di Tecnica Economica, University of Siena. His main research interest is in the economics of technological innovations.
Andrzej Przestalski (Poland)	Assistant Professor of Sociology at the Adam Mickiewicz University, PoznaA. His interests include the theory of social differentiation and history of sociology. He is a co-author of *Ownership and Society* and co-editor of *On Social Differentiation* (1992) and *Krytyka rozumu socjologicznego* (*Critique of Sociological Reason*) (1997).
Georg Ruhrmann (Germany)	Studied sociology and biology at the University of Bielefeld with habilitation in the field of risk communication at the University of Muenster. Since 1998 he has been Professor for Communication Science and Media Effects at the University of Jena. His main research areas are television, new media and risk communication.
Maria Rusanen (Finland)	Department of Biochemistry and Biotechnology, University of Kuopio. She has studied animal sciences and biotechnology. She specialises in issues connected with animal biotechnology.
Timo Rusanen (Finland)	Department of Social Sciences, University of Kuopio. He has studied social sciences, public health and education. His research focuses on social issues about biotechnology and public assessment of insecurity and risks in society.
George Sakellaris (Greece)	Biologist with a PhD in biochemistry. His postdoctoral research was in molecular genetics. He is Senior Researcher at the National Hellenic Research Foundation, Institute of Biological Research and Biotechnology (since 1986) in Athens, and since 1998 he has been responsible for the Office of Science Communication and Bioethics in the same institute. He is a committee member of the Hellenic Biotechnology Society.
Michael Schanne (Switzerland)	Head of a private office for applied social research. Researches science journalism and risk communication.
Franz Seifert (Austria)	Studied biology at the University of Vienna and, as a postgraduate, political science at the Institute of Advanced Studies, Vienna. He is currently a researcher at the Institute of Technology Assessment, Austrian Academy of Science, Vienna, and a collaborator in the Austrian project on 'Biotechnology in Austrian Public Discourse'.
Carla Smink (Netherlands)	Studied food science at Wageningen Agricultural University. She started working as a researcher at SWOKA in 1993 and became project manager in 1997. Her main fields of research are consumer attitudes towards food production and biotechnology.
Angeliki Stathopoulou (Greece)	Qualitative research manager and vice-president of Metron Analysis SA, a Greek research agency specialising in social and political research. She is a research consultant at the Department of Communication and Mass Media, University of Athens, and works on topics related to biotechnology, political behaviour and immigrants' socio-economic integration in Greece. Her special interest lies in developing qualitative research methodology.

Ester Stevers (France)	Doctorate student in political sciences and researcher at Paris II. She specialises in EC regulation and organisation science.
Bolesław Suchocki (Poland)	Assistant Professor of Sociology at the Adam Mickiewicz University, PoznaΔ. His main interests are methodology and social statistics. He is co-author of *Techniki pomiaru w socjologii* (*Techniques of Measurement in Sociology*) (1985).
Arne Thing Mortensen (Denmark)	Professor of Philosophy in the Department of Communication, Education and Computer Science, Roskilde University. His main areas of expertise are epistemology, philosophy of language and philosophy of mind. Since 1984 he has worked on topics related to science communication and public perception of science.
Helge Torgersen (Austria)	Studied biology at the University of Salzburg and did basic research at the Institute of Molecular Biology and the Institute of Biochemistry at the University of Vienna. He works at the Institute of Technology Assessment, Austrian Academy of Science, Vienna, and his research interests are the risk assessment of deliberate releases of GMOs and the social impact and public perception of genetic engineering.
Tomasz Twardowski (Poland)	Professor of Molecular Biology and Biotechnology at the Institute of Bio-organic Chemistry, Polish Academy of Sciences, PoznaΔ, and the Institute of Technical Biochemistry, Technical University of Łódź. His main research interests are regulatory mechanisms of protein biosynthesis and legislation on biotechnology. He has authored numerous articles on the function of protein biosynthesis machinery and the correlation between the development of biotechnology and legislation. He is editor-in-chief of the Polish biotechnology journal *Biotechnologia*.
Wolfgang Wagner (Austria)	Professor at the Department of Social and Economic Psychology, University of Linz. His research interests span everyday cultural and social thinking, social representation theory, distributed and shared cognition, group processes, and problems of the transfer and application of theory in professional practice.
Mercy Wambui Kamara (Denmark)	Master's student in the Department of Environment, Technology and Social Studies, Roskilde University. Her fields of interest include the relationship between policy-making, science and politics, with particular reference to biotechnology.
Hans-Peter Wessels (Switzerland)	Works at the Research department of the Swiss Federal Institute of Technology (ETH), Zürich. He is especially interested in biotechnology and policy.
Atte von Wright (Finland)	Microbiologist at the Technical Research Centre of Finland, studying food and safety. He specialises in the toxicology of food. He is a docent in general microbiology at the University of Helsinki and in microbial genetics at the University of Kuopio.

Acknowledgements

This book is the first volume to be produced from three years' hard work by an international team of some 50 researchers. Our research has been supported by the European Commission through the European Community Concerted Action, 'Biotechnology and the European Public', Contract No. BI04-CT95-0043; and it has been further supported in order to embrace Poland through the INCO programme of the European Community, Concerted Action, 'Biotechnology and the European Public', Contract No. IC20-CT96-0030. Our work has embraced a series of separately funded national studies, together with comparative assessments at the European level. It is a great pleasure to acknowledge the European Commission for supporting this work throughout. In particular, we thank our first Project Manager in the Commission, Andreas Klepsch, whose advice at the outset that we should concentrate above all on the conduct of high-quality research was an inspiration; and Maurice Lex, who most ably assumed responsibility for the project in mid-term.

In London, the successful coordination of such a large project would have been impossible without the extraordinary diligence and (on occasions) diplomacy of our Project Manager Eleanor Bridgman. Eleanor's contributions extended far beyond mere management, and included a substantial contribution to the research itself. Other researchers in London who contributed importantly to our work were Miltos Liakopoulos, Armin Helic and Hildur Hardardottir. This book was produced against an extremely tight programme thanks to the support of Giskin Day, Ela Ginalska and Jane Gregory. Graphic support was provided by Marco Crosta and Jerry Fowler.

Finally, we should like to thank all of our research colleagues in each of the participating national teams. It is a daunting challenge to attempt the coordination of a research effort embracing around 50 researchers in more than a dozen different countries. From the outset, however, our colleagues brought an enthusiasm and a commitment to the common purposes of the project that made our task not only possible but positively pleasurable. All of the European contributors to this volume have worked hard to produce a whole that is genuinely greater than the sum of its parts. Since their work is not included in this volume, we should like to single out for separate thanks here Torben Hviid Nielsen in Norway, Edna Einsiedel and colleagues in Canada, and Jon Miller and colleagues in the USA. Their time will come, in a second volume of findings that will be published in due course.

John Durant
Martin W Bauer
George Gaskell
Coordinators, 'Biotechnology and
the European Public' Concerted Action

Part I
Introduction

The representation of biotechnology: policy, media and public perception

George Gaskell, Martin W Bauer and John Durant

Introducing a strategic technology

Modern biotechnology is the third strategic technology of the postwar period, following nuclear power in the 1950s and 60s and information technology in the 1970s and 80s. These have all been identified as strategic technologies in their day, in the sense that they have been seen to carry the potential to transform our future life. Like other strategic developments, modern biotechnology has given rise to societal visions of both a utopian and dystopian nature. On the one hand, there are the claimed benefits: new diagnoses and therapies to eliminate diseases, new crop varieties to eliminate world hunger, new technologies for environmental remediation, and so on (for a striking recent example, see: http//:www.monsanto.com). But on the other hand, there is an equally dramatic list of claimed horrors: the re-emergence of eugenics; threats to biodiversity and ecological integrity, and a Brave New World in which genetic information is a key agency of supervision and control (for an equally striking example, see Mae Wan Ho).[1] Where biotechnologists and their supporters tend to see enormous opportunities for progress through the 'appliance of science', opponents tend to see biotechnologists either 'playing God' or 'opening Pandora's Box', with frightening consequences for the future.

From the very outset, modern biotechnology was viewed as having extraordinary and potentially far-reaching implications.[2] Unusually, the development of recombinant DNA (rDNA) technology in 1972–73 – the defining breakthrough in modern biotechnology – led to doubts being expressed by prominent representatives of the nascent biotechnological community. Concerns as varied as the health and safety implications of rDNA research and the long-term consequences of rDNA technology for the evolutionary process led to a self-imposed moratorium in 1974–75 on key types of research, pending the clarification of potential risks. Following the establishment of laboratory safety regulations by the US National Institutes of Health and a number of European countries, research progressed rapidly to the point where by the end of the 1970s the technology's economic benefits had moved to centre stage. In 1979, an EC report, *The Biosociety*, identified biotechnology as one of the key components of economic competitiveness into the twenty-first century.[3]

In recent years, this EC projection has been at least partially fulfilled. Modern biotechnology has yielded medical developments as varied as new pharmaceuticals, new forms of genetic testing, genetic fingerprinting, the first clinical trials of gene therapy, and the prospect of the first clinical trials of xenotransplantation. At the same time, agricultural biotechnology has begun to produce its first real contributions to the food industry, with new plants and products such as 'vegetarian' cheese (early 1990s), the 'Flavr Savr' tomato (1995); 'Roundup-Ready' soyabeans (1996), and 'BT maize' (1996). The accelerating pace of biotechnological research and development has been accompanied by a burgeoning of coverage in the mass media, increasingly intense public debate, and not a little activist mobilisation. Public interest has been further heightened by the cloning of Dolly the sheep (1997), as well as by continuing controversy over the growing of both experimental and commercial genetically modified crops. These developments have in turn provoked a host of regulatory initiatives across Europe. Policies have been framed at both the national and the European levels. Following the passage of the EU directives on contained use and deliberate release in 1990, policy debate has proliferated as different branches of biotechnology have matured.

The scientific community, industry, national governments and international institutions are all directly involved in different ways in the development, implementation and regulation of technological innovation. The role of various publics (as tax payers, consumers, patients, interest groups, individual citizens, etc.) is less direct but nonetheless vitally important. It may be hypothesised that the influence of these publics on the processes of technological change has grown in recent years. Public debate and public criticism of particular technological developments have become increasingly commonplace; and in response, policy processes appear to have become increasingly sensitive to 'public opinion'. Thus, some conventionally technocratic policy-making institutions have found places for non-expert ('lay') as well as expert representatives; and new forms of public consultation have been developed with a view to the establishment of 'socially sustainable' technology policies.[4]

There are a number of debates in the literature about the social control of technological change. Do controversies about new technologies arise out of 'irrational' public perceptions prompted by 'sensationalist' media coverage, or are they the result of legitimate political disagreement about alternative technological futures? Is public acceptance of new technologies best achieved by countering 'ill-informed' resistance to change, or by the facilitation of constructive dialogue between an increasingly concerned and involved public and representatives of science and industry? How is public opinion represented within democratic societies, and when is it most influential? At what stages in the innovation cycle does public opinion enter the debate? Is it true, as one recent study suggests, that public opinion has gained importance at progressively earlier stages in the development of new strategic technologies?[5]

Introducing the study: conceptual framework

This volume is the first major output of an international research project whose aim is to address questions such as these. The book brings together a set of key empirical findings that serve to document public discourses about and representations of modern biotechnology in Europe from 1973 to 1996. The research on which these findings are based has its origins in an EC-funded three-year Concerted Action, under the title 'Biotechnology and the European Public'. The Concerted Action brought together research teams from 13 European countries: Austria, Denmark, Finland, Germany, Greece, Italy, France, the Netherlands, Norway, Poland, Sweden, Switzerland and the United Kingdom. In addition associate members joined from the United States of America and Canada. Each participating team brought national funding to the Concerted Action. The European funding made possible a collaborative project by supporting the coordination, design and analysis of the major multinational studies featured in this book.

The immediate context for the present research was provided by two previous EC 'Eurobarometer' survey studies of public perceptions of biotechnology in Europe (Eurobarometer 35.1 in 1991 and Eurobarometer 39.1 in 1993). These surveys provide a unique resource for national and comparative international studies of public perceptions of biotechnology. However, survey data of this kind seldom provide a sufficiently detailed picture from which to adequately interpret either national or international trends. Ideally, survey research should be carried out alongside complementary contextual studies. By contextual studies here we have in mind such things as detailed qualitative research intended to explore people's understandings and images of the new technology, longitudinal media analysis designed to reveal significant patterns of reportage, and policy studies documenting significant features of the political and regulatory systems that are responsible for public policy.

The conceptual framework for the research draws on insights from social systems theory[6] and social representation theory.[7] We consider biotechnology as an emerging scientific–industrial complex – a complex of research, development and implementation. By this, we do not mean to imply that biotechnology is a unified field, complete with a single, hierarchical mechanism of command and control; rather, we regard it as a heterogeneous coalition of many different actors, institutions and interests engaged in a power game over the control of this complex. The biotechnological complex evolves alongside other established systems (economic, legal, mass media, political, etc.) which collectively constitute its environment. Change occurs in part through 'challenges' of one system upon another, and responses to these challenges. For research purposes, any particular system may be foregrounded as the focus of attention. For our purposes, it is not biotechnology itself but certain aspects of the political systems more usually regarded as parts of its systemic environment that are foregrounded in this way. In particular, we are interested in the way in which old and new structures, formal and informal, within the 'public sphere' shape the development of biotechnology.[8]

The economic, legal, political and media environments, among others, each give more or less attention to biotechnology at different times, and all have 'eyes' on other issues. As each turns its gaze to biotechnology, it constructs a different representation of the object. Thus, for the financial market, the representation is likely to emphasise investment, risk and profits; while for the media, it may consist largely in the 'news value' of particular developments. In this sense, the environment of biotechnology is a set of observers with different levels of attention and different ways of viewing. But these are more than merely passive observers. Their gaze is an active process that may facilitate and/or constrain the development of biotechnology in particular ways. The presumptions underlying this research are that, in the course of its 25-year development, first, biotechnology has regularly challenged observers within the public sphere, and second, these observers have at times responded with counter-challenges that have shaped the continued development of biotechnology itself.

By systematically observing the observers of biotechnology in the public sphere, we aim to document the influence of the public sphere upon the emerging technology. Our research gains access to the public sphere through what we term 'public discourses' about biotechnology. For present purposes, a public discourse is simply speech or writing within the public sphere that embodies one or more representations of biotechnology. The study focuses on three arenas of discourse and their interrelations:[9] the policy context, media coverage and public perceptions. Public discourse in each of these arenas has its own logic, partly depending on the national context. Our objectives as social scientists are, first, to understand the ways in which particular public discourses embody particular representations of the emerging system of biotechnology as a whole, and second, to explore the systematic interrelations of these discourses over time.[10]

A word of caution over matters of definition is in order. Throughout this study, the term biotechnology is generally used to mean 'modern biotechnology', i.e. those processes, products and services that have been developed on the basis of interventions at the level of the gene. In the literature, modern biotechnology thus defined is generally contrasted with 'traditional biotechnology', i.e. those processes, products and services that have been developed on the basis of interventions at the level of the cell, tissue or whole organism. While these are justifiable distinctions from the point of view of the biotechnologist, it is important to note that, in any particular social situation, they may or may not accord with particular representations of biotechnology in the public sphere. Since the public sphere is our principal object of interest, it is vital that we acknowledge what the public understands by biotechnology, irrespective of whether this lay definition complies with any particular professional definitions.

Organisation of the research

The research has been concerned with the collection of internationally comparable data in three arenas of public discourse: public policy, media coverage and public perceptions. The research has been organised in the following way. First, each participating country has conducted a longitudinal (historical) analysis of the development of public policy for biotechnology over the period 1973 to 1996 (a similar longitudinal study of policy developments at the European level has also been undertaken). This period was chosen to embrace the entire history of modern biotechnology from the discovery of rDNA technology up to the point at which the research project actually commenced. Second, each participating country has conducted a longitudinal analysis of media coverage of biotechnology in the opinion-leading press, also from 1973 to 1996. Third, all participating countries have contributed to the development and analysis of a representative sample survey of public perceptions of biotechnology. This

survey was fielded in 1996 in each member state of the EU, together with Norway and Switzerland. (Similar surveys were also developed and fielded in Canada and the USA, but the results of these North American studies are not included in this volume.)

Clearly, the total quantity of empirical information involved in these three parallel investigations is inevitably very large. For purposes both of condensation and of clarity it is presented here in the following way. Part II contains a series of national profiles. These provide synoptic accounts of the main results in each participating country. Each national profile describes the broader context of the development of biotechnology, the evolution of policy debates and regulations, the way in which biotechnology has been presented in an opinion-forming media outlet, the nature of public attitudes, and an integrative summary. Part III provides an overview of the results of the Eurobarometer Survey on Biotechnology. Selected findings are described, giving both the European and the national perspectives; and these are followed by a series of tabulations of the results for each question. Finally, Part IV provides a short integrative conclusion. This volume is the primary sourcebook for the results of the research. A second, companion volume of analytical essays is in course of preparation. Finally, in the Appendices, detailed information on the methodology and results of the media analysis and of the Eurobarometer survey are provided.

From the outset of this research, methodological issues have been to the fore. Our aim has been to study public policy, media coverage and public perceptions in each country in ways that would facilitate meaningful comparisons both between the three arenas and across the participating countries. This has necessitated the establishment of agreed procedures for the conduct of each part of the research. In the case of public perceptions, where the chosen method was the social survey, the task was relatively easy: survey methodology is a highly formalised and well documented field. By contrast, there are no equally well-developed and generally accepted methodologies for the analysis of policy or media coverage. For this reason, considerable attention has been given to the development of protocols for these parts of the research. In the remainder of this chapter, we outline the principal features of these protocols as an essential aid to the proper interpretation of the remainder of the book.

Public policy

Rationale

Public policy is an important expression of public aspirations, attitudes and values. Public policies may have many different explicit aims: they may seek to promote public goods (for example, through the encouragement of innovation), or to prevent public harms (for example, through the imposition of health and safety regulations); they may seek to protect the interests of producers (for example, through the patent laws), or those of consumers (for example, through product labelling requirements); and they may seek to reconcile conflicting ideals or interests (for example, in the provision of guidelines for the acceptable conduct of research on human embryos). Public policy is the outcome of activities in political forums. In pluralistic democracies, these activities are necessarily multiple and multivalent. In other words, at any particular time no single actor or interest group is likely to dominate the policy-making process to the exclusion of all others. Instead, different actors and interest groups vie for influence in a political process (part private, part public) involving competition, cooperation, coalition making and breaking, and compromise. In the end, what transpires may be something that no single actor or interest-group originally intended. In the EU, policy-making for biotechnology takes place at both the national and the European levels. To the complex of actors and interests operating at the level of the individual nation-state must therefore be added a second complex of actors and interests (including the EC, the European Council and the European Parliament) operating at the level of the 15 member states.

For the purposes of this study, the aim has been to characterise public policy developments in relation to biotechnology over the period 1973–96. Needless to say, this is an extremely ambitious task. A full-scale historical reconstruction of policy developments in all areas of biotechnology, across the entire history of biotechnology, and across all participating countries, was well beyond both our resources and our needs. Our primary concern was not with the provision of a meticulous historical record but rather with the characterisation of key events in the policy debate as they have both shaped and in turn been shaped by the larger public sphere for biotechnology in Europe as a whole. Throughout, we have kept our sights firmly on the goal of informing our analysis of public perceptions and media coverage at both the national and the European levels. For this reason, we have chosen to chart the evolution of the debates in the legislative arena with an emphasis on the changing 'frames' of the debate and their origin in particular sponsors/constituencies.[11] By a frame, we mean a particular dimension in which policy-making (debate, disagreement, legislation, etc.) takes place. Typical frames are: economic development, health and safety of workers, environmental impact, ethics and consumer choice (labelling). By sponsors and constituencies we mean particular institutionalised interest groups such as political parties, consumer organisations, environmental pressure groups, religious groups, animal rights groups, industrial working parties, agricultural organisations and third-world interest groups. We characterise constituencies in relation to particular frames in the debate, and in terms of their contributions to policy, their position in the debate, their publicity activities and impact, and their credibility with the public.

In reviewing the history of biotechnology policy-making, we have concentrated in the main on formal policy-making processes; that is, on the institutionalised activities by which official public policy has been established. However, wherever possible we have also paid attention to informal influences (such as lobbying, and the oppositional activities of non-governmental organisations) on the formal sector. We have been interested in questions of three main types: those that concern the 'framing' of biotechnology within the policy field; those that concern the mechanisms by which policy has been framed; and those that concern the relationships between individual nation states and the EU. In the first category, we have considered questions such as: which issues have been debated? How have these issues been framed by the selection of themes? Which have been the principal sponsors/constituencies of particular themes? How has the policy process dealt with opportunities and risks in relation to biotechnology? Have policy-makers concentrated on the control of processes or products?[12]

In the second category, we have considered questions such as: what have been the distinctive 'policy cultures' for biotechnology in Europe? What have been the principal mechanisms for generating biotechnology policy in Europe? What has been the influence of public opinion upon policy-making? Have policy-making processes tended towards the 'technocratic' or the 'participative' mode?[13] Is there evidence of 'institutional learning' as policy-makers develop new instruments and forums in light of previous experience?[14] In the third category, we have considered the timing of policy processes: how early or late do particular countries become engaged with biotechnology policy-making? How far do particular countries 'lead' or 'follow' in particular areas of policy-making? What are the relations between national and European initiatives on biotechnology?[15]

Methodology

We turn next to the methodology developed for the policy study. We began by developing a chronology of key policy developments in each country for the period 1973–96. This chronology was based partly on published sources and partly on interviews with key actors from different arenas, including government, industry and non-governmental organisations. Each chronology took the form of a brief tabulation of events, together with a series of 'identity cards' providing more detailed information about events judged to be of prime importance. The aims of the chronologies were: to document concisely the most important policy initiatives in each country, to provide a baseline of data for comparison with the media study (see below), and to provide a baseline of data for purposes of international comparison. The chronologies were complemented by a review of the 'policy culture' in each country. By policy culture in this context, we mean the prevailing norms and styles of policy-making at the national level. These were judged to be vitally important for the interpretation of national similarities and differences.

Once the chronologies had been completed, it became clear that, even at this level of data compression, the amount of empirical information available for purposes of international comparison remained uncomfortably large. For this reason, it was decided to engage in a further process of data compression by converting each chronology into a single 'regulatory profile'. The purpose of the profiles is to provide in a standardised form a concise summary of policy developments in each country over almost a quarter of a century. By a process of trial and error, policy developments were classified into ten areas: reproductive technologies, gene therapy, genetic screening, transgenic animals, genetically modified food, GMO releases, GMO contained use, health and safety, research and development policy, and intellectual property rights. These categories are empirically derived; they reflect the ways in which for the most part European policy-makers have dealt with biotechnology. In each area, the profile allows us to record a limited range of policy developments by means of standard symbols. Key information contained in the profiles includes: the creation of policy-making bodies, the establishment of formal policy review processes, the production of formal reports (discussion documents, government white papers, etc.), and the creation of both statutory and non-statutory (voluntary) regulations. The timing of each event is provided in the profile itself, and the nature of key events is described in numbered footnotes.

The national profiles in Part III of the book include the profiles together with short narrative accounts of key policy developments in each country. In practice, it has proved difficult to squeeze a wealth of detailed information into the straitjacket of the profile; undoubtedly, a considerable amount of potentially useful information has been lost in the process. However, this is a price worth paying for the ability to stand back from the detail and assess the overall pattern of activity in each country. For it is only in this way, we suggest, that significant patterns at the international level can be discerned.

Media coverage

Rationale

The mass media constitute a major forum of the public sphere in modern societies. There is general agreement in the literature that the mass media are enormously influential, but much less agreement about the exact nature of this influence. It is variously argued that the mass media serve to 'frame' issues in the public domain, that they serve an 'agenda-setting' role, and that they serve to lead and even to shape public opinion. For our purposes, the mass media are best viewed as an important source of representations of biotechnology in the public sphere. Thus, the media function both to explain and legitimate formal policies ('top-down'), and to signal issues and themes arising from informal political forums ('bottom-up'). In these senses, the mass media may be viewed as 'channels' of communication both from government to the public and from the public to government. Throughout, it is the complex interrelationships between media discourses, policy discourses and public perceptions with respect to biotechnology that are the focus of our attention.

The comparison of media coverage across all of the participating countries and in eight different European languages is an opportunity to map the cultivation of a new technology across different publics. By design the project has elective affinities with 'cultivation analysis'.[16] However, it is distinct because of its focus on the opinion-leading press and its multicultural contexts over time. For this reason, we refer to the media analysis as a 'cultural indicator' analysis which contributes to a better understanding of the emerging representation of genes and genetics in late-twentieth-century Europe.[17] The media analysis comprises two elements: a time-series analysis of the coverage of biotechnology in general, and a cross-sectional qualitative analysis of the media coverage of one or more particular issues. It is the time-series analysis with which we are principally concerned here. This time-series analysis has two immediate aims: first, to establish an indicator of the intensity of coverage over time, and second, to characterise this coverage in a longitudinal content analysis.

Because we were interested in comparisons with the policy analysis, the same time frame was adopted in both studies. First, we established an indicator of the absolute intensity of media coverage by estimating the number of all relevant articles on a year by year basis. Although it was clear from the outset that new genetic technology does not have the same news value across all European countries, nothing systematic was known about any national differences: previous studies are limited both in scope and time.[18] Second, we established a relative indicator for each country of the growth of 'genetics' as scientific news. For each country, we determined the high and the low points for coverage; then we defined the peak year for each country as the index = 100 and recalculated the relative figures for each year. On the basis of the these two indicators, and in conjunction with the content analysis, we then defined a 'phase structure' of the media debate in each country. Phases were defined on the basis of relative peaks and troughs in media coverage in each country.

The final element in the media study was the creation of a corpus of media material in each country for purposes of comparative analysis. The method indicated for this purpose is classical content analysis. We chose this type of analysis from a multitude of textual analysis techniques because: first, it allows for systematic (i.e. replicable and valid) comparisons on the basis of a common coding frame; second, it can cope with large amounts of material; and third, it is sensitive to symbolic material, albeit through a process of simplification. Content analysis treats media texts as the 'objectified' traces of the complex communication process from senders to receivers in each country. The aim of the analysis is to deliver a systematic and comparable interpretation of these traces. Crucial in this process is the coding, which constitutes a set of coder questions which the text is answering and in light of which the coders establish, more or less reliably, an agreeable interpretation of the materials on a case by case basis. Contrary to some historical claims,[19] we do not consider this type of interpretation to be the final objective reading of the material. Rather, we view the media content analysis as a resource best understood in combination with the results of the policy and public perception modules of the study. In this sense we are following the programmatic model of 'parallel content analysis'.[20]

Methodology

Ideally, we would have liked to assess the whole of the media coverage of biotechnology. However, taking into account the diversity of the media systems in the various countries and the time period chosen, a selective procedure was the only practical way forward. The method adopted for the selection of material is specified in the guidelines that were negotiated among the participating teams (see Appendix 3). These guidelines specify: the choice of media outlet; the definition of what was to count as a media report of biotechnology; and the sampling strategy to be employed. As we are dealing with the emergence of a new technology in the public sphere, we have selected the opinion-leading press for study. By 'opinion leading', we refer to outlets that are read by decision makers for information and by other journalists for inspiration. The reasons for adopting the opinion-leading press are: first, that the function 'opinion leader' provides a functional criterion for selection independent of other characteristics such as circulation or quality (occasionally, as in the case of the UK study, this criterion led to the use of more than one source within a single longitudinal study); second, that in most participating countries national newspapers still act as opinion leaders; and third, that newspaper materials make relatively convenient and accessible sources for purposes of data collection and analysis. We assume that for each country we can identify one or two press outlets that stand as proxies for the nature and intensity of media coverage more generally.

To establish a comparable sampling frame for media analysis across 12 countries is a challenge in itself. Our strategy was to establish functional equivalence with the help of guidelines for the project. Some newspapers offer an historical index of articles. This constitutes a self-classification of journalists. For the early years, several of us relied on this entry point, although we were aware that this classification was not necessarily exhaustive. Such indices were checked by manual scanning, under a protocol according to which the number of issues scanned was inversely proportional to the amount of relevant material they were expected to contain (the smaller the number of expected articles, the greater the number of issues that need to be checked in order to establish a reliable intensity index). With on-line resources such as FT-Profile or CD-ROMs from certain newspapers, the complexity of sampling is reduced to the question: what is relevant material? To answer this question, the project defined a core set of keywords translatable into each of the eight languages. These were: 'biotech*', 'genetic*', 'genome' and 'DNA'.

The fact that the project dealt with eight different languages led to some semantic uncertainties. The denotation of 'biotechnology and new genetics' has to be recovered from a changing lexicon of words and phrases both across languages and over time. It became clear, for example, that as the technology has developed so too has the vocabulary that denotes it. In English, for example, the term 'recombinant DNA' was current in the 1970s but disappeared later on; the term 'biotechnology' was not commonly used until around 1980; and terms such as 'genetic engineering', 'genetic manipulation' and 'genetic modification' all appear and disappear later on, in what seems to be a complex game played for control of semantics in the public sphere. Our research strategy has been to offer each national research team a choice of functionally equivalent suggestions for manual sampling, index sampling, on-line sampling and associated quality checks (see Appendix 3).

Finally, our project sampled material for two purposes, to measure the intensity of coverage, and to establish a corpus for content comparisons. Teams have chosen a different approach to this bifurcation. In some countries the material in the corpus is proportional to the coverage using a fixed sampling ratio, while in other countries, not least for reasons of statistical comparison of periods with small numbers, the corpus is not proportional to the coverage but rather uses a fixed sample size across certain periods. Hence, the frequencies in the data corpus are suitable only for relative comparisons of contents. The comparison of intensities has to rely on the data that is collated in the table of estimates taking into account the different sampling procedures (see Appendix 6, Table 2 for absolute intensity estimates).

The basic idea of the content analysis was to establish a basis for international comparisons of the coverage of 'biotechnology and new genetics'. Inspired by Lasswell's pioneering but failed project of a 'world attention survey',[21] we undertook more modestly to characterise the attention given in 12 European countries to this new technology over the period of the study. The coding frame constitutes the set of comparative core variables. Each country developed further analytical questions to suit their particular research purposes. The basic idea of the coding frame is to deliver a characterisation of coverage in terms of frames, thematic structure and evaluation of biotechnology. Frames, themes and evaluations are further differentiated (see Appendix 5 for coding frame). The unit of analysis is the 'single press article', which is read by the coders and interpreted in the light of the questions posed by the coding frame. As most coders were highly educated members of the national research teams, their readings are likely to reflect the subcultural features of those that have produced the articles and thus to constitute in this sense a valid, albeit not a universal, reading.[22]

For each article, the coding frame assesses: the section of the newspaper in which it appears, the size of the article as an indicator of news importance, the format of the article, and whether the article appears as controversy. The news event is also characterised by authorship, the actors identified with biotechnology, the theme, its location, and its attributed consequences in terms of risks and benefits. An indication of the evaluation tone is gained by convenient 'negativity' and 'positivity' ratings. A key feature of the coding is the identification of 'frames' of coverage. As in the policy analysis, so here a frame constitutes the dimension on which disagreements are elaborated. For our purposes, we adopted the set of frames observed by Gamson and Modigliani[23] and added further frames identified in the course of pilot work, including 'nature/nurture', 'ethics' and 'globalisation'.

The national profiles provide media profiles which condense the results of the media content analyses, much as the regulatory profiles condense the results of the policy chronologies. The media profiles provide a phase structure defined on the basis of the peaks and troughs of coverage over time. Different countries report different numbers and timings of phases on this basis. Across each phase, the profiles compare the relative structure of coverage on five key variables: frames, themes, actors, locations, and risk and benefit. These allow us to characterise and compare changes in the cultivation of biotechnology across different countries. For each variable, a 5% threshold is introduced to show the major shifts over time. Multiple codings for themes, actors and locations are integrated into single frequencies.

It has been a major concern of the research to ensure data quality cross-nationally. Quality management measures adopted have included careful negotiation of sampling and coding proce-dures, familiarisation with the sampling and coding procedures in the context of local constraints, revision of the coding frame to take account of local pilot work, and formal reliability checks for both within-country and cross-country consistency (see Appendix 7 for reliability measures). Within-country consistency was assessed on the basis of a sample of 10 per cent of the coded material in some countries; and cross-country consistency was assessed on the basis of two English articles which were coded by all 12 teams. For within-country comparisons, the average reliability between coders measured as percentage of agreements was $r = 0.72$, ranging from 0.61 to 0.87; and for cross-country comparisons the equivalent figure was $r = 0.65$, weighted by the number of values in each category, ranging from 51% to 82% depending on the formula used.

Public perceptions

Rationale

The third module of the research is concerned with direct measurement of public perceptions by means of random sample social survey. We use the term public perceptions because most of the obvious alternatives – public opinion (with its connotations of political opinion polling), public attitude (with its connotations of exclusive interest in personal dis-positions) and public acceptance (with its connota-tions of the public relations industry) – are unsatis-factory. In particular, the related terms 'public acceptance' and 'acceptance research' are both semantically and pragmatically flawed. Semantically, 'acceptance' makes an unhelpful distinction between technology and the public – as if we had to deal with two sharply separate entities, when in fact the two are part and parcel of the larger social system of biotechnology; and it implies that the public has only a single choice (either to accept or to reject) when in fact the public is a constitutive part of the processes that serve to shape the technology. At best, 'acceptance' is a euphemism for a mixed bag of people's enthusiasms, discontents, worries, imaginations, fears, anticipations and resistances. In order to achieve a more realistic grasp of the dynamics of public perception and its effects on new developments, we need to stop attributing deficiencies to the public in terms of 'lack of acceptance'.

In order to avoid the pitfalls of the concept of 'public acceptance', we prefer the more neutral notion of 'public perceptions'. By 'public perceptions' in this context we mean all of the ideas that people may have concerning biotechnology. As such, the term embraces interest in, understanding of and attitudes towards biotechnology; but also, it includes the images, hopes, fears, expectations and even forebodings that people may experience when they think about biotechnology. The term 'perception' includes the processes that lead to a particular structure of imagination about a new technology at any moment in time. The importance of imagination lies in its capacity to go beyond a given reality and to anticipate a negative or positive future that inspires present actions.[24] In the case of biotechnology – literally, the technology of life itself – the importance of individual and collective imagination in shaping public perceptions can scarcely be exaggerated. The cultural resonance of key phrases – 'test tube baby', 'genetic engineering', 'cloning', etc. – has as much to do with their metaphorical and mythopoeic powers as with their scientific and technological significance.

So far as biotechnology is concerned, public opinion has been described as the 'major uncertainty for investors and government alike'.[25] Cantley finds that experts who are inclined to dismiss public opinion as irrelevant are liable to the moral charge of anti-democratic elitism. Furthermore, he suggests that those who disregard public opinion ignore the empirical correlations between public perceptions of biotechnology, trust in various players, and the perceived need for rules and regulations. Whoever wants to constrain the regulatory machinery of the state for economic or ideological reasons, needs to take into account public opinion and to consider the trust bestowed upon various actors. With the realisation that the provision of public information does not necessarily lead to public acceptance, trust has become a key concern of the 1990s. Significantly, this concern has arisen at a time when many political scientists and pundits have acknowledged a growth of public scepticism about institutionalised authority across the democratic world.

Methodology

From the outset, our research was organised around the opportunity to field a social survey through the Eurobarometer Office of the EC. Following extensive qualitative research using individual and focus group interviewing in the spring of 1996, a survey instrument was designed and pilot tested. Following necessary modifications, the Eurobarometer on Biotechnology (46.1) was conducted during October and November 1996. The survey was conducted in each member state of the EU, using a multistage random sampling procedure providing a statistically representative sample of national residents aged 15 and over. The total sample within the EU was 16,246 respondents (i.e. about 1000 per EU country). In addition, similar samples were achieved in Norway (1996) and in Switzerland (1997). The survey drew on the questionnaire employed in two previous Eurobarometer surveys on biotechnology (35.1 in 1991 and 39.1 in 1993). Where possible, questions were repeated for purposes of trend analysis; but changes both in biotechnology itself and in the public debate about biotechnology dictated the need for a number of new question sets in the survey instrument. The revised questionnaire embraced items on the following topics:

- optimism/pessimism regarding technological developments, including biotechnology;

- knowledge of elementary scientific knowledge relating to biotechnology;

- beliefs about the role of nature and nurture in the development of human attributes;

- specific attitudes on six applications of biotechnology measured on four dimensions: usefulness, risk, moral acceptability and overall support;

- general attitudes to the monitoring and regulation of biotechnology;

- confidence in different agencies to regulate biotechnology;

- confidence in different institutions to tell the truth about biotechnology;

- future expectations about the contribution of biotechnology to society;

- the importance of the issue, sources of information and attentiveness to the issue;

- political attitudes; and

- socio-demographic characteristics.

The results of the Eurobarometer survey are presented in full in Part III of this book (technical information concerning the survey is provided in the Appendices 1 and 2). In addition, the national profiles describe key features of public perceptions in each of the participating countries as represented in the survey. Here, we wish simply to raise a number of issues that should be borne in mind when interpreting the results of the survey. First, the survey was conducted in November 1996 (exept for Switzerland: June 1997); that is, just as the issue of genetically modified soyabeans was beginning to be discussed in Europe, and a few weeks before the surprise announcement of the birth of Dolly the sheep. It is important to recall, therefore, that the survey provides a snapshot of public perceptions at a key point in time. Since then, further significant developments have taken place both in biotechnology and in the public debate about biotechnology. These developments do not invalidate the survey results, but they do oblige us to be cautious before attempting to extrapolate from them to the present. Partly for this reason, the percentages of respondents agreeing or disagreeing with particular propositions in the survey are generally of less interest to us than is the underlying structure of the answers, as revealed for example in the relations between responses to different questions about biotechnology, in the relations between responses to questions about biotechnology and socio-demographic characteristics, and in the trajectory of opinion over time as shown by the time-series questions.

A second word of caution relates to the inevitable limitations that accompany any cross-national study of this kind. The Eurobarometer questionnaire was fielded in no fewer than ten different European languages across all the member states of the EU, Norway and Switzerland. Inevitably, such an exercise raises difficulties to do with the accurate translation of key terms. All of the key terms in this study – biotechnology, gene technology, etc. – have linguistic and other connotations that are partly culture-specific. While every attempt has been made to ensure semantic equivalence in the choice of terms, it should be remembered that unavoidable semantic differences may occasionally affect results. Furthermore, it is always possible that

other, non-linguistic cultural differences (including differences in the way that respondents report knowledge and ignorance, attitudes and non-attitudes, etc.) may obtrude. Here, as in so many other areas of scholarly research, we are obliged to use data that are as good as they can be but not always as good as we would wish.

A final caveat on data quality

As will be apparent to the reader, the longitudinal and comparative analysis of public discourse about biotechnology in three domains across many countries involves the integration of a very large body of empirical materials. Throughout the project, we have been particularly concerned with data quality. This is no easy undertaking when addressing empirical sources of such different types. Some may be inclined to feel that we are too dependent on empirical sources, while others may worry about the robustness of particular measures. In our view, however, the principal need in this area is for informed interpretation; that is, for understandings of the public representations of biotechnology that are firmly rooted in the real worlds of public debate and decision-making. We believe that the distinctive contribution of this volume lies in the range and richness of the empirical resources that it brings to bear on topics which are too often the subject of fact-free surmise and speculation. An overly fastidious approach to issues of data quality would probably have prevented our embarking on – let alone completing – this research project; but at the same time, a cavalier approach to data quality, while obviating the need for all of the detailed methodological work of the past three years, would have risked substituting personal prejudice for social reality. We and our research colleagues have attempted to survey an extremely broad field using techniques which, though necessarily less than perfect, were considered to be sufficient for the purposes required. It is for others to judge how far we may have succeeded.

Notes and references

1 Mae Wan Ho, *Genetic Engineering – Dream or Nightmare? The Brave New World of Science and Big Business* (Bath: Gateway Books, 1998).
2 Berg, P, Baltimore, D, Boyer, H W, Cohen, S N, Davis, R W, Hogness, D S, Nathans, D, Roblin, R, Watson, J D, Weissman, S, Zinder, N D, 'Potential Biohazards of Recombinant DNA molecules', *Science* 185 (1974), p303.
3 EFB, 'European Federation of Biotechnology and Dechema Biotechnology in Europe', *Fast Occasional Papers no. 59* (Brussels: EEC, 1982).
4 Joss, S and Durant, J (eds), *Public Participation in Science: The Role of Consensus Conferences* (London: Science Museum, 1995).
5 Bauer, M, 'Resistance to new technology and its effects on nuclear power, information technology and biotechnology', in Bauer, M (ed), *Resistance to New Technology, Nuclear Power, Information Technology, Biotechnology* (Cambridge: Cambridge University Press, 1995).
6 Luhmann, N, *Social Systems* (Stanford: Stanford University Press, 1995).
7 Moscovici, S, *La psychanalyse, son image et son public* (Paris: PUF, 1961); Wagner, W, 'Queries about

social representation and construction', *Journal for the Theory of Social Behaviour*, 26 (1996), pp95–120; Bauer, M, and Gaskell, G 'Eine forschungstrategie für soziale repraesen-tationen', in Witte, E (ed), *Social-psychologie der Kognitionen: soziale Repraesen-tationen, subjektive Theorien, soziale Einstellungen* (Lengerich: Pabst, 1998).

8   Habermas, J, *The Transformation of the Public Sphere* (London: Polity Press, 1990).

9   Hilgartner, S, and Bosk, C, L, 'The rise and fall of social problems: a public arena approach', *American Journal of Sociology*, 94 (1988), pp53–78.

10 Jaspers, J M, 'The political life cycle of technological controversies', *Social Forces*, 67 (1988), pp357–77.

11 See Gamson, W and Modigliani, A, 'Media discourse and public opinion on nuclear power: a constructivist approach', *American Journal of Sociology*, 95 (1989), pp1–37.

12 Jasanoff, S, 'Product, process, or program: three cultures and the regulation of biotechnology', in Bauer, M (ed), *Resistance to New Technology – Nuclear Power, Information Technology, Biotechnology* (Cambridge: Cambridge University Press, 1995).

13 See, for example, Hisschemoller, M, and Midden C J H, 'Technological risk, policy theories and public perception in connection with the siting of hazardous facilities', in Vlek, C and Cvetkovich, G (ed), *Social Decision Methodology for Technological Projects* (Dordrecht: Kluver Academic, 1989).

14 See, for example, Evers, A and Novotny, H, *Ueber den Umgang mit Unsicherheit: die Entdeckung der Gestaltbarkeit der Gesellschaft* (Frankfurt: Suhrkamp, 1987); Beck, U, *Die Erfindung des Politischen* (Frankfurt: Suhrkamp, 1993).

15 Wheale, P and McNally, R, 'Biotechnology policy in Europe: a critical evaluation', *Science and Public Policy*, 20, (1993), pp261–79; Wald, S, *Biotechnology in the OECD Committee for Scientific and Techno-logical Policy: Evolution and Main Events 1980–1993* (OECD, June 1993).

16 Gerbner, G, 'Toward "cultural indicators": the analysis of mass mediated public message systems', in Gerbner et al (eds), *The Analysis of Communication Content* (New York: John Wiley & Sons, 1969), pp123–32.

17 Bauer, M, 'Science in the media as cultural indicators: contextualising surveys with media analysis', in Dierkes, M and v Grote, C (eds), *Between Understanding and Trust: Science, Technology and the Public* (Reading: Harwood, in print); Bauer, M and Gaskell, G, 'Eine Forschungsstrategie fuer Soziale Repraesentationen', in Witte, E H, (ed), *Sozialpsychologie der Kognition: Soziale Repraesentationen, subjektive Theorien, soziale Einstellungen* (Lengerich: Pabst, 1998).

18 Ruhrmann, G, 'Genetic engineering in the press: a review of research and results of a content analysis', in Durant, J (ed), *Biotechnology in Public: A Review of Recent Research* (London: Science Museum, 1992); Hansen, A, 'Journalistic practices and science reporting in the British press', *Public Understanding of Science* 3 (1994), pp111–34.

19 Berelson, B, *Content Analysis in Communication Research* (Illinois: Free Press, 1952).

20 Neuman, W R, 'Parallel content analysis: old paradigms and new proposals', *Public Communication and Behaviour*, 2 (1989), pp205–89.

21 Lasswell, H D, 'The world attention survey: an exploration of the possibilities of studying attention given to the United States by newspapers abroad', *Public Opinion Quarterly*, 5 (1941), pp456–62.

22 Eco, U, *Apocalypse Postponed* (Bloomington: Indiana University Press, 1994), Ch. 2.

23 Gamson, W and Modigliani, A (1989).

24 Boulding, K E, *The Image: Knowledge in Life and Society* (Michigan: The University of Michigan Press, 1956).

25 Cantley, M F, 'The regulation of modern biotechnol-ogy: a historical and European perspective', in Brauer, D (ed), *Biotechnology – Legal, Economic and Ethical Dimensions* (Weinheim: VCH, 1995).

Part II
National profiles

Austria

Wolfgang Wagner, Helge Torgersen, Franz Seifert, Petra Grabner and Sigrid Lehner

The economy and biotechnology

Biotechnology became a political issue in Austria only recently, but it must be considered one of today's most controversial technologies. However, awareness of biotechnology remained slight for a long time, although public perceptions of techno-logical development, as well as the political status of innovation in general, have changed considerably since the 1970s.

Research and development

Until the 1970s Austria suffered from a significant lack of R&D policies. Despite high rates of economic growth, the post-war decades were characterised by imitation and technology imports, resulting in short-term positive effects. In the long run, however, the effects were negative: a growing technological dependency and a mentality of avoiding the risks and costs of R&D. The early 1970s saw a shift towards a somewhat more innovative strategy, and during the 1980s there were some attempts to initiate a catch-up modernisation. However, tech-nology policies more or less failed due to inherent structural weaknesses, and Austria has only recently managed to increase its R&D rate to the EU (and OECD) average.

Biotechnology remained of little economic significance. Although in 1980–81 the government tried to start developing an Austrian strategy in biotechnology policy by formulating research priorities and programmes to promote applications, budgetary restrictions and organisational problems prevented the development of a coherent political strategy.[1] However, these programmes helped some university and industrial research institutions to improve their international reputations and to establish a world-class laboratory through a joint venture. There are also a couple of pharmaceutical companies engaged in the development and production of drugs based on recombinant DNA techniques, such as the recombinant bovine soma-totropin.

It would however be an overstatement to claim that those cutting-edge labs and firms represented the general standard in Austria. A recent survey by Ernst & Young shows that Austria has relatively few biotechnology companies compared to other European countries.[2] No single big firm is based here, most of the more innovative companies are branches of multinationals, and there are very few new companies starting up. Consequently, the political weight of the Austrian biotechnology industry is much less than in neighbouring Switzer-land or Germany, especially in agro-biotechnology where there is no big Austrian player. This sheds some light on the delay in release applications and on the relatively little notice the government took of industry's arguments about competitiveness when a strong public opposition emerged.

Agriculture

Austrian agriculture has been subject to considerable changes during recent decades: compared to the 1950s, today about one-quarter of the workforce produces roughly double the amount of goods on an area reduced by 15%. Although there has been an enormous push for modernisation, and lower prices and incomes resulted in migration from the rural areas, Austrian agriculture is still small-scale and mostly family-run. The 1990s have seen a shift in the official agrarian policy (under the heading of 'eco-social market policy') as the former production function has been supplemented by an ecological dimension.[3] This led to the current high number of about 23,000 organic farmers in Austria, which is more than the total in the rest of the EU. In an EU comparison, the *Independent* reported: 'Generous taxpayer subsidies, concerned consumers and the Alps have made Austria the developed world's leading organic farming nation. Just over 11% of its farmland receives no chemical fertilisers or pesti-cides'.[4] The equivalent value in the UK is only 0.3%.

Political system

The post-war Austrian political system can be described in terms of continuity and change: **1** in the party system; **2** in social movements; **3** in the role of bureaucracy; and **4** in Austria's corporatist modes of decision-making.

1 The party system has shown a remarkably high degree of continuity during the post-war period. At its heart are the Austrian Social Democratic Party (SPÖ) and the conservative Austrian People's Party (ÖVP), with a grand coalition of these two being the most frequent form of government. Until 1970 the ÖVP was the dominant force, and after a series of electoral defeats the SPÖ became the leading party.

While the fact that, by 1998, the SPÖ had been in government for 28 years shows the continuity of the Austrian party system, developments in the other parties exemplify its changes. The transformation since 1986 of the formerly insignificant Austrian Freedom Party (FPÖ) into a populist right-wing party has put considerable pressure on the government. As a consequence of the FPÖ's move to the right, its liberal wing broke away in 1993, constituting the Liberal Forum and thus adding a fifth party to the Austrian Parliament. More pertinent for the present context, however, was the arrival in Parliament in 1986 of a Green party.

2 The success of the Greens in the 1980s reflects profound changes in Austrian civil society at that time, notably the emergence of new, ecologically oriented social movements. Unlike other countries, 'the key event for political mobilisation leading to new forms of political conflict was not... the student unrest of 1968, but the conflict over nuclear power.'[5] This struggle culminated in a referendum in 1978, when a slight majority rejected the use of nuclear energy in Austria, thus preventing the opening of a nuclear power plant already built in Zwentendorf. In the aftermath of this conflict, a broad and differentiated scene of new social movements emerged in the 1980s. The conflict over energy resurfaced in 1984, when activists prevented the construction of a hydro-electric power plant in Hainburg on the Danube. The spectacular events in this context came to be seen as a turning point in recent Austrian history. The government, confronted by an unexpected nationwide mobilisation and by the tabloid media supporting its opponents, finally had to cancel its plans. The emergence of new social movements during the 1980s has clearly challenged and – to some extent – changed features of the Austrian political system, if only by adding new actors and issues, and establishing new forms of political conflict.

3 Perhaps more than in other countries, Austrian bureaucracy is a factor of continuity. Perpetuating a political culture that has its roots in the paternalism of the late Habsburg monarchy, its influence often brings about an attitude of fragmentation. The tendency is to consider the public not as an independent actor but as being dispersed in a variety of client relation-ships. This hardly furthers public accountability and responsiveness. Thus, socio-political change often takes the shape of a 'top-down revolution'. Austrian politics is extremely legalistic, and most government decisions have to take the form of a law, which implies that Parliament has to be involved. However, bureaucracy is actually more influential than the constitution, with its strong emphasis on popular sovereignty outlines, since government ministers generally depend on the cooperation of the hierarchically structured departments.[6]

4 Another feature is the corporatist mode of decision-making called 'social partnership'. Developed in the post-war period, the system served to prevent open clashes of interest and thus to guarantee acquiescence and social harmony. Social partnership implies that decisions can only be made unanimously after the essential political groupings have had their say in protracted negotiations behind closed doors. Consequently, few highly concentrated, centrally organised and quasi-monopolistic interest groups are involved in political decision-making. Until the 1980s, their cooperation and influence were steadily extended, but growing criticism has led to serious problems of legitimacy.[7]

Public policy

The political debate on biotechnology began in Austria only in the mid 1980s. Although very few took part in it, there was growing concern reflected in open discussions and enquiries in which genetic engineering and artificial reproduction were bracketed together. After the passing of the Act on Artificial Reproduction, discussions concentrated on the question of a legal framework for genetic engineering. Originally the need for a specific law was denied by industry and the scientific community. The Academy of Sciences set up a Commission for Recombinant DNA to adopt the NIH guidelines for Austria and to render legislation unnecessary. But when Austria started to align its laws to EU directives (the application for membership was submitted on 17 July 1989), a special law was inevitable. Also the pharmaceutical industry had concluded that a lack of regulation might lead to a loss in competitiveness. After exploring the existing legal framework, negotiations for a law started in 1991, and the Ministry of Health attempted to involve all relevant parties (including NGOs).

The Austrian Parliament also dealt with the issue: it chose the subject of 'Technology Assessment of Genetic Engineering' for its first ever Inquiry Commission in 1992. The goal of the expert hearings was to consider as many views as possible. In nine sessions the Commission tried to cover the whole range of issues related to genetic engineering, but it paid less attention to GMO releases than to other aspects. The final report recommended: the assessment of impacts on human health and the environment; the ethical and social sustainability of applications; mandatory labelling of products; and greater public participation should be mandatory. The Inquiry Commission was successful in reaching a broad consensus, but conflicts arose especially within the conservative ÖVP. Industry and scientists still favoured self-regulation. The Ministry of Health continued negotiations and presented a revised draft law before the Commission's report was issued.

The experiences of this first attempt at TA clearly reflected Parliament's weak position. It showed that instruments such as parliamentary inquiry commissions interfere with traditional patterns of the Austrian political system.

After significant changes the draft law was passed by Parliament in June 1994 and enforced in January 1995. It regulated the contained use and release of GMOs, genetic testing and gene therapy. It also contained a provision for the avoidance of 'social unsustainability' (*soziale Unverträglichkeit*) of products. Its meaning remains unclear and its practical significance is nil.[8]

The Commission did not achieve its explicit aim of promoting a broader public discussion, although surveys showed implicit public unease. The attempt to anticipate foreseeable conflicts via legal regulation also failed. Unlike Germany, where the debate on biotechnology faded with economic problems, in Austria public debate, conflict and polarisation set in with the first applications for GMO releases.

GMO releases and political debate

At the beginning of 1996, the Austrian Health Ministry as the relevant authority received three release proposals, and the Green Party and environmental organisations took up releases as a prime issue. A public research institute proposed a GMO release to do bio-safety research, commissioned by the Ministry itself with the tacit aim of paving the way for future releases. Public opposition arose immediately, and increased considerably when a release proposal for a herbicide-resistant maize was submitted. NGOs collected 30,000 objections with the help of the biggest tabloid newspaper. The Ministers of Agriculture and of the Environment publicly opposed the proposal, not least because genetically modified crops were seen as a threat to Austria's many organic farmers. The proposal was withdrawn. Another proposal initially met far less opposition, but eventually provoked a scandal as the company released the plants before permission was formally granted.[9] The Minister of Health announced a two-year ban on all GMO releases, and although the Federal Chancellor lifted it immediately, Austria has not seen any deliberate releases yet.

The companies' as well as the authorities' way of handling the issue added steam to the rising debate. Protesters exploited the political momentum and launched a people's initiative to ban genetically modified food, the release of GMOs and patents on genes.[10] There were hardly any protests against medical and basic research applications. The debate over whether genetically altered soya and maize could be imported (which EU law allowed), and how they should be labelled, added a European dimension, and Austria succeeded in maintaining its embargo on genetically modified maize. In 1997, after an emotional campaign supported by the tabloid press, the people's initiative became the second most successful in Austrian history, with about 1.23 million signatories.[11] The government came under pressure from two sides: on the national level NGOs supporting the people's initiative reproached it for following EU regulations against the people's will, and industry and the research community kept warning that Austria would fall behind in science and technology.

In a revision of the law, the government responded to some of the critics' concerns (but not those addressed in the people's initiative). Subsequent deliberate-release applications were again withdrawn after NGO actions.[12] In order to comply with domestic needs, Austrian representatives pushed for a more restrictive biotechnology policy on the EU level. After the crisis surrounding BSE,[13] the EC and some member states showed greater awareness of public opinion and adopted a broader definition of risk, as well as closer scrutiny of new products. The EU policy had thus slowly drifted in a direction that could be interpreted as acknowledging some elements of the Austrian stance.

Taken together, we can identify the following phases of biotechnology policy:

- The pre-history from Asilomar up to 1985 saw no attention given to biotechnology, due to delays in economic and scientific development and the pattern of technology import.

- The 'dawn' of 1985–90 is characterised by the first initiatives to support biotechnology research and by emerging regulatory debates, but little public debate.

- The 1991–94 period saw the opening and closure of the debate: future EU membership forced regulation; and the Health Ministry allowed initial broad participation in negotiations for a law. In 1992, the Parliamentary Inquiry Commission adopted a critical view and a conflict arose between Parliament and the Ministry, as the latter turned to traditional patterns of closed negotiations in the final phase of the regulatory process. There was still little public debate, though surveys showed public unease for the first time.

- Then 1995–97 brought an escalation as a consequence of the first release applications: NGOs discovered the issue and an illegal release provoked public outrage and debate, finally resulting in a successful people's initiative. The government found itself caught between EU regulations and public pressure. While the use of biotechnology in medicine was accepted, food and agricultural applications were hotly contested.

The fact that the evolution of the debate is not fully reflected by our profile (see Figure 1) is due to characteristics of the Austrian political system such as the tradition of negotiations ('social partnership') behind closed doors, and the lack of a strong tradition of committees, boards or other formal bodies, or of non-statutory regulation. Events during 1991–94 might suggest strong public awareness of biotechnology, but this was not the case at all. In fact, public awareness strengthened only in 1996 following the first release proposals, and peaked before the people's initiative in 1997.

With regard to its key issues, the first years of the debate were dominated by the dichotomy of classical attitudes to technology: as means of modernisation, and as a matter of ethical and ecological concern. Since the first release proposals for GMOs in late 1995, public attention in Austria has shifted towards genetically modified food. This has definitely been influenced by the BSE scandal, and by the debates on genetically modified soya and maize within the EU. Austrian politicians have also been facing increasing discontent since Austria joined the EU. One of the dominant questions concerns the space for national state actions within the EU. One of the crucial points already in the debate over membership was (under the question of environmental standards) the situation in Austrian agriculture. Since the debate about genetically altered food has broadened, general distrust in industrialised agriculture has grown remarkably, and organic agriculture has looked like an economically feasible market-niche alternative.

Media coverage

We selected the weekly *profil* and the daily *Die Presse* to be scanned for articles on biotechnology. These influential Austrian newspapers are both opinion leaders, and both show continuous reporting on biotechnology since the 1970s. *Die Presse* and *profil* also represent different poles of the Austrian ideological left–right spectrum, which allowed us to cover some political plurality.

The intensity of media coverage shows an almost exponential development (see Figure 2). From 1973 until the early 1990s, there was only a slight increase in the number of articles. With 1997 set as a 100% baseline, intensity oscillates around 6% during the 1970s and around 15% during the 1980s. From around 1994 onwards, there is a significant rise in media reporting about biotechnology.[14] This increase in intensity is accompanied by various changes: in the thematic structure, the predominant actors, and the general framing of biotechnology reporting. All these changes fit with a development characterised by a late and massive politicisation in Austria (see Figure 3).

Some topics covered remain more or less stable over time: basic research and medicine, for example,

figure in narratives of progress over the whole period. Similarly, ethical considerations accompany reporting and commenting on biotechnology from its beginnings until today. By contrast, politics came on stage rather late, in the early 1990s. It is characterised by a broadening of conflicting perspectives on biotechnology, which eventually become dominant. The evolution of the media discourse can be divided into five periods which have distinct patterns of content and intensity (see Table 1).

Phase 1, 1973–76: The early history of biotechnology in the Austrian media. There were very few news articles mentioning biotechnology. Their predominant themes were narrow in scope. Medicine and basic research were discussed as promising fields of progress. However, even at this early stage some voices expressed moral concern.

Phase 2, 1977–84: First diversification. A slight increase in intensity was accompanied by a thematic diversification. Economic and political topics were added to the hitherto dominant scientific and medical ones. Political actors (e.g. the government) were mentioned more frequently, but as yet there was no indication of political controversy. Reports on biotechnology rarely referred to Austria itself. Foreign countries, particularly the USA, were the main site of events. Biotechnology was still depicted in positive terms even if some criticisms were voiced. Health and environmental risks were mentioned in an ethical context.

Phase 3, 1985–90: The economic perspective. During this period the economic perspective on biotechnology became pre-eminent. The economic frame climaxed with biotechnology being constructed as an object of high economic potential. Accordingly actors from the sphere of economics and industry featured centre-stage.

Political topics become more diversified than before, because of the new regulatory problems. Political argument peaked in the third quarter of the 1980s. As actors from the civil society, like NGOs, came on stage, biotechnology in the context of democracy became an issue. Also, biotechnology reports were referring increasingly to Austria and less to the USA.

While not yet revealing the massive entry of political argument into the media discourse, this phase prepared the ground for the 1990s. Political discourse in media coverage was neither as heated as would be expected in an actual political debate, nor did we find a high portion of articles of controversial content. Biotechnology was still evaluated positively, albeit decreasingly so. Sceptical articles added objective risks for health and the environment to their arguments, which finally superseded the ever-present ethical concerns.

Phase 4, 1991–94: Politics centre-stage. This phase saw a clear beginning of the truly political debate on biotechnology. Its characteristics were a maximum of texts concerning regulation, out-weighing even progress in medicine and basic research, which – unlike economics – were always relevant. In relation to politics the significance of economic arguments decreased sharply. Ten per cent of biotechnology news concerned implications for democracy (see Figure 3). Political actors gained further importance and for the first time constituted the same proportion of actors as those from science. Also, almost 50% of articles referred to Austrian issues – a maximum since the inception of the discourse.

The fact that 'biotechnology policy' now meant dispute and conflict, rather than mere administration and governing, was clearly indicated in the number of controversial topics and articles. Besides the government, political parties now played an important role as political actors. Media coverage of biotechnology increased sharply until 1997, as did the negative tone of the reporting.

Phase 5, 1995–97: Political controversy on biotechnology. The last phase was entirely dedicated to political conflict on biotechnology. The trends in the preceding phase continued and intensified, as did the participation of civil actors and political parties in the dispute. Intensity quadrupled compared to its former level. The frame 'democracy' became the most dominant. The risk debate was enriched by the 'risk to the consumer' argument.

In fact, this last phase coincided with the illegal field release of genetically modified potatoes by a private enterprise in spring 1996. The media and politics were driven by enormous public pressure which culminated in the people's initiative against certain aspects of gene technology in April 1997. The media – particularly the tabloid press – played a major role in the mobilisation of opponents of biotechnology. After this event, biotechnology remained on the public agenda until the end of our media sample.

The quarter-century covered by the media content analysis was characterised by a number of indicators which illustrate its specific evolution. After long phases of relatively little media awareness – indicated by low intensity during the 1970s and 1980s – the debate gradually became strongly political during the 1990s and finally burst in a heated media hype in 1996 and 1997.

Given that the political process began several years earlier, it is still surprising, however, that politics and its related discourse peaked so late in Austria. Two crucial political events, the (unique, for Austria) establishment of a Parliamentary Inquiry Commission on biotechnology in 1992 and the passing of the Genetic Engineering Act in 1994

were merely beginnings whose relevance was not recognised by the media or the public. Only when faced with proposals for experimental releases of genetically modified crops in Austria in 1995–96, and when shortly afterwards the question of soyabeans and maize imports emerged, did the public and the media force politicians to react.

Public perceptions

According to our survey, Austrians are, on average, the most critical of modern biotechnology in Europe. Whatever the specific reasons for this resistance – a general scepticism towards new technologies, historical experiences, and so on – it is embedded in a unique pattern of attitudes and social characteristics when compared to other European nations. Austrians had a very high level of contact with the topic at the time of the survey (November 1996): 79.1% had either heard or talked about biotechnology.[15] This was most likely due to the high degree of attention the media paid to release experiments of genetically modified crop plants and the European policy towards modified maize and soyabean imports in the autumn of 1996.

Survey results: attitudes

Figure 4 shows the effects of contact with the topic of biotechnology upon optimistic vs pessimistic attitudes. Respondents are called 'optimist' if they agreed that biotechnology would have a positive effect upon life in the future and 'pessimist' if they believed the opposite. More of the people who reported contact were either optimistic or pessimistic; fewer of them responded 'don't know'. Contact, then, is accompanied by familiarity with the issue and polarises Austrians' attitudes. This polarisation is more pronounced with women than with men.

Contrary to expectations, the high exposure reported by Austrians in 1996 does not mean greater factual knowledge about biology. Together with Portugal, Greece, Spain and Ireland, Austria belongs to the lowest third of all European nations in terms of basic knowledge. Compared to a similar survey in 1994, the Austrian knowledge score has increased only very slightly (1994, 2.49; 1996, 2.55 on a five-point scale of knowledge items).[16] A question asking whether only genetically altered tomatoes have genes and naturally grown ones do not, or whether both have genes, was answered correctly by 33.5%, which is slightly less than the European average (35.1%). Denying genes to naturally-grown fruit and plants reflects a non-factual understanding of biological processes and gene technologies. If genes are perceived as not to be part of nature, any genetically manipulated organism is contaminated by something artificial; just as chemicals may contaminate our food, technologically introduced genes are seen as contaminating natural organisms. This social representation[17] is further corroborated by analyses of interviews and focus groups. Newspaper headlines reading 'Keep Austria Gene Free' during the mobilisation phase of the 'Anti Gene Technology' people's initiative in spring 1997 come as no surprise.[18]

Figure 4 also shows the polarisation effect of knowledge about biology. Fewer respondents with some basic knowledge resort to 'don't know' than do respondents with low knowledge, and instead they join the ranks of either the optimists or pessimists. More knowledge does not simply increase technological optimism: instead, it has a polarising effect.

Resistance to biotechnology in Austria is fuelled by two sources. This is shown by the 'natural experiment' of two people's initiatives which were fielded simultaneously during the spring of 1997. One initiative was to ban mainly agricultural applications of gene technology in Austria, and the other concerned women's rights. The former initiative was signed by 1.23 million people; the latter by 645,000. Approximately 43% of all signers signed both initiatives. Those 57% who signed the gene technology initiative but not the women's rights initiative mostly supported conservative parties such as ÖVP and FPÖ. Those who signed both initiatives mostly supported the Ecological (Green) Party or the Liberal Party (LIF). Potential Social Democratic voters were divided about these issues.[19] This segmentation roughly indicates that opponents to biotechnology are split into two halves: one associated with a more conservative political approach, and the other with a more progressive political outlook. This pattern resembles findings about a 'green' (environmental) and a 'blue' (conservative – 'black' in the Austrian case) form of resistance to new technology.[20]

In an open question respondents had to spontaneously produce ideas, words or sentences which came to mind when thinking about biotechnology. Whereas pessimists' responses were in general clearly negative, optimists were split into roughly equal thirds, producing negative, positive or neutral responses. More pessimists than optimists made moral statements (i.e. their utterances contained prescriptions and warnings), more of them expressed fear, and more referred to an idealised image of nature which should not be spoiled (see Tables 1 and 2).

Respondents with less factual knowledge tended to produce rather more ideas with a moral component (18.5%) than those with a higher level of knowledge (13.9%). These latter respondents referred more to concrete applications of biotechnology in medicine (11.1% vs 17.2%), food production (16.3% vs 25.3%) or agriculture (6.3% vs 11.7%). They also talked more about the economic consequences of biotechnology (6.7% vs 11.3%).

When it comes to specific applications of biotechnology, Austrians, like people in other countries, encouraged medical applications (production of medicines, genetic testing) significantly more than food and agricultural applications such as crop plants (see Figure 5).

The Austrians' attitudes towards these applications were strongly determined by usefulness and moral acceptability in a series of multiple regressions (average $\beta_{usefulness}$ = 0.38, average β_{moral} = 0.54). The more an application is seen as useful and the more it appears morally acceptable, the more people tend to encourage it. There is virtually no influence of risk perception upon encouragement.

A path analysis[21] reveals that the more the Austrians perceived the risks of biotechnology, the more they preferred traditional breeding methods (0.28) and the less they shared in a general technological optimism (–0.26). On the other hand, biotechnology was more morally acceptable to respondents with more factual knowledge (0.31), to those expecting more positive consequences (0.37), and to those who were more relaxed about its control (0.28).

More Austrians prefer traditional breeding methods to genetic manipulation than any other Europeans (yes, 69.2%; no, 17.0%; don't know, 13.8%). They think that the hereditary characteristics of plants and animals should be produced in the traditional way, i.e. by selective breeding. This preference holds true even for 68.0% of the 28.8% of Austrians who consider traditional breeding methods to be less effective than biotechnological methods (as effective, 47.9%; less effective, 28.8%; don't know, 23.3%).

Survey results: regulation, trust and media

In general, Austrians have the strongest opinions in Europe, i.e. the fewest 'don't know' responses (12.0% vs 24.3% for Europe) about regulating biotechnology. More than in any other country they think that current regulations are insufficient 'to protect people from any risks linked to modern biotechnology' (67.3% vs 51.81% for Europe). Accordingly, only 16.6% (Europe, 27.4%) 'would buy genetically modified fruits if they tasted better.' Some 69.7% (Europe, 57.5%) say that they would not buy such fruit.

Austrian pessimists trust both national public bodies (22.3%) and international organisations, such as the UN and WHO (27.9%). Optimists clearly prefer international bodies (national control, 10.6%; international control, 37.8%). On average, trust in national public bodies (16.9%) is higher than in Europe (11.1%). Fewer Austrians (12.3%) than other Europeans (22.3%) would entrust regulation to scientific organisations. Like other Europeans, however, only a minority would entrust regulation to the EU and its public bodies (5.9%).

When asked about whose information about biotechnological issues to trust, more Austrians than other Europeans would rely on consumer organisations (22.4% vs 19.2% for Europe) and environmental organisations (19.4% vs 16.4% for Europe). Less trusted than these institutions (but still enjoying more trust than in the rest of Europe) are national (7.1% vs 4.0% for Europe) and international public bodies (6.5% vs 4.3% for Europe). There is comparatively less trust in information provided by universities (6.9% vs 11.% for Europe) and by the media (0.6% vs 4.84% for Europe). In this respect Austria resembles Germany, France and Denmark.

However, these findings must be qualified by the degree of contact with the issue, i.e. having read and talked about biotechnology, and by optimism vs pessimism: people who are at least minimally acquainted with the topic tend to trust consumer organisations (39.4%) more than unacquainted respondents (29.5%), who either don't know (16.2% vs 6.3%) or have more trust in public authorities (9.5%) than in consumer groups (4.5%). Likewise, optimists would rely more on consumer groups (43.7%) than on environmental organisations (24.1%), both of which pessimists trust about equally (32.2% and 31.5%).

Another consequence of Austrians' short-term exposure is their significant lack of opinion, positive or negative, about potential consequences of biotechnology. Their rate of 'don't know' responses tops the European chart with an average of 18.5% compared to an average rate of 11.1%. Only Greeks, Portuguese and Spaniards equal or top the Austrians' indecision, in a few areas. Biotechnology is an unfamiliar issue for Austrians which is seen to offer no new perspectives. More importantly, Austrians do not perceive clear threats either. This comes as a surprise given their strong rejection and high recent exposure to media reports.

Despite the Austrians' lack of trust in the newspapers (only 0.6% trust the media), there is some association between newspapers and their readerships' exposure to biotechnology, their average level of basic biological knowledge and their optimism vs pessimism regarding biotechnology's impact upon future life. On the one hand there is the group of readers of tabloid newspapers[22] who are more pessimistic (pessimists exceed optimists by more than 10 percentage points), a balanced mix of less and more biological knowledge (the percentages of people with low and high knowledge do not differ by more than 10 percentage points), and of whom more than 10% report never having read or talked about biotechnology. On the other hand, there is the group of higher-quality papers[23] whose readers are mostly optimistic, have higher knowledge, and of whom 90% report exposure to the topic.

It seems that it is not the quality press but the tabloid and similar newspapers that attract more opponents of biotechnology.

Survey results: Austria compared to other countries

The pattern of public attitudes and beliefs about biotechnology in Austria cuts across the patterns of other European countries. It combines a uniquely high negative evaluation with high exposure to the topic and little factual knowledge about biology. In these respects Austria resembles Germany. The reasons for this situation can be found in the recent history of how politicians have dealt with biotechnology, and the late but massive response of the media in general and the overwhelmingly negative publicity for biotechnology in the tabloid press in particular.

Like most other Europeans, Austrians maintain a more positive attitude towards medical applications than to applications in food production and agriculture. Their moral concerns far outweigh questions of risk and utility. However, medical applications – at least those known at the time of the survey – did not intend to cross the species barrier between and within the plant and animal kingdoms, as do some agricultural applications. On the one hand, medical applications are obviously useful to respondents: everyone gets sick, and the prevention and treatment of disease offer personal benefits. On the other hand, food and crop applications are far-fetched and do not seem to bear on individual well-being. Lacking personal relevance, genetically modified food appears superfluous, and its possibly improved quality seems highly dubious.

People with access to more information and more knowledge about biotechnology are not automatically supporters. They are more certain of their opinions, which may be negative or positive. More supporters tend to be male and educated above average. More opponents tend to be female and/or less well educated. Regulation is widely seen as insufficient, and there are few trusted information sources.

Commentary

According to the survey data, Austrian public perceptions of biotechnology are unique in Europe. Low support, low expectations and a 'menacing' image are combined with comparatively little factual knowledge. Austria is among the European countries where particularly agricultural applications of biotechnology receive the least support, and where the general image of biotechnology, or, rather, gene technology, is least favourable. Recent political events – a popular people's initiative against agricultural biotechnology and the Austrian import ban on maize at the EU level – are in line with these attitudes. This unique situation deserves an explanation. What is so peculiar about Austria that its people are more sceptical towards biotechnology than those in other countries? Which factors have contributed to the Austrian situation?

Explanations may lie in apparently unrelated historical and cultural factors. The early phases of the history of biotechnology policy in Austria were not so different from those in other countries. However, a general feature was the delay both in policy-making and especially in public debate. In the 1980s, biotechnology at first appeared as a proxy for cutting-edge high technology, but soon became controversial through its perceived association with reproductive technology. Neither image, however, reached the public, and press coverage remained insignificant.

In the early 1990s, the question of how to regulate biotechnology, if at all, was ultimately resolved by the wish to join the EU, which meant that Austria had to adopt the EU directives. The law-making process put the issue on the political agenda in the early 1990s. Biotechnology became the subject of some cautious and temporary political experiments, but was still of little interest to the public.

From a situation with hardly any public debate, biotechnology suddenly became a major political issue because of the first release applications, and public awareness grew. Soon, new food products coming in from the EU triggered a broad public debate. The subsequent people's initiative was considered very successful in terms of votes, but not politically. The issue remained on the public agenda and led to a policy gridlock. So we are left with two questions: **1** Where do the particular underlying attitudes come from? and **2** What led to the turn of events of 1996?

The source of 'Austrian attitudes'

It is obvious that public attitudes played a role in these events. However, we cannot explain the turn in politics by a fundamental turn in underlying attitudes leading to a sudden rejection. The previous Eurobarometer survey carried out in 1994 in Austria showed a rather similar picture then of low acceptance and low factual knowledge.[24] This implies that what we have observed is predominantly caused by a higher awareness which, however, was not accompanied by higher knowledge. Following the 'deficit' model of public understanding of science – that a lack of factual knowledge leads to a lack of acceptance – one might conclude that there had been a gut rejection.

However, a lack of factual knowledge cannot be *the* decisive factor, since support for biotechnology appears to be higher in countries with a similarly low level of knowledge. Also, the Austrian data show that

people with a higher factual knowledge have a more pronounced opinion and answer 'don't know' less often, but that nevertheless there are a considerable number of people sharply opposed to biotechnology especially in agriculture and applied to animals. Since in 1996 Austrians were considerably more aware of the power of modern genetic methods compared to traditional breeding than in 1994, we may speculate that there has been a learning process, not so much in terms of factual knowledge, but in terms of practical relevance. This is not linked to acceptance, since many respondents did not approve of these methods.

In conclusion, although knowledge may be important, we have to search for other factors that might have contributed to the particular Austrian set of attitudes, which refer to recent Austrian political and economic history.

1 Austria is lagging behind in general technology awareness. Technology has been on the government agenda several times, but has never become an issue of real importance.

2 Austrians feel particularly stricken by globalisation effects, and react in a technology-averse way. During the 1980s, there were economic developments with heavy impacts. For example, the formerly strong public sector with its unprofitable heavy industries got into trouble and, after investing a lot of public money, the government cut the public sector and privatised what was left. When the Iron Curtain was lifted in 1989, Austria found itself bordered on one side by countries in turmoil. It faced the threat of domestic companies leaving to take advantage of the low wages across the border, and of immigration of cheap labour that might contribute to rising (but still low) unemployment. Finally, EU membership became necessary because of the economic challenges. In a 1994 referendum, two-thirds of the Austrian population chose EU membership, but initial support had faded a year later as the promised cornucopia of cheaper goods and more jobs did not materialise.

3 Austrians' trust in their government and state institutions is decreasing. Apart from distinct events which clearly undermined this trust, we also find the phenomenon of 'weariness' of politics (*Politikver-drossenheit*), as in other West European countries. New social movements and demands for participation challenged the political systems, but the politicians reacted inadequately.

4 The influence of NGOs is particularly strong in Austria. Since the referendum on nuclear power and the successful protest against a hydro-electric power plant which gave rise to the Green Party, there has been a lively Austrian NGO scene which has succeeded in exerting some political influence.

NGOs are particularly active in environmental and siting issues, animal welfare and agriculture, but are less concerned with traditional consumer issues. However, up to the mid 1990s biotechnology was not really an issue. Although there had been a clear negative position within the Green Party, and there was a tiny organisation devoted solely to questions of biotechnology from the late 1980s, biotechnology was of little relevance to the more important NGOs. The issue only experienced a sudden upturn with the first release applications – there was an incident, and the topic was suddenly on the agenda.

Triggers of the events of 1996

In 1995, the Green Party and some big NGOs faced a lack of 'big issues', and were in need of such a topic around which to organise their efforts. Attempts to build on past successful campaigns had failed. The aim was to emphasise a particular problem as a means both of stating a clear identity and of stimulating public awareness of the environmental movement. The long-standing but hitherto underrated topic of biotechnology was finally chosen as the main focus of activity for 1996.

The first release applications were handled in an unprofessional way, giving rise to a media hype supporting NGO activism. The first application was commissioned by the Health Ministry itself, which the public felt was politically improper. Later, a company did not wait for the decision on its planned release, and went ahead without formal permission. This made good copy for the tabloids, and opened up space for coverage of biotechnology. Austria's biggest tabloid newspaper (*Kronenzeitung*) is very influential and reaches over 40% of the population. It has shown a sympathy for green issues on several occasions; for example, it strongly supported the movement against the hydro-electric power plant in 1984. Its support for the people's initiative undoubtedly contributed to its success. However, the choice of topics by a mass medium is not least a question of

adaptation to the readership's interests. Whether *Kronenzeitung* served or shaped these interests, the arguments fell on fertile ground.

The first Austrian release applications gave rise to a debate that was kept alive by the soya and maize cases coming up shortly afterwards. During the debate on releases in Austria the agenda was set: consumer issues and questions of general agricultural orientation were to the fore. These topics were pursued in the subsequent debates around soya and maize. Now aware of consumer and agricultural issues, Austrian public opinion forced the government to adopt a technology-critical role at the EU level, which in turn was in line with the anti-EU feeling in Austria. On the other hand, Austria appeared as a kind of technology-critical avant-garde, which in turn fitted in with its role in questioning environmental standards and especially with its anti-nuclear energy policy. In the aftermath of the BSE crisis, the EU was more open to demands for tighter regulation from various sources, meeting concerns that were also emphasised by Austria.

Summary

Shedding some light on the Austrian situation reveals a complex, multifaceted problem, both with respect to the sources of the underlying attitudes as well as to the triggers of recent events. The Austrian case may show that there is no such thing as a simple explanation for a particular set of attitudes towards biotechnology. Lack of factual knowledge cannot be held responsible for a 'lack of acceptance', nor does an increase in public awareness and debate automatically lead to an increase in acceptance. If we adopt a broader cultural perspective, the Austrian data may tell us something about general attitudes with regard to technology, agriculture and the kind of modernity that is seen to be worth pursuing in this country. If we stick to the simple question of which factor(s) account for an observed 'lack of acceptance', we may not get too useful an answer.

Acknowledgements

This research was funded by the Austrian Fonds zur Föderung der Wissenschaftlichen Forschung, number P11849-SOZ, granted to Wolfgang Wagner. Part of Franz Seifert's research was supported by the Institute for Techology Assessment, Austrian Academy of Sciences.

Figure 1. A regulatory profile for Austria

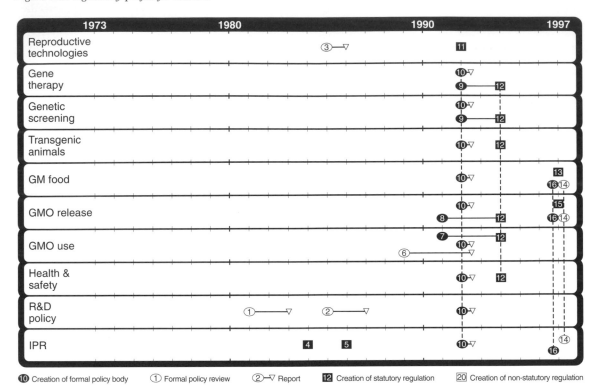

1 *1981–83*. Taskforce (Commissioned by the Federal Ministry of Science and Research) to take stock of the Austrian potential of biotechnology – report taken as the basis for new research priority: biotechnology and genetic engineering.

2 *1985–87*. Taskforce (commissioned by the Federal Ministry of Science and Research), named Biotechnological Concept. Its goal was to define future priorities for biotechnology R&D.

3 *1985–86*. Commission of the Conference of the Austrian Rectors. Report to Parliament – August 1986.

4 *1984*. Amendment to the Patent Act.

5 *1986*. Amendment to the Patent Act.

6 *1989–92*. Commission of the Austrian Academy of Sciences to translate the NIH Guidelines.

7 *1991–94*. Interministerial working group on industrial production.

8 *1991–94*. Interministerial working group on deliberate release.

9 *1992–94*. Interministerial working group on gene analysis.

10 *1992*. Parliamentary Enquiry Commission technology assessment on the example of genetic engineering.

11 *1992*. Act on Artificial Procreation.

12 *1994*. Genetic Engineering Act.

13 *1997*. Decree on novel food (anticipation of novel food directive).

14 *1997*. People's initiative.

15 *1997*. Ban on the import of genetically altered maize.

16 *1997*. Special Advisory Board concerning the People's Initiative on Genetic Engineering (September 1997 to April 1998).

Figure 2. Intensity of articles on biotechnology in the Austrian press

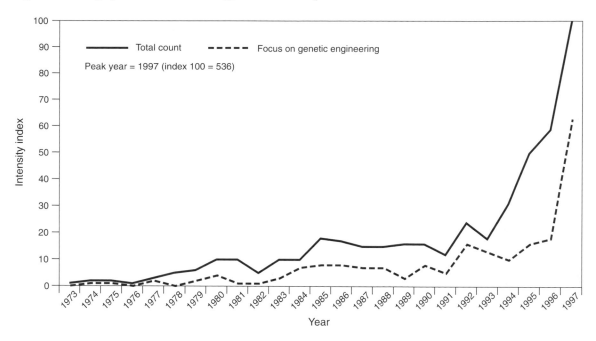

Figure 3. The development of frames in the media coverage of biotechnology in Austria

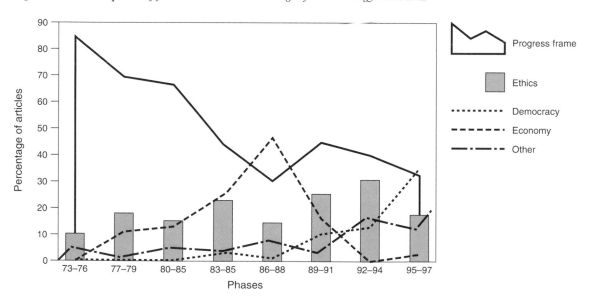

Figure 4. Attitudes to biotechnology, 1996

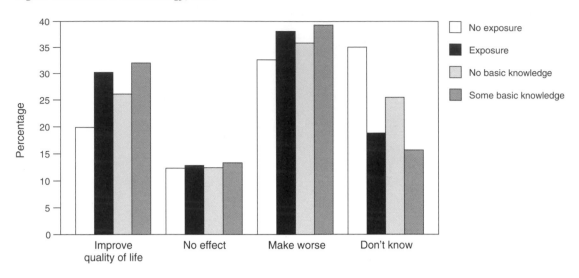

Figure 5. Attitudes to applications of biotechnology in Austria, 1996

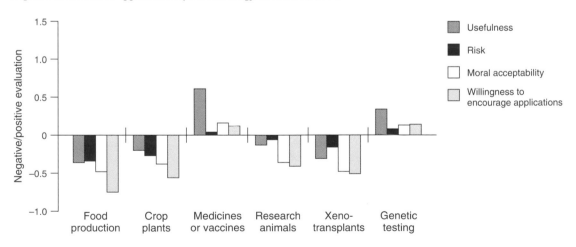

Table 1. Austrian media profile[a] (for an explanation of terminology please see Appendix 5)

Phase	Freq.[b] (%)	Frame (%)		Theme (%)		Actor (%)		Benefit/risk (%)		Location (%)	
1. 1973–76	2	**Progress**[c,d]	85	*Medical*	35	*Scientific*	70	*Benefit*	76	*Austria*	50
		Ethical	9	*Basic research*	32	Political	20	Neither	15	USA	25
		Nature/nurture	5	*Ethical*	12	Industry	10	Both	10	Other Europe	13
										Other countries	13
2. 1977–84	17	**Progress**	59	**Medical**	21	**Scientific**	62	**Benefit**	54	**Austria**	31
		Ethical	19	Basic research	15	Political	19	Neither	18	*USA*	28
		Economic	16	Economic	13	Industry	11	*Both*	17	Other Europe	18
				Ethical	10			Risk	11	*Germany*	9
				Political	7					World	10
3. 1985–90	27	**Progress**	36	**Medical**	18	**Scientific**	42	**Benefit**	65	**Austria**	39
		Economic	35	*Economic*	14	Political	23	Both	17	USA	18
		Ethical	20	Basic research	13	*Industry*	17	Neither	9	Other Europe	19
		Pandora's Box	5	*Political*	13	*Ethical*	5	Risk	9	*Other countries*	14
				Regulation	10	*International*	5			*Germany*	9
				Ethical	8	EU	5				
				Animals	5						
4. 1991–94	24	**Progress**	43	*Regulation*	32	**Scientific**	36	**Benefit**	41	**Austria**	48
		Ethical	30	Medical	19	*Political*	30	Neither	30	Other Europe	16
		Accountability	13	*Security & risk*	9	Industry	10	*Risk*	21	USA	15
		Globalisation	3	Ethical	7	*NGOs*	8	Both	9	*World*	13
		Runaway	6	Political	7	*Media/public*	7				
		Nature/nurture	4	Basic research	6	EU	5				
				Identification	4						
5. 1995–96	30	**Progress**	45	**Basic research**	24	**Scientific**	46	**Benefit**	59	**Austria**	38
		Ethical	23	Medical	21	Political	26	Neither	26	*Other Europe*	27
		Nature/nurture	9	*Regulation*	10	EU	7	Risk	9	USA	20
		Accountability	9	Political	9	Media/public	6	Both	7	World	7
		Runaway	5	*Agriculture*	7	NGOs	6				
		Globalisation	5	Economic	6						
				Security & risk	5						

a For reasons of comparability, the media profile covers the period 1973–96.
b Percentage of corpus in the period; total $n = 302$.
c Bold indicates highest frequency within phase.
d Italics indicates highest frequency within category.

Table 2. Categorisation of utterances by optimists and pessimists in Austria

	Pessimists (%)	Optimists (%)
Negative tone	55.8	29.3
Ambivalent tone	11.8	11.3
Positive tone	2.7	20.8
Neutral tone	18.0	21.9
Moral tone	22.3	12.0
Expression of fears	29.0	12.4
Mention of 'nature'	18.0	11.7

Notes and references

1 Gottweis, H, 'Biotechnologiepolitik', and Gottweis, H and Latzer, M, 'Technologiepolitik', both in Dachs, H, Gerlich, P, Gottweis, H, Horner, F, Kramer, H, Lauber, V, Müller, W C and Tálos, E (eds), *Handbuch des politischen Systems Österreichs* (Wien: Manz, 1997).

2 Ernst & Young, 'European Biotech 97: A New Economy', The Fourth Ernst & Young Report on the European Biotechnology Industry (Stuttgart: Ernst & Young, 1997)

3 Hofreither, M F, 'Agrarpolitik', in Dachs, H, Gerlich, P, Gottweis, H, Horner, F, Kramer, H, Lauber, V, Müller, W C and Tálos, E (eds), *Handbuch des politischen Systems Österreichs* (Wien: Manz, 1997).

4 *Independent* (London), 12 April 1996.

5 Gottweis, H, 'Modernization conflicts, new democratic culture and shifts of power in Austrian politics', *German Politics and Society*, 21 (1990), pp48–62.

6 Gerlich, P, 'Political culture,' and Müller, W C, 'Political parties', both in Lauber, V (ed.), *Contemporary Austrian Politics* (Boulder, CO: Westview Press).

7 Tálos, E, 'Corporatism – the Austrian model', in Lauber, V (ed.), *Contemporary Austrian Politics* (Boulder, CO: Westview Press).

8 Seifert, F and Torgersen, H, 'How to keep out what we don't want. An assessment of 'Sozialverträglichkeit' under the Austrian Genetic Engineering Act', *Public Understanding of Science*, 6 (1997), pp301–27.

9 The experimental release of a potato with modified starch was announced under the heading of 'renewable resources'. After an anonymous report to the authorities, the company had to collect up the seed tubers after one week.

10 A people's initiative (*Volksbegehren*) is a constitutional tool of direct democracy. It is a way of putting a bill before Parliament if it is supported by at least 100,000 voters; supporters have to sign the initiative in the presence of the authorities and all the signatures must be collected within one week. Parliament is then obliged to deal with the proposal, although there is no legal requirement beyond this. Between 1963 and 1994, 14 people's initiatives were launched and succeeded in raising the requisite number of signatures; most of them were organised or supported by one of the opposition parties. Three of them were successful in leading to the enactment of a law that met their demands.

11 In parallel, there was another people's initiative on women's rights, which attracted half as many supporters.

12 Grabner, P and Torgersen, H, 'Österreichs Gentechnikpolitik – Technikkritische Vorreiterrolle oder Modernisierungsverweigerung', *Österreichische Zeitschrift für Politikwissenschaft*, 1 (1998), pp5–27.

13 The scandal surrounding BSE in British livestock elicited deep concerns among European consumers, and also had major impacts on the beef market in Austria.

14 Note that part of this increasing number of articles in the 1990s may be due to more effective scanning methods by using on-line access.

15 Note that most of the questionnaire referred to 'modern biotechnology'. This term is often understood as signifying 'green', 'biological' or 'natural' methods of farming in Austria. Hence it has a traditionally more positive ring to it than the term 'genetic engineering'. Therefore all figures reported here are likely to be influenced by the 'bio' connotation.

16 We do not uphold, as does the so-called 'enlightenment' or 'public understanding of science' model, that the level of scholarly knowledge in the public provides a measure of competence on complex technological issues. For this model see Durant, J R, Evans, G A and Thomas, G P, 'The public understanding of science', *Nature*, 340 (1989), pp11–14. Public competence, or ignorance for that matter, cannot be easily assessed; however, it is surely not only reflected in factual knowledge about a technological development, but also and more importantly in the way the public is able to cope with new technologies in everyday life. See Irwin, A and Wynne, B, *Misunderstanding Science? The Public Reconstruction of Science and Technology* (Cambridge: Cambridge University Press, 1996).

17 Wagner, W, *Alltagsdiskurs – Die Theorie sozialer Repräsentationen* (Göttingen: Hogrefe, 1994).

18 Headlines like this one do not necessarily mean that the journalists responsible for them are ignorant about biology. However, among the less-informed public such messages confirm the widespread belief that genes are foreign to organisms.

19 Hofinger, C and Ogris, G, *Wählerstromanalyse des Gentechnik- und Frauenvolksbegehrens 1997* (Vienna: Institute for Social Research and Analysis, 1997).

20 Nielsen, T H, 'Behind the colour code of "no" ', *Nature Biotechnology*, 15(7) (1997), pp1320–1.

21 A LISREL structural equation model including the exogenous variables education level, sex, preference for traditional breeding, general technological optimism, age, trust in public institutions, importance of the issue, environmental concern, knowledge, religiousness, optimism in biotechnology and trust in NGOs, and the endogenous variables positive expectations from biotechnology, desire for control, level of contact with the issue, risk perception and moral acceptability. Encouragement could not be included in this model because of co-linearities. The figures reported here are total effects on risk and moral acceptability.

22 Such as *Kronenzeitung, Kurier, Kleine Zeitung, Täglich Alles* and *Die Ganze Woche*.

23 Such as *Salzburger Nachrichten, Der Standard, Die Presse, Oberösterreichische Nachrichten* and the weekly *profil*.

24 Torgersen, H and Seifert, F, 'Aversion preceding rejection: results of the Eurobarometer survey 39.1 on biotechnology and genetic engineering in Austria', *Public Understanding of Science*, 6 (1997), pp1–12.

Address for correspondence

Wolfgang Wagner, Department of Social and Economic Psychology, University of Linz, A-4040 Linz, Austria.

Denmark

Erling Jelsøe, Jesper Lassen, Arne Thing Mortensen, Helle Frederiksen
and Mercy Wambui Kamara

The history of Danish biotechnology

Economically and structurally, agriculture was for a long time the most important production sector in Denmark. Industrialisation came relatively late, starting only in the second half of the nineteenth century. The economic crisis in the 1870s and 1880s meant a transition in Danish agriculture from vegetable to animal products, and Denmark's producer cooperatives subsequently established strong positions in the world market, especially in the dairy and meat industries. Before the turn of the twentieth century, a strong technology base had developed in these sectors. Food production became one of the most important industries in Denmark, and included not only dairy and meat production, but also sugar factories, breweries and distilleries. The sugar factories and distilleries merged in 1989 to form Danisco, which is the leading Danish company in plant biotechnology.

A number of biotechnology companies were established on the basis of the agro-industrial development. The Carlsberg Breweries established its own laboratories very early on, and began a significant research effort that is still important for Danish research in plant biotechnology and brewing processes. Christian Hansens Laboratories started using calf stomachs to make calf rennet for cheese in 1874, and has achieved a leading position in the world market for rennet and starter cultures for food production, especially for the dairy industry.

In the 1920s when Banting and Best isolated insulin, newly founded Danish companies were among the first to initiate industrial production of insulin from pancreas. This was the origin of Nordisk Gentofte and Novo, which merged to form Novo Nordisk in 1990. In the postwar period Novo and Nordisk Gentofte expanded their expertise in and production of industrial enzymes and proteins for medical use, and during the 1980s they developed processes based on GMOs for producing insulin and other proteins.

Because of the late industrialisation and the small domestic market, Danish companies are comparatively small. Traditionally, industrial policy has been relatively general, providing infrastructure, a well-educated labour force, and so on. Large-scale, sector-oriented technology programmes, on the other hand, were not part of Danish industrial policy until the 1980s. After 1980, a number of such policy programmes were initiated, including three consecutive biotechnological programmes reflecting the relative importance of Danish biotechnological research and industry. A programme in food technology, including food biotechnology, was launched in 1991.

The Danish political system

The Danish political system is based on elections by proportional representation, and many parties are usually represented in the Danish Parliament. Since the 1920s the Social Democratic Party has been the largest party in Parliament, but no party, neither the Social Democrats nor any of the conservative parties, has had an absolute majority. Both social democratic and conservative governments have had to get support, and frequently partnership in government, from smaller parties. This parliamentary situation in which decisions are made through negotiation between parties has strengthened the role of consensus-seeking in Danish political culture.

In the 1980s, a particular situation arose in which a conservative minority government consisting of four parties was confronted with a 'green majority' of four other parties: the Social Democrats, the two parties to the left of the Social Democrats, and the small Radical Liberal Party. This 'green majority' used its influence from 1982 to 1988 to carry a number of decisions in Parliament against the wish of the government, especially regarding environmental and defence issues. It was also influential in relation to the Gene Technology Act in 1986.

The consensus-seeking element of Danish political culture has also influenced the policy processes of social interest groups. It has been expressed differently in different sectors: Danish labour-market policy has been characterised by strong traits of corporatism with three-part negotiations between the state, employers' organisations and labour organisations about the content of regulation as well as the administration of the rules. Similarly, in relation to agricultural policies, Danish farmers' organisations have been very influential. In areas such as environmental policy and physical planning, other practices based on public consultation with interest groups have been implemented. This is an important principle in the Danish Environmental Act, which was passed for the first time in 1971. The same principle was applied in the 1986 Act on Gene Technology and the Environment. The principle of public consultation implies a relative openness about the approval of production and products based on gene technology; this openness has also been maintained since the implementation in Danish legislation in 1991 of the EU directives on deliberate release and contained use.

Denmark joined the EC in 1972, even though there was widespread public scepticism about membership. This scepticism culminated in 1992 in the rejection in a referendum of Danish support for the Maastricht Treaty. During negotiations about the internal market, in the mid 1980s, EU sceptics argued that Danish environmental regulation would be undermined through the harmonisation. As a result of the negotiations the so-called 'environmental guarantee' was established, which is described in the following section of this chapter. Widespread scepticism towards the EC/EU was the background against which Denmark took part in the EC/EU policy processes related to new biotechnologies during the 1990s. Politically, there has been close parliamentary control of the Danish government's negotiations in the Council of Ministers through the Market Committee (since the beginning of the 1990s, the Europe Committee) in Parliament.

Since the end of the 1960s, several new grassroots organisations have emerged that focus on issues of environment, nuclear power and other technology-related issues. One of these was the environmental NGO, NOAH, which was formed in 1969. NOAH played a major role in the rise of the Danish environmental debate, and later, in the 1980s, it played a similarly important role in the debate about new biotechnologies. Another important NGO was the Organisation for Information about Nuclear Power (OOA), which was a driving force in the movement against nuclear power in the 1970s. This movement, and the popular resistance to nuclear power, eventually lead to the parliamentary decision in 1985 that nuclear power should not be part of Denmark's energy plans. Environmental controversies in the 1970s and early 1980s resulted, in several cases, in a lack of confidence in the environmental authorities, who had tried to conceal or understate environmental pollution. This experience of the nuclear and environmental debates contributed to the background for the biotechnology debate in the 1980s.

Public policy (Figure 1)

Six phases can be identified in the history of Danish policy-making and media coverage of biotechnology. These phases can be characterised as follows.

Phase 1, 1972–76. This was a phase of political quiet. The Asilomar conference did not provoke any immediate reaction in the Danish political system,

and after a short debate in the media, the biotechnology issue disappeared for some years.

Phase 2, 1977–83. This was a phase of non-statutory regulation. The first Danish policy initiatives were launched in 1976 in the shape of the Registration Committee concerning Genetic Engineering (RUGE), a non-governmental committee set up as a result of an initiative from the Danish Research Councils. Apart from serving as a voluntary registration and control body, RUGE aimed to follow international developments and to assess the need for regulation of biotechnology.

Around 1980, the first trials of IVF treatment at the national hospital in Copenhagen sparked a debate about reproductive and human gene technologies. Bjorn Elmquist, an MP from the Liberal Party, asked a number of questions of the Minister of the Interior, and subsequently made a proposal to the European Council about setting limits on the development of human gene technology.

Phase 3, 1984–86. The third phase was characterised by the shaping of the first Gene Act and the first industrial uses of manipulated organisms. The Minister of the Interior appointed the advisory group the Gene Splicing Committee in 1983, on the recommendation of RUGE and the National Technology Council.[1] The group was to consider the needs for public regulation of gene technology and, if considered necessary, to draft legislation. The result in May 1985 of this work on regulatory needs was a report that drafted no less than three new Acts.[2] However, because of conflicts within the conservative-led minority government, the group's report did not result in any immediate legislative initiatives.

Up until 1985, discussion about biotechnology in the Danish Parliament had been very sporadic, and was focused on ethical and human issues. This picture changed in 1985, after the two medical companies Novo and Nordisk Gentofte announced their plans to use GMOs for producing insulin and human growth hormone respectively. These applications, together with the as yet politically neglected report from *Gensplejsningsudvalget*, initiated both a debate and a political process focusing on safety and environmental issues. The government, however, did not react until an opposition MP from the Socialist People's Party raised a question in Parliament, which eventually led to the passing of the Act on gene Technology and the Environment, covering all three areas mentioned in the original proposals from *Gensplejsning-sudvalget*. This Act was carried unanimously and enforced in June 1986.[3]

Alongside the companies involved and the Ministry of the Environment, NOAH emerged as an important actor in 1985, filing complaints against the two approved products, and later, in 1986, as a critical voice in the political processes around the gene legislation. During the following five or six years, NOAH was the only important NGO opposing new biotechnology, and was, in the early years, one of the very few sources of information about genetic engineering.

When the Gene Technology Act was adopted in June 1986, Denmark became the first country to adopt a specific Act regulating biotechnology. The main principles in the Gene Act were: **1** a ban on deliberate release; **2** the Minister of the Environment could bypass this ban under 'extraordinary circumstances'; **3** possible permission for the contained use of GMOs after approval by the Minister of the Environment; and **4** all foods containing or consisting of genetically modified material must be approved.

After the adoption of the Act, occupational health related to gene technology also became a regulated area by means of a statutory order in 1987. Reproductive technologies and human gene technology were dealt with in a report by a committee appointed by the Ministry of the Interior in 1984.[4] The report was followed by a lengthy debate in Parliament in April 1985. Further negotiations about legislation in this field resulted in the passing of the Act in June 1987 to establish an ethical council and to regulate certain biomedical trials. The Danish Council of Ethics played an important role in debate and assessment of new biotechnologies affecting humans.

Phase 4, 1987–91. The fourth phase was characterised by some minor adjustments in the Gene Technology Act and the implementation of EU directives, but it was mainly an era of industrial application of biotechnology. The years after the adoption of the Act were marked by an intense debate about both the deliberate release and the contained use of GMOs. For its part, the biotechnological industry demanded an easing of regulation. This was expressed in a publication from the newly formed Association of Biotechnological Industries, which compared Danish regulation with international regulation.[5]

New on the scene was De Danske Sukkerfabrikker (DDS), the Danish sugar factories (from 1990 a part of Danisco), and its seed producer, Maribo Frø. Until the planned revision of the Gene Act in 1988–89, the debate mainly focused on contained use as an issue related to the medical and enzyme industry in Denmark, whereas the debate on deliberate release was influenced to a great extent by events in the USA. The result of the revision of the Act was the relaxation of regulation of industrial large-scale trials, and the revocation of the delaying effect of complaints.

The first governmental programmes to support research activities within modern biotechnology appeared in 1982, and were followed by some minor programmes up to 1986, when the first biotechnological research programme of 500 million DKR (about UK£50m) was initiated. Although all of these programmes were focused on basic biotechnological research, technology assessment projects had already been initiated within the first programme in 1982.

The biotechnological research programme provided the background for many of the debates and social-science research activities that took place in the following four years. One of the key players in these activities was the Danish Board of Technology,[6] which was appointed to administer a grant of 21 million DKR (about UK £2m) for technology assessment and information. This allocation was the result of the political influence of the green majority in Parliament on the 1986 research programme. In the subsequent second and later biotechnological research programmes, these activities were dampened down.

Together with the Danish Council for Adult Education,[7] the Danish Board of Technology was also appointed to administer the money allocated for information activities within the grant. The activities designed to stimulate debate in this period consisted of a series of public meetings and consensus conferences, most of which were arranged by the Board or NOAH, and of a large number of local discussion meetings.

In December 1988 Maribo Frø submitted an application for small-scale field trails of sugar beet resistant to Roundup® herbicide. Since this was the first case of deliberate release of GMOs in Denmark, the application was subject to a review and a parliamentary debate, as the Minister for the Environment had promised during the negotiation of the Gene Act in 1986. There was no reference to the clause in the Gene Technology Act demanding 'extraordinary circumstances' as a precondition for approving deliberate release, neither during the debate in Parliament nor in the administrative handling of the application in the National Environmental Administration. After the approval had been given, Maribo Frø carried out the field tests as planned, during the summer of 1990. One result of these efforts by DDS to produce a manipulated sugar beet, and of the subsequent political processes, was that the focus of the debate about deliberate release changed and acquired a Danish angle, as had the debate about contained use some years earlier.

The other major political event between 1988 and 1991 was the negotiation of the two EEC directives on contained use and deliberate release (90/219/EEC and 90/220/EEC), and the implementation of the directives in the Danish Act on Gene Technology and the Environment.[8] This meant that the so-called ban on deliberate release ended. A special feature was a provision stating that if there was any danger to the environment or health in Denmark, the general approval principle could be bypassed (this was known as 'the environmental guarantee').

Phase 5, 1992–96. After the rather hectic earlier phases, the fifth phase was characterised by a fall in public debate about biotechnology. Contained use did not appear as an issue of either political or public interest. One explanation for this could be that the level of regulation satisfied most actors; but it could also be because of a general (and remarkable) absence of critical actors in this phase.

As genetically manipulated food products approached the market, the opposition in Parliament started a discussion about the labelling of these products. A debate in 1993 showed a general agreement that all products based on genetically engineered processes (except additives) should be labelled individually on the product packaging. This attitude was reiterated in a parliamentary decision in 1994, which stated that the government would work nationally and in the EU to seek labelling of all products produced using GMOs. The government followed this line in its attempts to influence the regulation of novel foods in the EU.

Reproductive technologies and gene therapy were also important issues that Parliament dealt with almost continuously throughout this period. Thus, the Act concerning an ethical committee system for science was passed in 1992 and revised in 1996. The Act on artificial fertilisation was passed in June 1997 after both a lengthy parliamentary reading and much public debate. In 1995, following a report issued by the Danish Health Agency and hearings arranged by the Danish Council of Ethics, the Minister of Health made a statement to Parliament about gene therapy, and announced the establishment of a gene technology council in Denmark. This idea was later abandoned, however, because of resistance from the medical profession.

Other issues in the political arena in this period were genetic screening, on which an Act was passed in 1996, and the patenting of biotechnological inventions. A decision on this type of patenting was taken by Parliament in 1991, in order to give the government a binding mandate in relation to EU negotiations about the issue.

Phase 6, 1996 onwards. This sixth phase is characterised by a revival of the biotechnology debate. Despite Danish resistance, maize and soyabeans were the first products approved for marketing in the EU, and thus in principle in Denmark too. When the first cargo of soyabeans arrived at Aarhus, they prompted a lively political debate about labelling, and a more general debate about genetically manipulated food products. The result was a provisional labelling guideline, which stood until the novel food regulation took over in 1997 and eased the labelling requirements.

The other major issue in this phase was cloning. Shortly after the arrival of Dolly the cloned sheep, Danish researchers announced their plans to clone a calf. In the following months, cloning became an issue for both the public and the politicians. The Danish Board of Technology arranged a public parliamentary hearing on cloning in April 1997, and in May, Parliament carried a decision on cloning. It allowed research involving cloning and the insertion of the cloned egg into an animal, but required an abortion afterwards. There was general agreement in Parliament to ban cloned animals in husbandry.

Media coverage

Pluralism

Coverage of biotechnology in the Danish media is pluralistic in the sense that people who are interested in biotechnology (opinion leaders) normally have access to and use several important sources of printed information.

The rate of daily newspaper consumption is generally high. More than 80% of the population read a newspaper every day, and about 30% read more than one newspaper. There is a tradition of local diversity in the newspaper supply. In the past, local regions would have had four independent newspapers, each representing one of the dominant political parties: conservatives, agrarians, liberal 'radicals' and social democrats. Thus a kind of local dialogue was a part of opinion-making. Local newspapers are increasingly under economic pressure: typically, only one survives in each region; but many people would subscribe to it as well as to a national newspaper representing their preferred political viewpoint.

There is also a tradition of associations or societies that form organisations around common interests – economic, professional, political or recreational. Often, a person will receive periodicals from several of these associations: for example, monthlies published by a union (such as *Ingeniøren*), a trade association (*Dansk Landbrug*), a consumer organisation (*Samvirke*) or a 'grass-roots' organisation (*NOAH-bladet*). There are about 600 such periodicals. The biggest 21 of these are read by 28% of the adult population. This percentage is higher (about 50%) for people with higher education. Some of these periodicals have a special interest in and viewpoint on biotechnology, and have high credibility for their readers.

Biotechnology has been a topic of interest for the Danish public, as the number of registered articles in newspapers and other periodicals testify. A search in Artikeldatabasen for the period 1971–96 gives 934 articles. The following is the breakdown among national newspapers.

Berlingske Tidende (circulation 190,000): 136 articles
Børsen (40,000): 29 articles
Aktuelt (36,000): 54 articles
Information (22,000): 233 articles
Jyllandsposten (170,000): 76 articles
Kristeligt Dagblad (15,000): 92 articles
Politiken (143,000): 159 articles
Weekendavisen (57,000): 65 articles

Considering the pluralism of the media coverage of biotechnology, it would be inappropriate to designate any of these newspapers as opinion leaders. We have selected *Information* and *Politiken* as indicative of the most important content of the debate in relation to political and economic decision-making. Other newspapers and periodicals may have been as influential; but these two newspapers, although not the biggest, covered biotechnology amply during the period, and they reflected fairly well what was happening. In Table 1, all figures are based on codings of 300 of the articles in *Information* and *Politiken* (89 articles were omitted to make the sample homogeneous over time).

Continuity in the debate

The public debate during the period 1972–96 can be seen as a sequence of five phases. These phases were not very different, but each was characterised by dominating frames and themes. The differences should not overshadow obvious continuities in the debate: for example, ethical concerns have been important through all phases, and an examination of the arguments reveals that most of the ethical questions were formulated during the first phases and were repeated in the context of later discussions.

Another continuity is the recurring effort to popularise the basics of molecular cell biology, especially recombinant DNA technology and its application in medicine, industry and agriculture. All newspapers have published, on suitable occasions, informative articles that explain what the Flavr Savr tomato is, how GMOs may modify other organisms, how cloning is done, and so on. People interested in the debates would have read articles like these, for example as follow-ups to television programmes; the articles could also have functioned as appetisers to the numerous pamphlets and books where more coherent information can be found in popular forms. In this way, newspapers have played a role in what could be seen as the continuing education of the public.

However, this role seems to have changed through the period. In the 1970s, when microbiology was first brought to public attention, the proportion of informative articles was high. In the 1990s, general presentations of DNA research declined to become just one among several almost equally weighted themes. Now, newspapers seem to presuppose a generally well-informed public which can be approached without reiterating the basic explanations needed 20 years ago.

Changes in the debate

While these continuities were apparent, each phase also had its distinct features. Until 1976, biotechnology was mainly heard of in reports about potentially epoch-making progress in the scientific study of genes, and in reports of responsible scientists' concern for risky consequences of their work. Later in the 1970s (Phase 2, 1977–83), when the self-imposed moratorium was cancelled and new results and visions began to appear in scientific journals, the Danish public debate turned into a clearer confrontation between at least three positions:

- technological optimists who requested extra Danish R&D effort for the benefit of Danish science, industry and agriculture;

- technological optimists who saw possibilities for solutions for environmental and social problems in new 'sagacious' technologies; and

- critical experts who warned of environmental consequences as well as of trespassing ethical borderlines that might endanger social and political values.

The debate was mainly defined by people who could claim expertise: scientists in public and private positions as well as so-called 'counter-experts' – people with a scientific background and with experience from the intense Danish debate in the mid 1970s about nuclear power. As already mentioned, the Danish national hospital carried out the first trials with new reproductive techniques (IVF) in 1980. The moderate peak at 1980–82 on the intensity graph (Figure 2) reflects the media debate following this. From the early 1980s politicians took an active interest in the debate, and they began preparing new legal regulations of the area in 1983.

Phase 3 (1984–86) was characterised by Danish industrial and agricultural involvement in gene technology production, and by media confrontations between its advocates and a growing public opposition. Two of these media debates serve as examples: in July 1985 the Danish medical firm Novo was granted administrative permission for a fermentation plant designed for the GMO-based production of insulin. The permission was given on condition that production conform to forthcoming regulations about the contained use of GMOs (passed as *Lovom Genteknologi*, 1986). In the media, Novo was criticised for its secretive, uninformative attitude towards the public. The second case concerned DDS, which was refused permission in 1986 to run an open-field experiment with a genetically modified sugar beet (made resistant to Roundup®). This decision created a rather pointed debate in the media between DDS spokespeople and supporters of the new regulations.

Phase 3 ended with a political fait accompli: the Gene Technology Act securing expert assessments of environmental risks as well as a procedure for public hearings before GMOs could be used in industry or agriculture. The law was described in the media by its proponents as a positive support to research as well as to industrial and agricultural applications of gene technology. With its clear regulations and the ensuing public acceptance, the law would create the best economic and political conditions for investments in biotechnology, they said. As a kind of confirmation, an R&D programme was launched including money for research and for an extraordinary public information campaign about biotechnology (20 million DKR over the four-year span of the programme).

Phase 4 (1987–91) was the four-year period of this biotechnology research programme. In the media, confrontations about regulation and deregulation still played an important role. One of the issues was EU regulation vs Danish regulation, a theme that came up as a political question related to other political discussions about the EU.

This was also the period when the role of 'ordinary people' came into focus. Novo announced a new attitude to the public, appreciating the modus vivendi brought about by statutory regulations. And the 20 million DKR for technology assessment and information was spent by the Danish Board of Technology on projects whose primary aim was to support and initiate democratic participation. The best known of the Board's initiatives were the consensus conferences – meetings between lay people and experts that resulted in a final document written by the lay participants and addressed to politicians and to the media.[9] The first of these conferences took place in April 1987, and concerned 'gene technology in industry and agriculture'.[10] The final documents from the consensus conferences were reviewed in the media, and were considered to be interesting because of their high quality as expressions of what could be called 'informed public opinion'.

Phase 5 (1992–96) was characterised by a diversification of the media debate. Very few articles dealt with gene technology or biotechnology in general. There were still sensational stories, for example about transgenic animals, which played on fears or offended ethical positions, but gene technology as such was not an issue. It seemed to be accepted as a fact – perhaps deplorable, like so many other features of the world, but something that, in any case, was here to stay. What made news now were specific questions about, for example, EU

regulation, the behaviour of American multinationals like Monsanto, and the patenting of life.

Related to this diversification is an interesting curve of intensity (Figure 2), which seems to have characterised the debate in the 1990s. After some years of decline, from 1994 the debate again moved into an intensive phase, producing a comparable number of articles to that of the mid 1980s (Phase 6, 1996 onwards). One explanation for this is that applications of biotechnology which were planned or prepared in the 1980s were now a reality. They raised ethical problems about applications on human beings, as well as problems and perspectives concerning agriculture and industry. Earlier, for example, cloning was discussed as a hypothetical application of biotechnology. Now, the realisation of cloning as exemplified by Dolly the sheep reopened the old questions with renewed energy and, to a great extent, with the same basic arguments. The first significant example was the debate about Monsanto's Roundup-Ready soya that started in Denmark at the end of 1996 and gave rise to a pronounced increase in the number of articles. In the media, the soya was followed by Novartis's maize and cloning in early 1997.

Public perceptions

Public perceptions in the 1980s

In the Danish political debate about biotechnology there has been a clear conception of public perceptions as an incentive for regulation and public control. This idea can be traced back to the early 1980s, when industrial interests in biotechnology were first expressed.

A few years later, in 1984, when Novo and Nordisk Gentofte made their first applications to the Danish authorities for approval of production facilities for fermentation with GMOs, public opinions directly against the new biotechnologies were expressed for the first time. NOAH, the leading NGO in the debate about genetic engineering, initiated a petition against Nordisk Gentofte's plans, and a significant number of local people signed it.

These events gave support to an impression that public perceptions would be an important factor in political decisions about biotechnologies. No attempts were made, however, to survey public perceptions of biotechnology until after the Act on Gene Technology and the Environment had been adopted in 1986. At the beginning of 1987 the National Food Agency, which was in charge of the administration of the Act at that time, wanted to start an information campaign about gene technology. A survey, which was run by Jensen and Thing Mortensen, was carried out as a series of qualitative interviews with focus groups and a few individual interviews. One of the most important general results of the interviews was a critical attitude towards the authorities. 'We are more sceptical than the authorities,' said one interviewee. In their report, the authors concluded that behind much of the negative or sceptical attitude was a fundamental distrust in the will and ability of the decision-makers to regulate and control.

This survey was carried out before the Biotechnological Research Programme's information activities started, and public experience was still influenced by the controversies about the applications by Novo and Nordisk Gentofte. Once the Biotechnological Research Programme had been launched, a larger investigation by Borre was funded by the Danish Board of Technology. It consisted of four consecutive surveys between September 1987 and May 1989.[11] The results indicated that there was an increase in the number of supporters of gene technology from September 1987 to May 1988, and that there was a less polarised situation with respect to attitudes about gene technology in 1989 than in the years before. However, this decreasing polarisation that Borre saw may be disputed for methodological reasons: his classification led to a very large group with 'neutral' opinions, which change only very slightly.

The survey results also showed a quite critical attitude to certain aspects of gene technology. Forty per cent of respondents in April 1989 were against manipulation of plant genes, and 68% were against manipulation of the genes of cows and pigs. However, few were against manipulation of the genes of bacteria, or for carrying out amniocentesis. Furthermore, results showed little trust in the government (67% had little or no trust), and much higher trust in researchers and environmental movements.[11]

Despite the methodological problems, these results were seen by politicians and others responsible for funding the information activities as confirmation that the strategy behind the broad popular debate had been correct.

Consensus conferences

Another activity that was more directly initiated by the Danish Board of Technology was the consensus conference. These conferences are not opinion measurements in the usual sense, and with a small panel of lay people who have volunteered as participants, they do not give a representative picture of public perceptions. A unique feature of consensus conferences, however, is that they serve as a model of the social process of informed opinion development, which may be an important reason for the great attention that the consensus conferences have attracted from the media and among decision-makers.

An outcome of the first conference was Parliament's decision not to fund animal gene-technology projects under the first Biotechnological Research Programme (in the second programme from 1990 this was changed, however). Critical attitudes expressed both in the final document of the first conference and in later conferences, such as the one in 1992 on transgenic animals, have contributed to confirming the public's idea that restrictive rules about biotechnology are a precondition for accepting new biotechnologies in Denmark.

The Eurobarometer surveys results

In the 1990s the Eurobarometer surveys provided comprehensive sets of quantitative data that enabled us to make comparative analysis for all of the EU. In all three surveys – 1991, 1993 and 1996 – Denmark was ranked first or second among the EU countries with respect to the average number of correct answers to questions measuring knowledge about biotechnology. At the same time, Danes have low expectations of improvements in their way of life due to biotechnology and genetic engineering, both compared to the EU average and in absolute terms.

As a more general conclusion, Denmark fits into a pattern according to which the countries in northern Europe have the highest degree of measurable knowledge and the lowest expectations of possible improvements of our way of life. From 1993 to 1996, Denmark also followed a general trend of slightly increasing knowledge and a small change to more critical attitudes towards biotechnology and genetic engineering. There was a pronounced fall in the number of 'don't know' and 'no effect' answers for increasing numbers of correct answers to the knowledge questions both in 1993 and 1996. Increasing knowledge is associated with a greater tendency to form an opinion, either positive or negative. Accordingly, there was a low percentage of 'don't know' answers in Denmark compared to the EU average.

Survey results: risk perception

In 1991 and 1993 risk perception was very high in Denmark compared to most other EU countries for all six applications of modern biotechnology included in the survey (Figure 3). Regarding applications related to use of microorganisms and to medicines and health, there was relatively high support. Applications on plants spanned a middle position, and applications related to food and to animals generated negative responses. This differentiated pattern was also seen in the EU as whole, but it was especially pronounced in Denmark. A similar pattern was found in Borre's survey from 1989. All in all, these results indicate that the genetic engineering on microorganisms that gave rise to heated debate in the middle of the 1980s was generally accepted after the adoption of the Gene Technology Act. The support for medical applications is more likely to be related to the perceived benefits of these applications.

Interestingly enough, risk perception in the 1993 survey is least negative for applications on animals, which at the same time is the application that has by far the lowest support. This indicates strongly that the most important reason for rejecting these applications is not the perceived risk but ethical considerations.

In the 1996 Eurobarometer survey there is a similar differentiation of the answers, with positive responses to medical applications and negative responses to applications related to foods and animals. In particular, the negative responses to food-related applications are characteristic of Denmark. This survey makes possible an analysis of different predictors of willingness to encourage the various applications. The results are similar to those found for other EU countries. The most important predictor for this willingness is, for all applications, moral acceptability. This is true even for food production and crop plants, for which the answers at first glance seem to indicate that risk should be an important predictor. This is shown in Figure 3. Multiple regression analysis shows that the predictive value of risk is about the same for food production and xenotransplants ($\beta = 0.086$ for food production and 0.098 for xenotransplants). By contrast, the predictive value of moral acceptability is very high ($\beta = 0.522$ for food production and 0.615 for xenotransplants).

Influence of debate on attitudes

Evidence for critical attitudes towards genetic engineering on foods was also found in two surveys published in a leading conservative newspaper, *Berlingske Tidende*, in December 1996 and December 1997. The first survey was undertaken when the debate about Monsanto's genetically engineered soya was at its height in Denmark, and the second one was undertaken a year later as a follow-up to the first. In the first survey 68% of respondents said that they thought genetically engineered foods should be prohibited. In the second survey 63% wanted the foods prohibited.

The respondents were also asked two questions in both surveys about buying such foods. They were asked if they would buy a genetically engineered tomato if it had a longer keepability and a better taste. In the first survey 74% said 'no' and 20% said 'yes'. The second survey produced almost identical results. When asked whether the production of genetically engineered foods should be increased if it

could reduce food shortages in the world, 53% said 'no' and 34% said 'yes' in 1996, and 57% said 'no' and 34% 'yes' in 1997.

In the 1996 Eurobarometer survey, respondents were asked if they would buy genetically modified fruits if they tasted better: 78% in Denmark said 'no' and 15% said 'yes'. Furthermore, in the Eurobarometer survey 86% wanted labelling of genetically modified foods. The same question in *Berlingske Tidende*'s survey in December 1996 indicated that 95% wanted labelling. The 1996 Eurobarometer survey was conducted in the first half of November, before the debate about the soya had become a big issue in the media. Not surprisingly perhaps, the debate led to even more massive support for labelling, which was one of the key issues in the political discussion.

The fact that the results of the two *Berlingske Tidende* surveys were almost identical is an indication of the intense debate about genetically engineered foods during 1996–97 that started with the soya case and included genetically engineered maize, the adoption of the EU Novel Foods order, the cloning of Dolly and the EU directive on patenting of biotechnological inventions. Not all of this is about foods, but parallel to the media coverage and debates about genetic engineering there have been almost equally intense debates, as mentioned above, about several issues related to food and agriculture.

Commentary

In the 1970s modern biotechnology was barely an issue for the Danish public. There were few articles in the media, and almost no debate. In the same period, other public debates about technology-related issues attracted much attention, notably the debates about nuclear power and environmental pollution. Before 1976, nothing about modern biotechnology happened on the political scene either. The only political initiative of the 1970s, the appointment of RUGE, came from the Danish research councils, not from politicians. The influence of the USA is clear: researchers apparently wanted to anticipate a debate like the one that followed the Asilomar conference in the USA.

The need for regulation

At the beginning of the 1980s the first signs of applications of new biotechnologies in Denmark sparked debate and demands for regulation. The first trials of IVF triggered the first real media debate about gene technology around 1980. Further debate arose when industrial interests emerged. This led to an increasing awareness among some politicians of the need for regulation. Experience of the nuclear debate also played a role: it became increasingly clear to the actors in the biotechnology debate that the consequence of the public resistance to nuclear power would be that it would not be introduced in Denmark. Given the relative importance of Danish biotechnological sectors, this was an undesirable perspective for those interested in developing new biotechnologies.

Despite this, and despite the absence of regulation, initiatives came slowly, apparently because of resistance from biotechnological researchers and industrialists, and reluctance within the conservative government. It was the applications from Novo and Nordisk Gentofte in 1984–85 that led to the dramatic increase in the level of debate, and at the same time provoked opposition MPs to propose the Gene Technology Act. Also crucial for the initiation of the debate were the activities of NOAH.

The period 1984 to 1986 was one of controversy. The debate in the media and elsewhere was very polarised, with biotechnological researchers and industrialists on the one side and the environmental lobby on the other. The outcome of this period – the Gene Technology Act – was the result of several circumstances working together:

- experience of the debates about nuclear power and the environment, which had revealed both the power of public resistance and that the credibility of the authorities was important in public perceptions;

- the fact that Danish economic interests in biotechnology are significant;

- the heated debate provoked by the Novo and Nordisk Gentofte applications, signifying that gene technology would be a controversial issue if no political initiatives were taken; and

- the special situation in the Danish Parliament with a conservative minority government and a 'green majority'.

When the Act was adopted, it was the world's first specific Act on gene technology. This, together with the apparently very restrictive regulations, had a strong symbolic value. The policy initiatives following the adoption of Act were in many respects a means of establishing a social consensus about biotechnology in Denmark. The broad social debate as well as the first consensus conference led to increased awareness and to qualification of the arguments. At the same time these initiatives signalled that the authorities welcomed public debate.

The minor revision of the Gene Technology Act in 1989 was the result of severe criticism from the biotechnology industry of the lengthy and troublesome procedures of approval. Novo and Nordisk Gentofte publicly threatened to move their research and production facilities abroad if the terms of the

Act were not relaxed. Further to these events, and once the Act was functioning in a manner acceptable to the biotechnology industry, the industry began to change its attitudes to the Act and to the strict regulation. A kind of social consensus about the regulation of biotechnologies in Denmark developed. The results of the surveys of public perceptions of biotechnologies in this period, as well as the outcome of the first consensus conference, were all seen as confirming that the policy in this regard had been correct.

From the beginning of the 1990s, policy activities became increasingly oriented towards EC/EU regulation of biotechnologies. Parliament's role in these regulations became one of following the negotiations in the EC/EU. Thus, the parliamentary decisions in 1991 and 1994 about patenting of biotechnological inventions and novel foods respectively were taken to give the government a binding mandate in the negotiations. In the public debate about the regulations, a critical attitude developed towards the EC/EU. It is reasonable to assume that the more general and widespread critical attitudes in Denmark against the EU both influenced and were influenced by the critical debate about the regulation of biotechnology.

Public debate

Only the regulation of biotechnologies concerning humans is still a national concern. In this area there have been public debates since the beginning of the 1980s. The medical researchers have been against regulation of research into gene technology in humans, and in the first half of the 1980s some medical researchers showed an arrogance and a condescending attitude that probably had the opposite effect to the one intended on the political initiatives that eventually led to the establishment of an ethics council in 1987. According to the 1996 Eurobarometer survey, it seems that the establishment of the council and the Act on an ethical committee system for science has fostered much public confidence in these bodies. In the first half of the 1990s, human reproduction and ethical issues were the two most important themes in our media sample, and gene therapy was also important. That is an indication of the continued critical attention to and debate about biotechnology in humans, which is also reflected in the extensive policy activities.

Despite all the policy activity in the first half of the 1990s, public debate was less intense than in the second half of the 1980s. It seems that there were no major changes that could shake the broad consensus, and that the only industrial applications of any importance were production based on contained use and microorganisms, which were broadly accepted, according to the surveys. Danish industry, and

Novo Nordisk in particular, repeatedly stated that it was satisfied with the regulation, which ensured acceptance and politically stable conditions for production. This satisfaction was stated in explicit opposition to repeated criticism, especially from the German biotechnology industry that wanted revisions of the EU directives.

The Eurobarometer surveys have shown an increasing differentiation of the attitudes to the various applications of new biotechnologies. Attitudes to medical applications gain support, and gene technology related to food and animals provokes negative attitudes. The differentiation of attitudes is paralleled by a diversification of media coverage and debate.

When Greenpeace campaigned against Monsanto's soya in 1996, the Danish debate experienced a strong revival. As when the debate arose in the 1980s, the fact that concrete applications were on their way triggered the heated debate. This time the reaction was also reflecting the more general and critical debate about food and agriculture that had taken place in Denmark in recent years. Continued attention brought about by Novartis's maize, the cloning of Dolly the sheep, and the patenting directive has made 1996–97 the period of probably the most intense debate since biotechnology became an issue of public interest. Media attention has been extensive, and even the conservative papers have chosen a rather populist angle in their coverage, which probably reflects the widespread popular resistance to genetically modified foods.

The social consensus that was established in the years after the Gene Technology Act has lost its basis, at least in the area of food and agriculture. By 1998 there are clearly conflicting interests on the biotechnological scene. While large groups in the population are against gene technology in food and in animals, the food industry increasingly sees gene technology as an area of strategic importance.

Acknowledgement

The Danish project 'Biotechnology and the Danish Public' has received financial support from the Danish Social Science Research Council.

Figure 1. A regulatory profile for Denmark

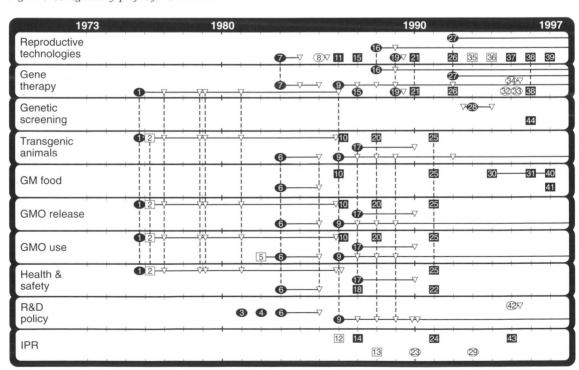

1 *June 1976–86.* Registration Committee concerning Genetic Engineering (RUGE) established.

2 *June 1976.* Voluntary control and registration of research involving genetic engineering by RUGE.

3 *May 1981.* National Technology Council subcommittee on gene splicing.

4 *February 1982.* Subcommittee on the biomolecular techniques of the research councils.

5 *October 1982.* Gene Splicing Committee established under the Ministry of Industry.

6 *1983–85.* Gene Splicing Committee undertakes a policy review to investigate the need for legislation and make proposals for new legislation. Report finalised in May 1985.

7 *1983.* The Commission on Ethical Problems in relation to egg transplantation, artificial insemination and embryo diagnostics is established under the Ministry of the Interior is established.

8 *March 1985.* Account to Parliament by the Minister of the Interior on ethics and medical technology (21 March); debate in Parliament on the account (27 March).

9 *1986.* Parliamentary Technology Board established.

10 *June 1986.* Gene Technology Act.

11 *May 1986.* Danish Health Agency approves IVF treatment at the Danish National Hospital (27 May), and IVF approved as an ordinary treatment on Danish hospitals (3 June).

12 *September 1986.* Danish Patent Office's Circular no. 11 of 1 September 1986: microbiological procedures and products of such procedures.

13 *November 1988.* Account to Parliament by the Minister of Industry of a proposal for the Council's Directive on Biotechnological Inventions, *KOM(88)496* (23 November); debate in the Parliament (1 December).

14 *December 1987.* Act on new plant varieties, no. 866 (23 December).

15 *June 1987.* Act on the establishment of an Ethics Council and regulation of certain biomedical trials, no. 353 (3 June).

16 *1988.* Establishment of the Danish Council of Ethics in pursuance of the Act on the establishment of an Ethics council etc., 1987–90 Annual reports 1988–96 (the Council is still functioning). Report: *Protection of Human Gametes, Fertilized Eggs and Embryos*, The Ethics Council, 1989.

17 *April 1987.* The Gene Technology Council appointed by the Minster of Education (1987–90).

18 *September 1987.* Statutory Order by the Ministry of Labour, no. 579 of 1 September 1987, on the register for gene technology.

19 *October 1989.* Committee under the Ministry of Health on biomedical research projects reports on research involving human subjects. Ethics/law, no. 1185.

20 *June 1988.* First revision of 1986 Gene Technology Act.

21 *May 1990.* Revision of the Act on ethical council and regulation of certain biomedical trials, no. 315 (16 May).

22 *October 1991.* Statutory Order no. 684 on gene technology and work environment (11 October).

23 *1990.* Research Policy Council: *Biotechnology and Patent Law. An investigation of current legislation in the food and agricultural sector in and outside Denmark and some considerations about adaptation of the legislation to modern biotechnologies*, by Mads Bryde Andersen and Mogens Koktvedgaard (Copenhagen).

24 *January 1991.* Parliamentary decision specifying five conditions for Danish acceptance of the EU proposal for a directive on legal protection of biotechnological inventions (24 January).

25 *October 1991.* Second revision of the Gene Technology Act and implementation of EC Directive 90/220.

26 *June 1992.* Act on a science ethical committee system and treatment of biomedical research projects, no. 503 (24 June).

27 *1992.* Establishment of science ethical committees in pursuance of the Act on science ethical committee system etc.

28 *June 1993.* Parliamentary committee on gene tests is established.

29 *1993.* Report on the patenting of human genes by Danish Council of Ethics.

30 *December 1994.* Parliamentary decision that all genetically modified foods should be labelled (6 December); the decision is taken to direct the government to follow this decision in negotiations with the EU about the Novel Foods Order.

31 *December 1996.* Instruction issued by the Danish Food Agency on the labelling of foods and food ingredients made on the basis of genetically modified soyabeans.

32 *May 1995.* Proceedings from a debate about gene therapy (3 May), issued by the Bioethical project GRAN, the Parliamentary committee on the Ethics Council.

33 *April 1995.* A report on the application of gene technology on humans (*Genes, Diseases, Humans*), issued by the Danish Health Agency.

34 *January 1995.* Parliamentary decision on current and future reproductive technologies, implementing among other things a ban on gene therapy on gametes (25 January).

35 *December 1993.* Circular on reporting about IVF treatment, etc. by the Danish Health Agency (22 December), issued with an instruction on reporting of IVF treatment.

36 *June 1994.* Circular no. 108 (13 June), issued together with an instruction on the introduction of new types of treatment in reproductive technology, no. 109.

37 *January 1995.* Parliamentary decision on current and future reproductive technologies (25 January).

38 *June 1996.* Revision of the Act on a science ethics committee system and regulation of certain biomedical trials, Act no. 499 (12 June).

39 *June 1997.* Act on artificial fertilisation, in connection with treatment by doctors, diagnostics and research etc., Act no. 460 (10 June).

40 *1997.* Novel Foods Order (EU 258/97) of 27 January. In force on 15 May.

41 *September 1997.* Soya and Maize Order (no. 1813/97) of 19 September, by the EU Commission.

42 *December 1995.* Review by the Ministry of Industry on the needs for and possibilities of biotechnology.

43 *July 1994.* Order on Protection of Plant Varieties in Europe (EU 2100/94) of 27 July, in force on 27 April 1995.

44 *July 1996.* Act on the use of health information in the labour market.

Figure 2. Intensity of articles on biotechnology in the Danish press

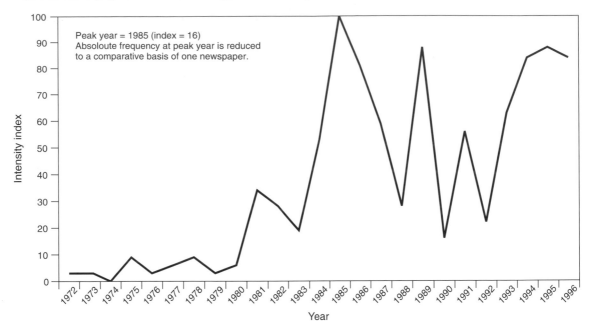

Figure 3. Attitudes to applications of biotechnology in Denmark, 1996

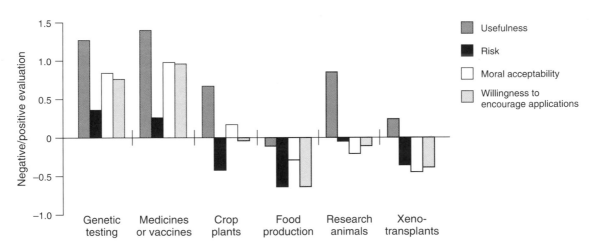

Table 1. Danish media profile (for an explanation of terminology please see Appendix 5)

Phase	Freq.[a] (%)	Frame (%)		Theme (%)		Actor (%)		Benefit/risk (%)		Location (%)	
1. 1972–76	2									Denmark	42
										USA	25
2. 1977–83	8	**Runaway**[b]	35	*DNA research*	21	*University*[c]	28	**Neither**	26	Denmark	40
		Progress	21	Reproduction	10	Industry	14	Risk only	26	*USA*	40
		Ethics	15	Ethical	8	Media	9	Benefit only	0		
								Both	0		
3. 1984–86	25	*Runaway*	50	DNA research	16	University	20	*Risk only*	73	Denmark	39
		Pandora's Box	5	Legal regulation	11	Industry	19	Neither	27	USA	20
				Economic	8	Media	11	Benefit only	0		
				Ethical	8	Military	9	Both	0		
				Microorganisms	7						
4. 1987–91	26	**Runaway**	42	**Political**	13	University	20	*Risk only*	73	*Denmark*	44
		Ethical	30	DNA research	11	Government	14	Neither	27	USA	20
		Accountability	13	Economic	8	Industry	13	Benefit only	0		
		Globalisation	3	Ethical	7	Parliament	8	Both	0		
		Runaway	6	Microorganisms	6						
		Nature/nuture	4	GMO release	5						
5. 1992–96	36	**Runaway**	36	**Reproduction**	13	**Media**	17	**Risk only**	66	Denmark	41
		Progress	11	Ethical	8	University	16	*Neither*	34	USA	17
		Pandora's box	10	DNA research	8	Government	16	Benefit	0		
				Diagnosis	8	Ethics		Both only	0		
				Gene therapy	7	committees	8				
				Microorganisms	6						
				Plant breeding	6						
				Human genome	5						
				Animal breeding	5						

a Percentage of corpus in the period; total $n = 300$.
b Bold indicates highest frequency within phase.
c Italics indicates highest frequency within category.

Notes and references

1 Teknologirådet, The National Technology Council, formed a committee in 1981 with the object of promoting commercial and social interests in biotechnology.

2 Indenrigsministeriets Gensplejsningsudvalg, *Genteknologi og sikkerhed* (Betænkning nr 1043: Copenhagen, 1985).

3 Miljøministeriet, *Lov om miljø og genteknologi*, Lov nr 288, 4 Juni 1986.

4 *Fremskridtets pris, Rapport fra indenrigs-ministeriets udvalg om etiske problemer ved ægtransplantation, kunstig befrugtning og fosterdiagnostik* (The Price of Progress: Report from the Commission of the Ministry of the Interior on Egg Transplantation, Artificial Fertilisation and Embryo Diagnostics) (Copenhagen, 1984).

5 Fink, K, and Terney, O, *Sådan reguleres genteknologi – Praksis og erfaringer* (Copenhagen: Forningen af Bioteknologiske Industrier i Danmark, 1988).

6 *Teknologinævnet*, later *Teknologirådet* (the Danish Board of Technology), was a body directed by Parliament to stimulate public debate about technology, assessing technologies and advising Parliament.

7 Dansk Folkeoplysnings Samråd (the Danish Council for Adult Education) administered a grant of 6.35 million DKR for information and local debate. In 1988–89 alone, DFS supported 350 meetings all over Denmark.

8 Miljøministeriet, *Lov om miljø og genteknologi*, Lov nr 3, 1991.

9 Grundahl, J, 'The Danish consensus conference model', and Klüver, L, 'Consensus conferences at the Danish Board of technology', in Joss, S and Durant, J (eds), *Public Participation of Science: the Role of Consensus Conferences in Europe* (London: Science Museum, 1995), pp31–40 and 41–51.

10 In addition to the conference in 1987, there have been four other consensus conferences about biotechnology: the human genome project (1989), transgenic animals (1992), infertility (1993) and gene therapy (1995).

11 Borre, O, *Befolkningens holdning til genteknologi II,* *Kommunikation og tillid* (The Attitude of the Population to Gene Technologi II, Communication and Trust) (Copenhagen: Teknologinævnets rapporter 1990/4, 1990); and Borre, O, 'Public opinion on gene technology in Denmark 1987–1989', *Biotech Forum Europe,* 7(6) (1990), pp471–7.

Address for correspondence

Erling Jelsøe, Department of Environment, Technology and Social Studies, Roskilde University, P.O.Bos 260, DK-4000 Roskilde, Denmark. Tel. +45 46742000, fax +45 46743041. E-mail EJ@teksam.ruc.dk

Finland

Timo Rusanen, Atte von Wright and Maria Rusanen

Research and development

Compared to the main industries of forestry and metals, modern biotechnology has made only a small economic impact in Finland. It is of minor importance even within the R&D sector, which is mainly represented by telecommunications research, for example, on mobile phones and telephone cables. But Finland is nevertheless interested in trying to use modern biotechnology as a tool for the industrialisation of its R&D.[1]

In 1996 there were about 50 companies applying gene technology in Finland. Most of these companies are small high-tech enterprises closely connected to universities. A few larger companies apply biotechnology in the food or pharmaceutical industries. Relatively large-scale biotechnological production ranges from diagnostic tests and industrial enzymes (which are significant exports) to biological pesticides. By 1998, the pharmaceutical industry and companies producing diagnostic tests were using far more gene technology in R&D than the food manufacturing or the wood processing industries.[2]

The most prominent areas of research in gene technology at the universities and research institutes in Finland are as follows. In Helsinki and Espoo there are research programmes on fermentation technology and plant biotechnology – Finland achieved the world's first transgenic barley. Research in Turku concerns bioreceptors; and in Oulu there are programmes in medical applications and diagnostics, such as for hormone-dependent cancer, osteoporosis and collagen, and fermentation technology. At Kuopio, research is concerned with transgenic animals, such as the 'erythropoietin cow' and transgenic fish, and with gene therapy for treating cardiovascular diseases and brain tumours.

Political constitution and culture

Finland is a constitutional republic with a parliamentary form of government ratified by the Constitution Act of 1919 and other constitutional laws. Supreme power is vested in the people, represented by the single-chambered 200-member Parliament. Parliament's functions include passing legislation. In practice most of the parliamentary decisions are based on governmental proposals by the Council of State (the Cabinet). The President of Finland also has considerable power in comparison with leaders of parliaments in other countries, including a suspensive veto and the right to dissolve Parliament in mid term. The President is the leader of foreign policy. The President appoints the Council of State, and can dismiss from the Council any of its individual members, if the Council so requests. The government must, however, enjoy the confidence of Parliament. Judicial power in Finland rests with independent courts of law.[3]

History partly explains the fact that there is usually no contradiction between the people and the law. The Finnish legal tradition is older than Finland as an independent state. The roots of the Finnish legislature date back to the time when Finland was a part of the kingdom of Sweden. In 1809 Finland became an autonomous Grand Duchy in the Russian empire, and had its own Parliament. At that time Finland could secure her threatened autonomy by means of her own constitution and legislative process. Legality became a penetrating ideology, which was consolidated by the fact that the administration was patriotic and not corrupt.

In 1906, parliamentary reform resulted in a single-chamber parliament based on universal suffrage. After this change, political parties became the very foundation of Finnish political life. They were also influential in the press and other media. If anything, this tendency has become stronger since Finland gained independence in 1917. Consensus and national unity have been important parts of the Finnish strategy for coping with its turbulent history during the twentieth century. Consequently attempts to influence political decision-making by means other than the ballot box have been regarded with suspicion, and associated with certain semi-fascist or ultra-leftist groups.

The political parties at the core of Finnish political life are numerous and varied. The non-socialist parties have generally held a majority in Parliament. In spite of the political contradictions between the towns and the countryside, after 1966 the Agrarian Party (now known as the Finnish Centre) and the Social Democratic Party, together with the smaller Swedish People's Party, have been the main parties of the coalition governments. Most representatives belonging to the Communist Party participated in most of the coalitions. The moderate conservative National Coalition Party was continuously in opposition until 1987, but its opposition was quite mild. Since 1987 the Coalition Party has been the other main part of government, either with the Centre Party or with the Social Democratic Party.

Changes in government have not caused any major changes in politics. The consensus between political parties and their programmes is illustrated by the 1995 government, the so-called Rainbow Coalition, which includes the conservative National Coalition Party, the Social Democrats, the former communists' Left Alliance, the Swedish People's Party and the Green League.[4]

In Finland, other social actors such as NGOs are relatively weak and uninfluential in decision-making processes. Many of these organisations are old: some of them were founded as long ago as the nineteenth century.[5] They are, in 1998, close in organisational terms to the formal sector. In Finland, these traditional organisations concerned with peace, nature conservation and women's movements have also been channels for new types of political activity in areas such as animal rights and food safety. Lately the influence and voice of NGOs have been somewhat increased due to the public concern about the safety of foods, environmental issues, animal rights and so on. Their voice is also represented in Parliament, and currently also in the Cabinet by the Green League.

The trade union movement has noticeable power because of the corporate structure of the labour market. It exercises this power by lobbying leftist MPs and through the sophisticated labour market policy system.

In summary, because rule by law has had a decisive role in the history of Finland as the safeguard of the nation against both external and internal threats, political decision-making has become rather consensus-based, with people trusting the formal political procedures to a greater degree than in other Western democracies.

Public policy

Phase 1, 1973–82. The introduction of gene technology in 1973–82 took Finland somewhat by surprise. Although the country had long scientific traditions in genetics, microbiology, virology and biochemistry, there were relatively few research groups specialising in molecular genetics. The main emphasis of Finnish molecular biology was concentrated on the study of the structure and functions of biological membranes. Consequently the first reactions of the scientific community were rather confused. Opinions were voiced of gene cloning being in the sphere of 'big science' and altogether beyond the scope of the limited resources of a small country like Finland. However, universities and research institutes very soon started several publicly funded projects with a conscious effort to introduce recombinant DNA technology, and in the early 1980s there were several high-quality groups working in the field. The Finnish medical, food and wood processing industries were also involved in

the developments from early on.

In 1973–82, the government, academia, and different grant-awarding and financing bodies were actively encouraging the first research programmes in gene technology. Also, several key enterprises in the pharmaceutical, food, chemical and wood processing industries took steps to consolidate Finnish know-how in the field, both by establishing their own research projects within their R&D programmes and by supporting the research done in different research institutions and universities.

Phase 2, 1983–91. During the period 1983–91 modern biotechnology became firmly established in Finland. Biotechnology was also well represented by the Finnish universities, which aimed to build connections between academic research and industry and to provide a basis for different spin-off enterprises. Some applications, such as industrial enzymes or experimental diagnostic products produced by gene technology, were also introduced.

Phase 3, 1992–97. Up until the late 1980s, no immediate need for special guidelines or legislation was seen. Peer control had an important role in controlling biotechnological research. In practice, because of the small size of the country and of the scientific and business communities, the implementation of the laws and regulations, when they were finally introduced in 1995, was rather straightforward and efficient (see Figure 1).

Finnish legislation during recent decades has, in general, followed the traditional Scandinavian 'socially responsible' model. 'Frame laws' have been typical of the way in which laws are passed in Finland. A frame law legislates on a certain area of life, but leaves much room for more detailed regulations by the ministries and administrative bodies. These frame laws (on health and safety, drugs, worker's rights and so on) have also regulated biotechnology.

In itself, law-making on biotechnology has been no exception to the standard Finnish legislative procedures. However, Finland's 1995 Gene Technology Act is an example of a new type of influence of international organisations on Finnish legislation: the Act is a frame law, but it also fulfils the demands of the transnational EU legislation.

When the Gene Technology Act was formulated, public discussion took place among mainly formal actors in Parliament, the Cabinet and the administration. However, the proposal was extensively evaluated by ministries and governmental bodies, as well as a total of 71 different institutes and organisations such as universities, trade unions, consumer organisations, animal welfare groups and Finnish Greenpeace. It should be noted that this evaluation was encouraged by the official bodies involved in the preparation of the law, rather than having been actively sought by the institutions and organisations polled.

The Finnish Gene Technology Act is, in general, slightly more restrictive than the minimum level defined in the EU directives. The Act refers to the contained use and manufacture of genetically modified organisms, and to their deliberate release in the environment.[6] In the special statute defining the law, genetically modified plants are consistently placed in a higher risk category than the EU directives require. The law also pays more attention to workers' protection than do the EU standards. Thus Finland, together with Greece, was the last EU country to harmonise its legislation in this respect.

The Gene Technology Act has had a significant role in guiding the biotechnology activities in Finland. The law gives to the expert boards (the Board for Gene Technology and the Advisory Board for Biotechnology) key roles in controlling scientific and marketing efforts in the field of biotechnology. However, due to the small population of the country, peer control still has a significant role. In 1997 a Finnish dairy company managed to obtain approval for a field test for antibiotic residues in milk based on a genetically modified bacterium. This happened after the Gene Technology Act was introduced, and so can be regarded as a test case.

There is no special legislation on reproductive technologies, but since the law does not specifically allow for human genetic manipulation, it is considered to be prohibited. The few experimental gene therapy trials have been controlled by the regulations of frame laws and specifically by hospital ethical boards, and thus have been evaluated case by case. The Finnish Novel Food Legislation has just started to implement the corresponding EU regulations.

The major scientific issues concerning gene technology in Finland are gene therapy, genetic engineering of animals, novel foods, and the use of biotechnology in forestry and wood processing. These are also the most prominent areas of either actual or potential medical or industrial applications. Thus they also represent fields where additional special legislation, to complement the national adaptation of EU directives, might eventually be needed. So far no specific laws about biotechnology have been introduced, nor are there any in preparation.

Media coverage

Finland has traditionally produced a large number of newspapers: in 1992, the total number of newspapers published was 242. The circulation of Finnish newspapers ranks among the highest in the world – the circulation of dailies is third only to Norway and Japan.[7] In Finland 95% of papers are delivered to subscribers' homes early in the morning, and 98% of Finns of 15 years and older regularly read at least one newspaper; 86% of the readership reads at least

two papers.[8] Measured in terms of annual turnover, dailies are by far the most significant mass medium in Finland.

The traditionally high rate of literacy in Finland has supported the expansion of the culture of reading newspapers. Since the late 1800s, the spirit of enlightenment adopted by public officials and the clergy was boosted by the expanding network of regional and local papers. The papers tended to be very factual and pragmatic.[9] The modern, pragmatic world view as well as new innovations in agriculture and improvements in health care and hygiene were efficiently disseminated in newspapers to all corners of the sparsely inhabited agrarian country.

In a way, the same spirit still exists in the contemporary Finnish press. As an example of part of modern progress, biotechnology is presented as a prosperous new methodology (see Figure 2 and Table 1). Biotechnology appears in Finnish newspapers as more pragmatic and less discursive than in the newspapers of many other EU countries, even when the themes of the articles do not differ very much from those published elsewhere. Biotechnology is presented as a series of breakthroughs and achievements which will improve standards of living. The tone in the articles is almost exclusively positive. The main frame in the articles is progress – economic or scientific. Critical aspects or ethical fears are less frequently included than in most other EU countries. These aspects have remained constant in Finnish press since 1973.

Phase 1, 1973–82. In this phase, discussion in Finland about biotechnology was limited to mainly professional circles. The chief concern was to introduce gene technology, and to reach international standards of research in the area. Scientists were the key actors in the discussion, and their voluntary adoption of safety regulations may have served to keep public concerns at a low level. During this phase the media and the general public remained silent about biotechnology.

Phase 2, 1983–91. In this phase, on the whole, modern biotechnology was regarded as a mainly progressive technique with rather few associated risks. No public concern was apparent, although industries avoided taking any risks regarding the public perception of their work. For example, beer produced by genetically modified yeast with certain superior technical properties was developed but never introduced on to the market, because it was not known what kind of reception it would elicit from the consumers. Some concern was voiced over animal welfare, especially when pictures of a transgenic 'Beltsville pig', which contained the bovine growth-hormone gene, were published in the Finnish media. But until the early 1990s, there was very little public discussion in Finland about biotechnology.

Phase 3, 1992–97. During the period 1992–97, Finnish gene technology was highlighted in the media for the first time. Attention focused on some much-publicised breakthroughs achieved in Finland. The first transgenic calf was produced by a university research team, and this led to industrial cooperation with a Dutch biotechnology company with the aim of producing pharmaceutically valuable compounds using transgenic dairy cows. Also at this time, the first human gene therapy trials were introduced.

Some public concern in the 1990s was also due to genetically engineered soyabeans and other novel foods. These issues also raised a debate in the media. The public concern was often presented in connection with other risks, such as BSE. Genetically modified foods and BSE both represented social risks; they were conceived of as having in some way a connection either to natural sciences or to high-efficiency farming; and they were imported from abroad.

The pragmatic, optimistic perception of biotechnology or new forms of technology in general is a typical Finnish attitude, according to the polls – and it is reflected in the media (or vice versa). This view is also typical of media reporting of the regulation of biotechnology. As the research activities are presented almost exclusively in a positive tone in the media, relatively little criticism is directed at the legislative process in biotechnology or peer control by scientists.

The actors in the majority of the articles are scientists expressing their views on progress in science. This predominance of scientists may partly be due to their central role in developing and controlling biotechnology, and the low level of public participation in the issue. Politicians are the second most common category, representing a certain, more distinctive, critical aspect towards biotechnological progress, although the main tone in this category is also optimism. The ultra-leftist and Green politicians have expressed their scepticism towards the techno-positive view about biotechnology. The industry, as the third most common actors, naturally expounds the economic prospects of biotechnology. This view is connected to the aim of trying to use modern biotechnology as a tool for 'R&D industrialisation'.[10]

In Finland animals are often presented in the media in connection with biotechnology (for example, animals as bioreactors). Contrary to the public concern shown in the surveys about animal welfare, the articles in the media tend more towards presenting animals as useful for scientific progress.

The benefits presented in newspapers include mostly medical advances which have been the most common theme since 1973. The majority of breakthroughs in Finnish biotechnology have been concerned predominantly with diagnostic and medical applications. By 1998 gene therapy was the main media issue. Risk perception is rather seldom presented in the media in connection with medical applications. When risk is presented, it is most often to do with ethical concerns.

Public perceptions

Finland represents a unique case of public perception among EU countries. As with the other Nordic countries, the level of knowledge about biotechnology and gene technology is rather high. This awareness of the issue is not, however, reflected by the relative pessimism or scepticism seen elsewhere in Scandinavia. Instead, acceptance of biotechnology applications is high and the risk perception is low (see Figure 3). In that respect the Finnish response is similar to that of southern European countries.

When considering genetically modified crop plants or genetically engineered foods, only a few Finns find these applications risky, while very many regard them as useful. The same tendency is almost as clear in the case of medical applications and transgenic animals. Medical applications are considered to be useful and not risky. They are seen as ethically sound ways for improving the health and well-being of humankind.

In the case of gene technology applied to animals, Finns are rather concerned about animal welfare. Animals are seen as defenceless creatures depending on human goodwill. Finns have also traditionally maintained close bonds with their domestic animals in the typical small Finnish homesteads.

These opinions reflect a certain kind of optimism towards biotechnological applications. However, we prefer to call this behaviour pragmatism, based on the agricultural past in a country with very difficult production conditions due to climate, geology and social history. In contrast to more conservative agricultural societies, Finns have been eager to embrace innovations in agricultural practices (new plant and animal breeds, machinery and so on), news of which has been brought to them by their own reading and propagated by progressive clergy at the local level and later on by teachers, consultants and local officials.

Attitudes to new technology

Finland was an agrarian society until the 1960s, and has urbanised extremely rapidly to become a post-industrial country. It can therefore be seen as a nation with a very pragmatic pre-industrial ethos combined with a post-industrial social structure. The people's experience of new technologies and the associated rapid rise in the standard of living has been overwhelmingly positive, and the disillusionment due to the disadvantages of uncontrolled technological development is only just starting to appear.

In general, Finns tend to adapt quickly to new

technology. The density of mobile phones in Finland is the highest in the world. According to the statistics, there were more than 2 million cellular phones in a country of 5 million inhabitants by the end of 1997. The Internet density is also among the highest in the world. It is obvious that Finns do not have a cultural resistance towards new forms of technology. The way of life in the sparsely inhabited countryside and the small towns favours the acceptance of new communications technologies. The public perception of biotechnology seems to share, together with electronics, the image of being at the cutting edge of modern technological progress.

It should be pointed out that there are no major genetically engineered crop plants that can be cultivated in Finland. The genetically modified barley developed in Finland is not being cultivated on a large scale. Soya and maize are imported into the country. Genetically engineered tomatoes are not yet marketed in Finland. So it is possible that the genetic engineering of plants is considered remote to Finnish everyday life.

To sum up, biotechnology has not, as yet, been a major issue in public debate in Finland compared with, for example, environmental problems. Various environmental issues have been debated by various actors with an intensity much greater than that accorded to discussions about biotechnology, even in recent years. These issues include the protection of primeval forests by the passive resistance environmentalists, the defence of animal rights by action groups, and, at the parliamentary level, the extension of the use of nuclear energy. Among the Finnish public, nuclear energy is a much more emotional topic than biotechnology, involving complicated political manoeuvring and industrial lobbying. The population in different polls has expressed much stronger opinions for or against nuclear power than it has about biotechnology.

In Finland, in the case of biotechnology, people trust experts to be responsible and truthful. The experts have inherited the same trust that public officers developed long ago in Finland's history. In contrast to people in some other European countries, the Finnish public feel that experts know and will inform people whether applications are safe or not.

It is impossible to say whether public opinion about the application of biotechnology will remain constant. Since Finland joined the EU as recently as 1995, Finland did not take part in the Eurobarometer surveys in 1991 and 1993, and so no comparisons are possible.

Commentary

In general the Finnish public has responded to biotechnology in a rather calm and dispassionate way, and the legislators have not felt any considerable pressure to control the applications of gene technology and biotechnology. For example, such emotional issues as human *in-vitro* fertilisation and other new infertility treatments have not caused a stir among the public. There has not as yet been any exceptionally sensitive issue that would have raised a notable public debate.

The public have – in comparison to the rest of Europe – very positive attitudes to biotechnology, and also great expectations of its applications. Risk perception is very low, and the application of genetically modified organisms is considered to be safe when passing safety regulations.

Due to the special political and economic history of Finland, the Finnish people in general apparently trust the political decision-making process and the administration in charge of the implementation of the laws and regulations. This fact is reflected in the development of the special Finnish biotechnology legislation and the public discussion connected with it. The role of NGOs and other informal actors has been rather limited, although their influence appears to be increasing.

In principle, due to the culture of reading newspapers and the high circulation of dailies, the press is a potential opinion leader in Finland. In the case of biotechnology, the press has by no means been the driving force in public discussion. Instead, it mirrors public opinion, and gives information on scientific achievements without dwelling on questions of risk or moral concern. As a rule, the media have reacted positively to biotechnology, presenting it to the public as an opportunity rather than a threat.

Because of the small population in Finland, the peer control offered by expert boards has played a significant role in controlling gene technology. The 1995 Finnish Gene Technology Act finally formalised the key role of the expert boards when controlling genetic research. The law gave authorities a tool with which to take action, if needed.

The law has existed for so short a time that it is too early to say anything definite about its consequences or effects on Finnish public perceptions of biotechnology. In research institutes the law apparently had very little impact on practice, which had already been regulated voluntarily before the legislation. For the industry, the law has probably meant clarification of rules and, maybe, somewhat boosted the development of biotechnological applications and products.

At the moment the gene technology law is apparently considered sufficient in scope for the foreseeable future. No calls for special legislation have been voiced. This is probably due in part to the fact that existing laws on foods, pharmaceuticals, occupational health and environmental protection are considered (as broadly formulated frame laws) also to apply to any new situations that might arise as a result of biotechnology.

Rusanen et al.

Figure 1. A regulatory profile for Finland

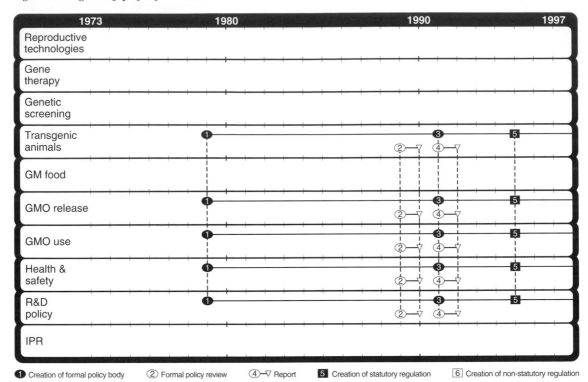

1 *1979.* Recombinant DNA expert group established by the National Board of Health.
2 *1989.* Expert group set up by the Ministry of Environment for the evaluation of a special gene law.
3 *1991.* Biotechnology Committee established by the Ministry of Social Affairs and Health.
4 *1991.* Committee established by the Prime Minister's Office to prepare a proposal for a gene law.

5 *1995.* The Finnish Gene Technology law and the establishment of the Board for Gene Technology. The law implements EU directives 90/219 and 90/220, covering deliberate release and GMO use. It does not address human genetic manipulation specifically, but this is considered to be prohibited under the legislation. The few experimental gene therapy trials that have taken place in Finland have been on a case-by-case basis.

Figure 2. Intensity of articles on biotechnology in the Finnish press

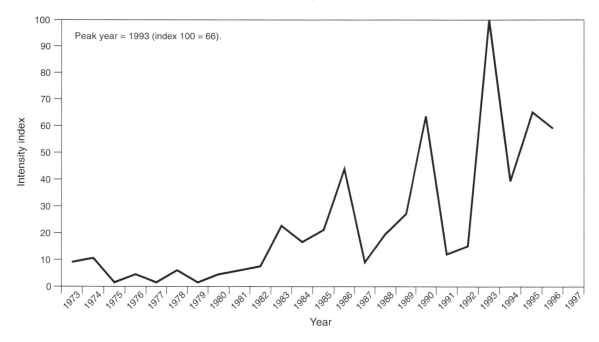

Figure 3. Attitudes to applications of biotechnology in Finland, 1996

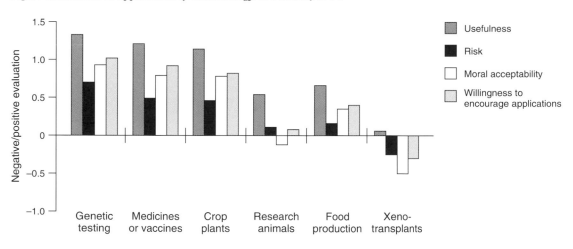

Table 1. Finnish media profile (for an explanation of terminology please see Appendix 5).

Phase	Freq.[a] (%)	Frame (%)		Theme (%)		Actor (%)		Benefit/risk (%)		Location (%)	
1. 1973–82	8	**Progress**[b]	34	*Medical*	37	*Scientific*	68	**Benefit only**	46	**Finland**	56
								Both	40		
2. 1983–91	38	*Progress*[c]	55	**Medical**	32	**Scientific**	60	*Benefit only*	51	*Finland*	71
		Economic	14	Economic	9	*Political*	17	Both	37	USA	11
		Ethical	8	Ethical	9	Industry	11	Risk only	6	Other Europe	8
		Accountability	8	*Political*	8						
				Animals	7						
				Regulations	7						
				Basic research	5						
				Agriculture	5						
3. 1992–91	35	**Progress**	45	**Medical**	18	**Scientific**	60	*Benefit only*	51	**Finland**	67
		Economic	20	*Economic*	12	*Industry*	14	Both	34	USA	12
		Accountabilty	11	*Animals*	9	Political	11	*Risk only*	8	*Other Europe*	9
		Nature/nurture	8	Political	7					World	5
				Regulation	6						
				Security & risk	6						

a Percentage of corpus in the period; total $n = 375$.
b Bold indicates highest frequency within phase.
c Italics indicates highest frequency within category.

Notes and references

1 Torgersen, H, Jelsoe, E, Lassen, J, Rusanen, T and Nielsen, T, 'Why Europe won't play ball. What is behind the European reluctance towards genetically modified food?', Paper presented to the AAAS conference, Philadelphia, February 1998.

2 Lampelo, S, *Biotekniikka laanin teollisissa klustereissa. Teollisuusprojekti* (Kuopio: Kuopion laaninhallitus, 1996), pp18–24.

3 Tiitta, A, *Find Out about Finland* (Helsinki: Otava, 1996), pp20–22; Council of State, *The Council of State and the Finnish Parliamentary System* (Helsinki: Information Unit of the Council of State, 1998), pp1–10.

4 Paastela, J, *Finland's New Social Movements* (Tampere: University of Tampere, 1987), pp7–11; and Tiitta, A, (1996), pp20–22.

5 Paastela, J (1987), pp7–11.

6 Komiteanmietinto, *Geenitekniikka. Ehdotus geenitekniikkalaiksi. Biotekniikkakomitean mietinto* (Helsinki: Ymparistominsterio, 1993).

7 Sauri, T, 'The mass media system in Finland: trends in development from the 1980s to the 1990s', in Sauri, T (ed.), *Joukkoviestinnän rakenteita 1980-luvulta 1990-luvulle* (Helsinki: Tilastokeskus, 1993), pp76–77.

8 Liikkanen, M, Paakkonen, H, Toikka, A and Hyytiainen, P, *Vapaa-aika numeroina 2. Kirjojen ja lehtien lukeminen, kirjastossa käyminen* (Helsinki: Tilastokeskus, 1993).

9 Tommila, P, 'Suomen sanomalehdisto 1700-luvulta nykypaivaan', in Nordenstreng, K and Wiio, O (eds), *Joukkoviestinta Suomessa* (Porvoo: Weilin & Goos, 1994), p50.

10 Torgersen, H, Jelsoe, E, Lassen, J *et al.* (1988).

Address for correspondence

Timo Rusanen, University of Kuopio, Department of Social Sciences, P.O. Box 1627, 70211 Kuopio. Tel. +358 17 162657. E-mail timo.rusanen@uku.fi

France

Suzanne de Cheveigné, Daniel Boy, Jean-Christophe Galloux, Anne Berthomier and Hélène Gaumont Prat

Economy

France is the largest country in the EU by area, and the second largest by economy and population. It is a very centralised country, and technologically advanced but with a top-down system: what one might call 'technological Jacobinism'. Whether it be Minitel, nuclear energy, Concorde or more recently biotechnology, development is decided in the upper echelons, then adopted more or less happily by the population. In recent years, the French economy has been characterised by slow growth, inadequate industrial investment and a high unemployment rate (more than 10% for the last ten years). The decline of traditional manufacturing led to a policy of restructuring manufacturing industries and an attempt to expand R&D activities in new technologies. A general feature of French science – unfortunately a feature shared by a number of European countries – is the poor contact between research and industry. High-quality research has trouble 'irrigating' industry.

France has a long-standing involvement in biotechnology because of its tradition of research and industry in the pharmaceutical and chemical fields, and because of the importance of its agriculture (France is the biggest agricultural producer in Europe). Most basic research in biotechnology is carried out by government research agencies active in the areas of agriculture, health and basic research (INRA, INSERM (Institut National de la Santé et de la Recherche Médicale) and CNRS (Centre National de la Recherche Scientifique) respectively) and by the Pasteur Institute, a private foundation. To give an example, the life-science department of the CNRS hosts 5700 researchers. But public funding of biotechnology R&D has been decreasing – it amounted to 360 million French francs in 1993, placing France fourth in Europe.

French research on biotechnology is strong in the areas of transgenic animals, monoclonal antibodies, genome mapping, extraction and purification, predictive microbiology, plant GMOs, automatic DNA sequencing, molecular probes, recombinant protein production, genetic therapy, vaccines, and sewage treatment.[1] But as far as industry is concerned, the only strong points have been vaccines and crop production of biomass. French industry has nevertheless been considered average in the following industrial areas: monoclonal antibodies, genome mapping, extraction and purification, plant GMOs, molecular probes, recombinant protein production, genetic therapy, vaccines, sewage treatment, and membrane separation.

The number of biotechnology firms has also decreased over the period 1990–95, from 70–90 down to 50–70. About two-thirds of these are large companies – little 'start-up' companies are quite rare owing to the difficulty of raising funds and of gaining the support of financial institutions or the stock markets. Among the large firms that use biotechnology, the main ones are Sanofi (the healthcare group of the huge oil conglomerate Elf-Aquitaine), Rhône-Poulenc-Rohrer (a chemical group that also owns Institut Mérieux – the world leader in vaccines) and Limagrain (a seed producer).

Political system and policy-making process

France is a founder member of the EEC. It is a republic governed by the constitution of 4 October 1958, headed by a President elected for a seven-year term. The system is a presidential one; however, with the recent political evolution since the defeat of the right-wing coalition in 1997, the role of the Prime Minister has been strengthened. The President oversees the operation of the state. The Prime Minister directs the activity of government. He or she is chosen by the President, and is usually from the political party that has been most successful in the general elections for the National Assembly. Although the government is, according to the constitution, led by the Prime Minister, he or she in fact acts under the direction of the President. The ministers (between 25 and 40, including Secretaries of State) are appointed by the President on the recommendation of the Prime Minister. Many are high-ranking civil servants: one of the peculiarities of the French political system is the close links between high administration, political leaders and economic decision-makers. Most of these belong to a small number of *grands corps* or groups whose origins often date back to their schooldays (at the prestigious École Centrale, École Nationale d'Administration, École des Mines, École Polytechnique or Conseil d'État).

The legislative branch of government, the Parliament, is composed of two chambers: the National Assembly and the Senate. The powers of the Parliament are rather limited under the constitution: the scope of legislative activity is narrow (the Parliament only votes on general texts and the laws put down only the principles – all applications are decided at government level) and its initiative is in reality diminished (the government is master of the agenda – very few laws are parliamentary initiatives). Consequently, Parliament seldom initiates policy.

France is traditionally a strongly centralised state. Even though the decentralisation laws of 1982 gave more powers to local governments, all policy and regulation in the field of biotechnology emanate from national authorities. French administration is relatively compartmentalised, very hierarchical and, politically, not very independent. Even though the main elements of public policy may be initiated within ministries, decisions and arbitrations are made at the level of the Prime Minister's Cabinet. This is due, on the one hand, to the fact that the borders between ministries are ill-defined (unlike in Anglo-Saxon administrative models), and on the other, to struggles of influence between different *grands corps* within administrative and political structures. These rivalries are often more important than political differences. In general, outside influences are only exerted at the ministerial level, where lobbying relies mainly on personal contacts. Of course, the political personnel (ministers and Cabinet) can also be influenced by outside elements that defy this administrative logic and obey a purely political one (the strategy of 'political risk'). But, on the whole, the administration is not very permeable to changes in political majority, especially in areas where the stakes are considered more technical than political.

Public policy

Reflection about policy in the area of biotechnology was initiated by the public authorities at the end of 1974, when they set up a group to examine security aspects. This was four months before the Asilomar Conference which was attended by several French scientists. In 1976, the report of the Research Commission of the Seventh Plan had already commented on the promise of biotechnology, but at no point was the question raised of its public acceptability.

Formally, the first control commission for genetic recombination *in vitro* was created in March 1975 within the Ministry of Research (DGRST). Thus, the security of GMOs was the first motive for developing public policy in the field of biotechnology. The initiative was taken by scientists mostly from the large public research organisations (at the time, few scientists had mastered the techniques of genetic recombination), and they met with a favourable response within the Research Ministry to which they were affiliated.

It rapidly appeared that policy concerning biotechnology in general and questions of security in particular could fit into the mould offered by the chemical industry: during the first few years of development, some of the control structures were shared and methods of analysis were similar to those applied to chemical products. Thus, up until the beginning of the 1990s, the Interministerial Group for Chemical Products was the main actor in the development of biotechnology policy. One can also find reference to this 'chemical model' in the area of industrial property and the protection of biotechnological inventions.

In a later period, policies appeared in parallel in the different technical areas opened up by biotechnology. As such, they can be called 'vertical policies', dealing with only one type of technical problem. GMO regulation is the only horizontal policy, crossing all fields of biotechnology, and therefore it plays the role of 'master policy'. In this sense, it can be said that there are not one but several policies on biotechnology in France.

Some fields have developed independently, because of their technical or juridical specificities, without really following the GMO 'master policy'. This is the case for industrial property and for pharmaceuticals. The definition of a true biotechnology policy per se only began in the late 1980s, following deliberations carried out at the European and world levels. For pharmaceuticals produced through biotechnology, technical regulations were rapidly set up at the European level, and were adopted without difficulty at the national level – a sort of adaptation to progress.

The GMO question became diversified as techniques developed. The original GMO policy branched out into policies concerning transgenic animals, novel foods and gene therapy. It should be noted, however, that the regulators did not always notice the interdependencies between certain policies: the biomedical area covers medicine, gene therapy and some genetic tests. The policies have nevertheless often been interpreted in different ways at different times.

To summarise, over the 23 years covered by this study, biotechnology has not been understood, as far as regulation is concerned, as a homogeneous and unified sector. This is apparent in the compartmentalisation of regulations in France, as well as, sometimes, in their lack of coherence.

Several stages can be distinguished in the evolution of public policies concerning biotechnology in France (see Figure 1).

Phase 1, 1974–92: the technocratic phase. There was no EC regulation in the 'master' area of GMOs. In France the authorities set up a true partnership with industry and research to control GMOs, on a self-regulatory basis, as in Anglo-Saxon countries. This voluntary and supple regulation contrasts with the usually restrictive character of the French normative framework. Politicians did not perceive biotechnology as a 'political' issue even if the need for regulation was felt at both technical and ethical levels. Even vocabulary, which was subject to a specific regulation (22 January 1986), had to be 'positive': the term 'genetic manipulation' (*manipulation génétique*) was to be avoided. The policy initiated in the mid 1970s, to expand R&D activity in the area of new technology in general and biotechnology in particular, was pursued even more actively from 1981, when the Left came to power. The stated goal was to have R&D expenditure rise to 2.5% of the GNP. At about the same time (1982), the government developed a 'mobilisation programme' for biotechnology with the aim of conquering 10% of the world market by 1990.

Phase 2, 1992–97: the political phase. Directives 219/90 and 220/90 were adopted at the EU level. They imposed more restrictive regulations, following the German model adopted two years earlier. The strict supervision they called for made French politicians wary. A number of events covered by the media (including the patenting of Harvard's transgenic mouse, and the Craig Venter patent) as well as worldwide discussions on biodiversity (such as the Rio Summit) brought biotechnology into the political field. Moreover, the discussion on the 'bioethical laws' (29 July 1994), which was mainly about *in-vitro* fertilisation, spilled over into biotechnology. Policies became less coherent within the narrow space left free by European directives for national initiatives. During this period, R&D programmes were drastically reduced, because of severe economic problems.

Phase 3, 1997–98. Changes in political majority had little effect on the evolution of public policy. Ecological preoccupations are undoubtedly more marked in left-wing governments, but they have never overridden a certain industrial realism. Texts and decisions made by one government have rarely been amended by the next. Two rare exceptions can be mentioned: the introduction of reservations in the ratification of the Rio Convention (negotiated and broadly approved by a left-wing government) and the decision to forbid the growing of transgenic maize taken by a right-wing government and revoked by a left-wing government. There is a wide consensus among officials concerning the policy that should be followed in the field of biotechnology, and the way this policy is implemented has not been radically changed by political alternation.

Actors and arenas

In France, the main actors defining public policy on biotechnology belonged to ministerial cabinets or to the industrial and research areas. Environmental groups, churches, trade unions and political parties were minor actors. Up until recently, no real public debate had emerged. These are global statements, of course, concerning a wide domain and a long period. They should be qualified, area by area, and policy by policy.

Discussions and decisions took place at a ministerial level, generally involving the Ministries of Agriculture, Research, Health, Environment and, sometimes, Justice. From the moment policies appeared in the area, in the mid 1970s, they were coordinated at the government level by interministerial structures, which were generally placed under the authority of the Prime Minister. This set-up allowed differences of opinion between ministries to be resolved (Industry, and to a lesser degree Research, were often opposed to Environment) and gave the Prime Minister and his Cabinet more efficient control over ministerial initiatives. Industrial organisations (Organibio, CNPF) and environmentalists ('The Greens', Greenpeace) exerted their influence at ministerial levels.

Industry was omnipresent in the defining of biotechnology policy, and was always served by the equation 'developing new technologies equals future jobs'. During the first phase (1974–92), regulation was founded on a form of self-regulation – a novel principle in France. Industry's influence undoubtedly decreased once public policy was essentially being decided in Brussels, where French industrialists have difficulty in competing with the lobbying from other countries. They receive little help from French civil servants delegated by their ministries, who, in keeping with their administrative tradition, are not accustomed to such techniques of influence. It should be noted that the field is not homogeneous: diverging interests have appeared between the pharmaceuticals industry, chemical industry, seed producers, etc.

Scientists have been omnipresent at all stages of decision-making. They have benefited from their own knowledge and expertise, which have been essential in an area which is largely open to prospective decision-makers and which, at the beginning, was obscure to political decision-makers. Public-funded research has been over-represented. It exerts a real influence over politicians and its public image is good – if somewhat tarnished by recent by scandals such as the one concerning HIV-contaminated blood, where some collusion was revealed between the areas of politics, science and business.

Established or emerging policies are generally backed by reports requested by the government, written either by scientific experts or by MPs with a special interest in the field. The former were called upon during the first phase (mainly Gros and Jacob, 1980, Bedossa, 1983 and Sautier, 1988), and the latter during the second phase (Chevalier, 1990). Experts have played a fundamental role in defining public policy in France. In the area of biotechnology, they mainly came from the large research organisations (INSERM, Institut Pasteur, Académie des Sciences), which gave their opinions the scientific legitimacy desired by those who commissioned the reports, or from the higher public administration (the Council of State or the Treasury). Experts were also found in ministerial committees or working groups; and later, as regulations developed and as regulatory bodies were set up, they occupied decisive positions in them. The importance of these experts in the definition of public policy has been all the greater because they are few in number (a few tens in all of France), their longevity has been remarkable (one of the participants in the first working group on GMO security in 1978 presides over the Genetic Engineering Commission 20 years later), and they have occupied multiple positions (A. Kahn was concurrently a member of both the National Ethics Committee and the Genetic Engineering Commission, and president of the Biomolecular Genetics Commission).

The influence of political parties has been weak because few of them have ever developed a real policy on biotechnology, as no real political stakes have been involved. Even the Greens did not develop a position on GMOs until 1992. Opportunism has been the rule, except when points of political doctrine were at stake, as in the links between biodiversity and north/south relations. This sometimes produces somewhat 'illogical' decisions, such as the ban on growing transgenic maize (but not on its import), which was seen as a trade-off by the Environment Minister of the time. Political parties have rarely initiated public policy in the field.

Although trade unionists have been associated with deliberative groups and with some regulatory bodies, unions do not appear to have played a major role in initiating or implementing public policy in biotechnology. On the other hand, except for agricultural unions (and even then only in areas where they are directly concerned, such as the intellectual ownership of new seed) they have not fought biotechnology. Recently, a small agricultural union, the Confédération Paysanne, opposed the import of Monsanto seed, mainly on the grounds of national independence.

The situation is the same for consumer associations, except that they are politically much weaker. Their influence has been increasing since foodstuffs produced by genetic engineering have begun to appear on the market. For example, the Federal Union of Consumers began a campaign against transgenic foodstuffs in 1994. The issue has developed since then, as consumer associations have become more and more dubious about novel foods. In the future, they could affect policy, particularly concerning labelling.

NGOs have had little influence in France.

Greenpeace does not seem to be independent of its European counterparts. They have not been as combative about biotechnology as they have been about nuclear energy, and their influence in the initiation or development of public policy has remained very limited.

The public are conspicuously absent among the actors. Although, throughout the 1970s, they were regularly mentioned in reports on biotechnology, the preoccupation was with informing and educating people. The reactions of the public or public participation in decision-making was not a matter for concern during the first phase, and was only slightly at issue during the second. Regulatory bodies (including ethics committees) are made up of experts or officials – they are not derived directly from the population in general. The organisation of a consensus conference in 1998 marked an evolution, though it came too late to modify policy tendencies in the field.

Parliament is now the main forum, at least when problems are submitted to it, although before 1992, the year of the first law on biotechnology, this was not the case. The Parliamentary Office for Scientific and Technical Choices played some role in analysing public policy in this area during the second phase. There were few written questions from MPs during the period under study. On the other hand, the discussions of the Bioethical Laws (1991–94), unlike those of the 1992 GMO Law, initiated a true national debate about biotechnology applied to humans.

Scientific meetings have also been important sites of debate, although they have remained largely in the hands of the same 'experts' for whom they provided an extra voice. The influence of the meetings was considerable, but they did not allow any public participation. Moreover, they have only considered topical questions, mainly about human applications.

The National Ethics Committee was created in 1983. Its composition reflected the image of the intellectual and scientific establishment. Its deliberations are secret, no dissident opinions are expressed, and its position is adopted by consensus. It is, in fact, less an arena than a regulatory body led by experts.

The growing importance of bioethics

Since public policy began to appear in the field of biotechnology in France, bioethical preoccupations have always accompanied the regulatory framework. They materialised with the creation of the National Consultative Ethics Committee, which is mid-way between a college of experts and a regulatory body. Politicians also took up the theme: numerous texts have been amended under the pretext of making them ethically consistent (in biomedicine, industrial property and biodiversity). Nevertheless, while political action so often refers to bioethics, the principles singled out by the Ethics Committee are not always followed in regulation. Such committees have become places of power, as can be seen in their quest for media coverage or in the activism of some of their more influential members. Such bodies are appearing at international levels (EU, UNESCO, EUROPABIO, Council of Europe), creating some competition. Whatever the case, the bioethical phenomenon has become indissociable from the development of biotechnology.

Media coverage

This policy-making process is reflected in *Le Monde*, the obvious choice as a daily opinion-leader in France. Its circulation is about 400,000, and it is read by about 2.2 million people (3.7% of the total population). It is a paper of reference, read by nearly every person in a decision-making position, by higher civil servants as well as by teachers and students – 57% of its readers are university graduates, compared to 17.5% in the general population and 40% among readers of daily newspapers. It has the largest readership of French dailies among the highest income groups. Politically, it is centre-left.

Other main national newspapers are the left-wing *Libération*, which has a younger readership and is important in particular in cultural areas, and the right-wing conservative *Le Figaro*, which is important for the economic sector. These two can be called 'quality papers', but do not serve as a reference in the same way as does *Le Monde*. The two main popular daily papers are *Le Parisien* and *France Soir*. Neither can be described as a tabloid. Regional newspapers are important in France: *Ouest France*, in Brittany, has the highest circulation of all daily papers. Circulation figures for the press in France are fairly low compared to those, for example, in the UK, Italy or Germany: 400,000 is a good figure for a daily national paper, and this is reached by *Le Monde*, *Le Figaro* and *Le Parisien*. *Libération*, a fairly recent creation (1973), has a circulation of around 200,000, compared with the nearly 0.5 million copies of other recent papers like *La Reppublica* in Italy or the *Independent* in the UK.

Our media study of coverage of biotechnology focused on the content of the opinion-leader *Le Monde*. The corpus was composed of articles selected as follows: by a manual search for 1970–75; using a newspaper index for 1982–86; and sampling one day in four (25%) with a complete text search for 1987–95 on a CD-ROM index. Search words were: biotech*, geneti*, ADN [DNA], transgen*, genique, genome, manipulation(s) + genetique(s), diagnostique + prenatal, diagnostique + pre-implantatoire, loi [law] + bioethique. The word gène

(gene) created a problem because accents are not recognised by the software, and gêne (trouble, bother, etc.) was a far more frequent occurrence. We therefore searched for 'gene(s) + maladie(s) [illness]'.

Different search methods give quite different results, so the trends in the figures must be interpreted with caution. Looking through the paper and using a newspaper index are similar in that only when the theme takes up a substantial part of an article will it be selected. On the other hand, a complete text search generates far more articles, including a new category where the theme is only mentioned in passing, used as a metaphor, etc. For example, in an article about the economy of a city or a country, biotechnology would be quoted, often along with information technology, in the context of future prospects. In another frequent case, 'genetic manipulation' (note the change in terminology) appeared in a film or book review, along with 'virtual reality' this time, to illustrate the way modern society is losing its way. These 'informal' references to genetic techniques are particularly interesting to observe: they illustrate the way 'biotechnology' is slipping into everyday use.

Figure 2 gives the relative index of articles per year (up until 1986, and extrapolated from the sample from 1987 onwards – the break in the curve indicates the change to a full-text search). It is interesting to note that we do not observe a continuous progression, but there was a depressed period, roughly from 1992 to 1994. This, not surprisingly, coincides with the decrease in industrial activity in biotechnology discussed previously. To characterise the coverage of the theme, we have chosen a phase structure, indicated in Table 1, related to the one for policy-making. The two isolated years, 1970 (only one article) and 1975, are kept separate. A break is made at the end of 1986 for methodological reasons, when the searching technique was changed. The break between 1991 and 1992 marks the drop in frequency, and also coincides with the beginning of discussions in the political arena around the implementation of the European GMO directives (see the policy phases). A final phase marks the renewal of interest in the theme.

In Table 1 we characterise the phases by indicating the percentages of all articles that were written in a given frame, together with given themes, actors, risk–benefit ratios and locations. Only items with a score above 10% are indicated. Since an article can mention more than one theme, actor, etc., the total percentage can be greater than 100. It is important to note that a number of the results in Table 1 can be related to the characteristics of the paper selected (*Le Monde*), and not necessarily to some typically French position. For example, *Le Monde* is more closely linked to the institutional areas of politics and science than to the economic sphere. This explains in part the greater prominence of the 'progress' frame in comparison to the 'economic' frame, and the presence of more scientists than industrial actors. If another more business-oriented paper like *Le Figaro* had been selected, the proportions would probably have been different. Frames are fairly constant, with a few exceptions: the economics frame was not yet present in 1975, and reached a high in the prosperous years of 1987–91. Those years were followed by a drastic loss of 'progress' in the early 1990s.

The themes are also fairly constant, to a first approximation. The medical theme was strongly present from the beginning of the 1980s onwards. The proportion of basic research (and of scientific actors) dropped steadily as applications, e.g. identification techniques, took up more space. Analysed year by year, the evolution of the economic theme, paralleled by the presence of industrial actors, is interesting. It was high during the 1980s – in 1987 industrial actors were present in a record 70% of the articles – and then it dropped to a low between 1990 and 1992 when less than 20% of the articles mentioned them. The first phase was a time of optimism about the economic prospects of biotechnology, but it also reflected a period of general prosperity for the French economy. The early 1990s were a time of disappointment in biotechnology companies, but this period was also one of economic stagnation. The trend since 1993, however, has once again been an increase.

The balance between risks and benefits also reflects the same drop in optimism. Benefits had been strongly present since 1982 – more so than in the pioneer days of 1975 when applications seemed a long way off. The proportion of articles reporting only benefits dropped steadily after 1987; those reporting only risk increased in the most recent phase; and there are more noncommittal 'neither risk nor benefit' articles.

Ethical preoccupations, either as a frame for stories or as a main theme, remain present throughout the period under study, in 10–20% of the articles – again an attitude that one would expect of *Le Monde*. Interestingly though, the ethical theme is in phase opposition to the economic one: it was relatively low when economic prospects were good and high in the pessimistic years. These ethical preoccupations, as we shall see, make up an important part of the public perception of biotechnology, in France as in the rest of Europe.

A look at the final phase gives an idea of present trends. Progress is back, and so are scientists. The EU has appeared as an actor. Themes have become more diverse as biotechnology penetrates new areas of public discourse. Yet, and this again is consistent with public perceptions, a certain wariness remains:

the proportion of optimistic, 'benefits only', articles has not increased, and 'security and risk' remains an important theme.

Public perceptions

The 1996 Eurobarometer results give us an interesting portrait of the French public's attitudes towards biotechnology – a technology with which they are not as yet very familiar, but about which they already have concerns.

The effects of various new technologies

The answers to the question on the expected effects of either modern biotechnology or genetic engineering (*les biotechnologies modernes/le génie génétique*) are similar to those of previous surveys. In France the technologies considered most promising are telecommunications (82% positive answers), new materials (74%) and information technology (71%). This hierarchy is close to the European one (respectively 81%, 64% and 77%). Solar energy, which is well supported in Europe, is less highly regarded in France: 25% of the population considers that its development will have no effect, compared to 15% in Europe. Finally, the social utility of space exploration is not convincing either to Europeans in general or to the French in particular: less than half of the people in the samples think that it will improve their lives.

In comparison, the image of biotechnology is much less certain. About a quarter of the sample expressed no opinion about it – much more than for the other technologies. Both in Europe and in France less than half (respectively 47% and 49%) think biotechnology will improve their life. Finally, and above all, because no other area of development provokes as many negative answers, the response 'things will get worse' scored 15% in France and 19% in Europe.

The use of the term 'biotechnology' in one half of the sample and 'genetic engineering' in the other produces different attitudes to the rest of Europe, the first expression being more frequently evaluated positively than the second (51% vs 43% positive answers). In France, however, the difference between the two is small (50% and 48% respectively). Compared to the results of the 1993 survey, the attitude towards 'genetic engineering' has evolved positively, whereas 'biotechnology' has regressed.[2] This evolution may be due to the fact that 'biotechnology' has recently been linked to crop applications, which are perceived negatively by the public, whereas 'genetic engineering' refers more to medical applications that are evaluated very positively, particularly in France.

Applications of modern biotechnology

Figure 3 shows the judgements made about the usefulness, risk and moral acceptability of applications of biotechnology, and about whether they should be encouraged. The French pattern is not very different from that of the rest of Europe. Medical applications come out on top both for usefulness and for acceptability. The risk factor is not to be discounted, though, since negative answers roughly balance positive ones. The appreciation of transgenic plants is a little lower: in France, the average scores are positive except for the risk factor, which is negative. As for animal applications, including xenotransplants and genetically modified laboratory animals, negative aspects are dominant in the judgements even though the French give them a positive score on usefulness. Finally, the last field, that of food products, is clearly the most criticised: even the usefulness of these applications appears to be doubtful. The usefulness score is 0 and the risk score is the highest at −0.61. These applications are not morally condemned, since the score of moral acceptability remains slightly positive in the final evaluation; but negative judgements dominate.

Compared to the averages for the rest of Europe, the scores given by the French are not very different. However, the small difference observed is in fact composed of two opposite effects. The French are more positive than other Europeans about medical and animal applications; on the other hand, they are more reserved about food. Although these differences are small, they fit the expected cultural traits: less preoccupation with the fate of animals is often considered to be characteristic of the south of Europe, and particularly of France whose national culture is supposed to be influenced by Cartesianism. Concern about transgenic foods may be due to the fact that the quality of food is very important in France.

Attitudes towards different applications vary according to social, cultural and demographic factors. In France, as in other European countries, men, the younger age-groups, people with scientific qualifications, or those whose knowledge about biotechnology is highest, have the most positive attitudes towards biotechnological applications. Nevertheless, some specifically French points should be noted. First, it appears that the differences observed for these different categories are weaker than in the rest of Europe (considered globally). The logical consequence is that we find fewer explanatory factors in France than in the rest of Europe. A linear regression for the other countries shows five factors significantly related to the general attitude score: sex, age, political opinion, knowledge level and the degree of importance given to these issues and to participation. In France, only the knowledge

level and the degree of importance produce significant differences. One interesting feature is that, in the French case, support for biotechnology is stronger on the left of the political spectrum than on the right, whereas the opposite is true for the rest of Europe. Finally, for the rest of Europe the relationship between knowledge level and attitude is strictly linear, whereas in France the curve begins to decrease at the highest levels of knowledge. A high level of knowledge about biotechnology tends, in France to slightly increase reservations about the development of this area of activity, perhaps due to a more precise perception of potential risks.

The final questions were about the regulations that the public wished to see applied to biotechnology, and about trust in various sources of information about the field. On the first point, Europeans tend to prefer regulation first by international organisations (36%), then by scientific organisations (22%). The French approximate this profile, although their confidence is lower in international organisations (29%) and a little higher in scientific organisations (25%). They are also more enthusiastic about ethics committees (16%) than are people in the rest of Europe (7%).

Concerning information about biotechnology, other Europeans trust consumer groups (33%) and environmental groups (24%); and here again, the French are not fundamentally different, although they have more trust in consumer groups (46%) and less in environmentalists (20%). For part of the sample, the same question was put distinguishing three areas of application: modern biotechnology in general, transgenic plants, and gene transfer to animals for xenotransplants. Europeans tend to modulate their answers according to the application, and one notes, for example, that environmental organisations are chosen more frequently for information on transgenic plants (24%) and the medical profession for xenotransplants (46%). French responses follow this rule, and vary in the same proportions according to the application. However, the French express a more frequent choice of consumer groups (for biotechnology in general and for transgenic plants in particular) and a higher rating of the medical profession in all three applications. This can be understood in the context of the preceding question, where we saw that the French tended to be more positive towards medical applications.

The question on associations with biotechnology

When asked an open question about what the word 'biotechnology' brings to mind, nearly one-fifth (19%) of the respondents declared themselves unable to give an answer. The other responses have been analysed with Alceste, a software package for textual analysis that sorts groups of words in context, thus identifying a number of themes evoked. Six profiles appeared, beyond the fraction of the respondents that frankly admitted their ignorance. One group (8%), which we call the echo group, replied invoking various new technologies – precisely the ones that were mentioned in the preceding question about the future consequences of telecommunications, new materials, etc. We add this group that clearly guessed to the 'don't know' group to get a total of 27% of the sample that does not know the term 'biotechnology'. This is consistent with the proportion that had no opinion on the subject.

Two generally favourable groups appeared, which we shall call agricultural and medical. The first spoke of agricultural applications (11% of the responses) in quite technical terms ('to resist', 'productive', 'cell', 'laboratory'), and the second (10%) referred to medical applications with a more affective vocabulary ('illness', 'to heal', 'to treat') and in a more enthusiastic tone ('to discover', 'possibility'). A number of the answers nevertheless express strong ambivalence.

Two groups are worried about biotechnology: the 'danger' group and the 'fear' group. Members of the first group speak in terms of danger (6% of responses), make judgements (using words like 'ethics', 'mankind', 'to manipulate'), and have wider ranging references ('Brave New World', 'the witch's apprentice', 'mad cow disease'). Their vocabulary is more sophisticated ('embryo', 'insemination') and they use more logical operators or subjunctive verbs. Members of the second group (14%) speak in terms of 'fear', rather than 'danger', and express less power (using the verbs 'to have to', 'to be able to'). The area of concern is closer (personal pronouns, 'babies').

Finally, a group (14%) speaks of biological applications, with an ambiguity that illustrates the limits of such an automated analysis. 'Biological', in French, can refer to pesticide-free food (the usual term is *aliments biologiques* – literally biological food). Reading through the answers, we find that a number of respondents in this group think that biotechnology is about the protection of the environment. Others have a far more precise idea of what is meant by the term, speaking of biological applications. A larger sample would be needed to sort out the two subgroups.

To summarise: if we estimate that about half of the last-mentioned group is mistaken, about one-third of the population does not know what biotechnology means, explicitly or not.[3] About 20% fear biotechnology, and 21% are generally in favour of it. (The balance of the answers were declared unclassifiable by the software.)

Beyond these global scores, the split between both the favourable and the unfavourable groups into those with more sophisticated references (the agricultural and danger groups respectively) and those with more of a gut reaction (the medical and fear groups) is interesting. We have related this classification to other variables in the Eurobarometer and have found marked differences between the groups. For example, men are over-represented in the agricultural group (55.7% vs 49.8% in the whole sample) and even more so in the echo group that guessed (61.6%). Women are over-represented in the medical (53.4%) and fear groups (58.2%), along with elderly people (27% vs 21%).

The less educated (under 15 years of schooling) are more strongly present in the fear group (24.8% vs 17.9% in the whole sample) as well as among the 'don't knows'. People with higher education are more strongly represented in the agriculture (37.4%) and medical (45.6%) groups, but even more so in the danger group (52.4%), than in the whole sample (33.3%). The same is true of people with a scientific education (agriculture 22.6%, medical 18.4%, danger 20.6%, vs 13.3% in the whole sample). Managerial and white-collar professions follow the same pattern. Finally people with low knowledge are over-represented in the 'don't knows' (46.1% vs 31.2% in the sample), and also in the fear group (38.3%). Those with a high level of knowledge are either favourable about biotechnology (agriculture 46.1%, medical 36.9%, vs 31% in the sample) or see danger (44.4%). Incidentally, they are also better at guessing: they form 41.1% of the echo group.

These results confirm those described above: higher education and knowledge do not automatically imply better support of biotechnology – they induce a strong split between positive and negative attitudes.

Commentary

Biotechnology was a topic mainly reserved for specialists in France up until the end of the 1980s. Policy was elaborated quite confidentially and media coverage was not very intense. Not surprisingly, therefore, we have found that biotechnology does not have a very clear image in public perceptions, with at least a quarter of the population not knowing what it is.

The situation is clearly evolving. Over the past year, a number of events – a conflict over the authorisation for growing transgenic maize, the cloning of Dolly the sheep, a number of criminal cases in which genetic fingerprinting has been used – have put modern biotechnology into the limelight. In fact, the public debate has been slowly increasing since the beginning of the decade, partly under the influence of the implementation of European directives. Public perceptions have become more complex – biotechnology is commonly related to other techniques or events such as *in-vitro* fertilisation, mad cow disease, cloning or AIDS. This is no doubt scientifically mistaken, but it reflects a growing preoccupation with techniques that touch on vital processes.

In this study, we have been able to follow the entry of biotechnology into the French public sphere. At the same time, we have seen French public policy and perceptions evolve, partly because of the influence of Europe, but perhaps also because of a 'coming of age' of the French public, who are now more ready and willing to make moral judgements about technical issues.

Acknowledgement

This study was funded by the Direction Générale de l'Alimentation, Ministère de l'Agriculture, the Institut National de la Recherche Agronomique, and the Centre National de la Recherche Scientifique, département des Sciences de l'Homme et de la Société.

Figure 1. A regulatory profile for France

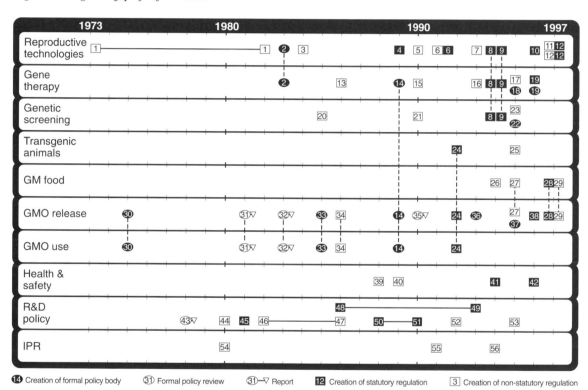

1   *1973.* The first Centre for Sperm Study and Storage (CECOS) established to support reproductive technologies development. These centres established the CECOS Federation on 23 March 1982.

2   *1983.* Creation of the National Consultative Committee on Ethics (CCNE), 23 February.

3   *1984.* Opinion no. 3 (CCNE) on ethical problems to do with reproductive technologies (23 October).

4   *1989.* Braibant's project of law on reproductive technologies and bioethics.

5   *1990.* Opinion no. 20 (CCNE) on organisation of gamete donation and its consequences (18 July).

6   *1991.* Opinion no. 24 (CCNE) on embryonic and foetal reduction (24 June); report by N Lenoir, 31 December.

7   *1993.* Opinion no. 40 (CCNE) on the post-mortem transfer of embryos (17 December).

8   *1994.* Law 94.654 (29 July) on reproductive technologies, prenatal diagnosis and human body products.

9   *1994.* Law 94.653 art. 16-4 of Civil Code protecting the human body (29 July) and prohibiting eugenic practices (OG 30 July). This law and Law 94.654 (see note 8) together regulate all aspects of reproductive technologies, prenatal diagnosis and human body. They are referred to as the 'bioethical laws'.

10 *1996.* Decree on regulation concerning medical health (12 November).

11 *1997.* Opinion no. 52 & 53 (CCNE) on collection of embryo tissue and cells (11 March).

12 *1997.* INSERM (National Institute of Medical Research (INSERM) report on *in-vitro* fertilisation (14 October). Decrees concerning studies of human embryos (27 May and 26 December).

13 *1986.* Opinion no. 8 (CCNE) on research and use of *in-vitro* human embryos for scientific and medical purposes and prohibition of germinal genetic therapy (15 December).

14 *1989.* Creation of the Genetic Engineering Commission on 11 May, decree no. 89/306 (OG May 13).

15 *1990.* Opinion no. 22 (CCNE) on gene therapy (13 December).

16 *1993.* Opinion no. 36 (CCNE) on the application of somatic gene therapy (22 June).

17 *1995.* National Academy of Medicine report on genetic diagnosis and gene therapy (28 March) .

18 *1995.* Creation of a Commission of Genetic Therapy on 28 March alongside the Agency for Medicines (OG 13 May).

19 *1996.* Law 96.452 (26 May), special statute on the products of cellular and genetic therapy. Establishment of the High Council of Gene and Cell Therapy (OG 29 May).

20 *1985.* Opinion no. 5 (CCNE) on problems raised by prenatal and perinatal diagnosis (13 May).

21 *1990.* Opinion no.19 (CCNE) on embryo research to achieve pre-transfer gene diagnosis (18 July), on which a moratorium was declared in 1986.

22 *1995.* Decree 95.558 of 6 May 1995 concerning the National Commission of Medicine and Biology on the reproduction and the prenatal dignosis and modifying the Health Code.

23 *1995.* Opinion no. 46 (CCNE) on genetics and medicine from prediction to prevention (7 November).

24 *1992.* Law 92.654 (13 July) concerning the use, control and the release of GMOs, and modifying Law 76.663 (19 July 1976) regulating industrial plants and the protection of environment.

25 *1995.* Decree no. 95.487 (28 April) implementing, for genetically modified animals, title III of Law 92.654 on the release of transgenic animals.

26 *1994.* Decree 94.46 of 5 January (OG 19 January) determining conditions for the deliberate release of GMOs used for human and animal consumption, other than plants, seeds, or plants or animals comprising products related to human or animal food.

27 *1995.* Order (18 July) determining the content of the authorisation register for the deliberate release and marketing of GMOs used as food for human and animals other than plants, seeds, or plants or animals contained in or being in contact with such food (OG 31 August).

28 *1997.* Ministry of Agriculture authorises the commercialisation of Ciba-Geigy's transgenic maize (4 February). However, on 12 February the Prime Minister forbids the culture of the transgenic maize in France.

29 *1997.* Advice published by the Ministry for consumer interests interpreting the legal provisions related to the labelling of GM food. On 17 June there was an Opinion of the National Council for Nutrition concerning the labelling of GM food.

30 *1975.* Creation of an administrative commission within the Ministry of Research, the National Classification Committee for In-vitro Recombinations, with the aim of controlling gene recombinations from a technical point of view (March).

31 *1981.* Royer report on the security of industrial applications in the field of biotechnology.

32 *1983.* Bodessa report on biotechnology and of security problems raised by GMOs.

33 *1985.* Creation of the Interministerial Group for Chemical Products (GIPC).

34 *1986.* Publication of reports on technical terminology for biotechnology: AFNOR NF X 42 000, NF X 42 001, and NF 42 002.

35 *1990.* Chevallier Report (OPECST) on applications of biotechnology to agriculture (December). Contains some analysis on GMOs and the risk of their release in the environment.

36 *1993.* Creation of CGB, decree 93.235 concerning the establishment of the Committee for the Study of Release of the Products of Biomolecular Engineering (23 February).

37 *1995.* Decree 95.487 implementing title III of Law 92.654 on GM animals (28 April).

38 *1996.* Decree 96.317 concerning the procedures related to the deliberate release of GM products and genetically modofied human body parts (10 April).

39 *1988.* Publication of several technical standards (AFNOR) related to security in biotechnology (NF X 42. 050, 051).

40 *1989.* Publication of a technical standards (AFNOR), guidelines for good practice for research and development in the field of biotechnology (NF X 42-070) (April).

41 *1994.* Decree no. 94.352 (4 May) concerning protection of workers against risks (*OG* May 6).

42 *1996.* Order on the protection of workers against exposure to biological agents in academic laboratories and industry (13 August).

43 *1978.* Report 'Gros-Royer-Jacob' commissioned by President Giscard D'Estaing, pointing out the importance of research in biotechnology.

44 *1980.* The Committee on Strategic Industries (Ministry of Industry) places biotechnology on the priority list.

45 *1981.* First governmental financial initiatives for the development of research in biotechnology.

46 *1982.* Publication of the foundation document on the Mobilisation Programme in Biotechnology, under regis of a national committee presided over by the Minister of Industry and Research (1982–86).

47 *1986.* Second version of the Mobilisation Programme is produced (1986–90).

48 *1986.* 'Nutrition 2000' Programme on biotechnology and food processing (1986–92).

49 *1993.* 'Tomorrow's Nutrition' Programme on biotechnology and food processing.

50 *1988.* Law 88.1138 (20 December) on the protection of individuals participating in biomedical research.

51 *1990.* Law 90.549 (2 July) modifying law 88.1138 (see note 50).

52 *1992.* Start of 'Biofuture' Programme (1992–97).

53 *1995.* The Ministry of Research set up 14 scientific committees responsible for carrying out projects in 14 strategic programmes in the life sciences, divided into seven priority areas (genetics, structural biology, etc.).

54 *1980.* Decree of implementation by France of the

Budapest Treaty on the international recognition of the need depositing of microorganisms for the purposes of patent procedure (25 November).

55 *1991*. Opinion (CCNE) on the commercialisation of the human body, recommending the non-patentability

of human genes (2 December).

56 *1994*. Law 94.653 on the human body, modifying article L.611-17 of the French intellectual property code, forbidding the patentability of the human body, human body parts and human genes (29 July).

Figure 2. Intensity of articles on biotechnology in Le Monde

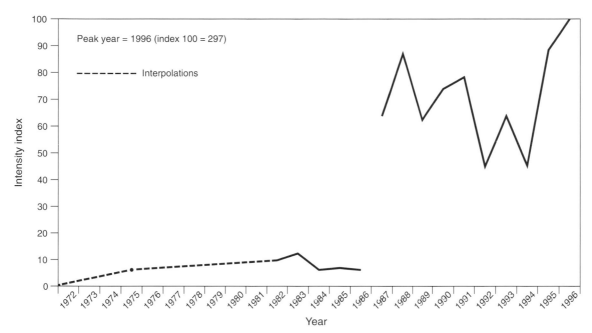

Figure 3. Attitudes to applications of biotechnology in France, 1996

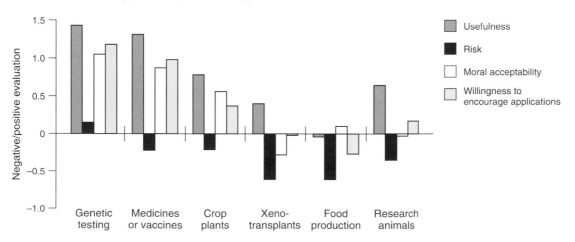

Table 1. French media profile (for an explanation of terminology please see Appendix 5).

Phase	Freq.[a] (%)	Frame (%)		Theme (%)		Actor (%)		Benefit/risk (%)		Location (%)	
1. 1970	1	**Progress**		**Medical**		**Scientific**		**Benefit only**		**Germany**	
				Basic research						USA	
2. 1975	17	**Progress**[b]	41	*Basic research*	53	*Scientific*	106	*Both*	47	*France*	71
		Pandora's Box[c]	12	Security & risk	47	Political	20	Benefit only	18		
				Ethical	29	Ethical	13	*Risk only*	18		
				Regulations	29	International	13	Neither	18		
						NGOs	13				
3. 1982–86	22	*Progress*	48	**Medical**	45	**Scientific**	72	*Benefit only*	48	*France*	55
		Economic	21	*Economic*	35	Industry	30	*Both*	26	USA	30
		Ethical	13	Basic research	30	Political	24	Neither	15		
				Ethical	21			Risk only	11		
				Political	18						
				Security & risk	14						
4. 1987–91	202	**Progress**	35	**Medical**	46	**Scientific**	55	*Benefit only*	48	*France*	70
		Economic	23	Basic research	27	*Industry*	39	Both	25	USA	16
		Ethical	10	Economic	24	Political	21	Neither	18		
				Ethical	14						
				Political	14						
5. 1992–94	143	**Progress**	27	*Medical*	48	**Scientific**	51	**Benefit only**	35	*France*	57
		Economic	20	Economic	18	Industry	26	*Neither*	30	USA	19
		Ethical	20	Ethical	18	*Political*	26	Both	20		
				Basic research	16			*Risk only*	16		
				Identification	13						
6. 1995–96	270	**Progress**	45	**Medical**	43	**Scientific**	63	*Benefit only*	36	*France*	50
		Economic	15	Basic research	24	Industry	21	Neither	26	*USA*	33
		Ethical	10	Economic	19	Political	14	Both	22	*UK*	12
				Security & risk	17	*EU*	11	Risk only	15		
				Agriculture	11						
				Ethical	10						
				Regulations	10						

a Percentage of corpus in the period; $n = 623$.
b Bold indicates highest frequency within phase.
c Italics indicates highest frequency within category.

Notes and references

1 Government report (1991), *Les 100 technologies clés pour l'industrie française – l'horizon de l'an 2000*.
2 In the 1993 Eurobarometer report, average scores are calculated by giving 1 point to the 'improve' answer, 0 to 'no effect' and −1 to 'things will get worse'.

For France, the scores for 1993 and 1996 were respectively 0.63 and 0.5 for 'biotechnology', and 0.27 and 0.41 for 'genetic engineering'.
3 This group consists of the 'don't know' group, the echo group and about half of the biological group.

Address for correspondence

Dr Suzanne de Cheveigné, Laboratoire Communication et Politique, 27 rue Damesme, 75013 Paris, France. Tel. +33 1 44 16 73 66 (or 64), fax +33 1 44 16 73 69. E-mail dechevei@gps.jussieu.fr

Germany

Jürgen Hampel, Georg Ruhrmann, Matthias Kohring and Alexander Goerke

The German political system

The German political system was configured by the experience of the collapse of the Weimar Republic. The idea of the control of political power, combined with mistrust in all forms of plebiscitary democracy, stem from this experience, which was decisive for the German Constitution, which allows only representative forms of public participation.

The federal structure of Germany is deeply rooted in its history. There are now 15 federal states (the Bundesländer), which participate in political decisions and have their own areas of regulation. Important internal affairs, culture and education (including schools and universities) are regulated mainly on the federal level rather than by the central state.

The legislative assembly consists of two parliaments: the Deutsche Bundestag which is elected by the people, and the Bundesrat which represents the interests of the federal states and whose members are delegated by the state governments. Both the Bundestag and the Bundesrat have to agree any law that affects the interests of the federal states.

In the 1960s and 1970s the German political system was in general a three-party system, involving the liberal-conservative Christian Democratic Union/Christian Social Union (CDU/CSU), the Social Democratic Party (SPD), and the liberal Free Democratic Party (FDP). In the 1980s, the newly established Green Party, which started as a single-issue environmental party, became a stable member of the party system. After reunification in 1990, the socialist Party of Democratic Socialism (PDS), the successor to the East German Socialist Unity Party (SED), was established. It remains, in 1998, a regional party with political importance mainly in East Germany.

In postwar Germany, a political model was implemented of a society that is as self-regulating as possible (this is the principle of subsidiarity). Whenever possible, associations should negotiate with the other associations in their fields. Thus the political system in Germany can be described as 'neocorporatist'. The state is in the position of guardian, and should become active only if self-regulation is not possible or if there are conflicts with other regulations.

The so-called *Tarifautonomie*, state-free collective bargaining as a way of handling labour conflicts, is a well-known example of this principle. Self-regulation can also be seen in other fields, such as medicine and science, that first became important in respect of genetic engineering. The medical profession has its own regulatory institutions, as does the research system.

From 1982 to 1998, the German government was a coalition between the Christian Democrats and the Liberals. The opposing Social Democratic Party, in coalitions with different parties, governed most of the federal states and therefore held the majority in the Bundesrat. Because the Bundesrat has to agree to laws affecting the interests of the federal states, the different majorities in the two parliaments led to a strong orientation towards consent in the German political system.

Modern biotechnology in research and industry

Biotechnological research in Germany is very competitive. The German world share of inventions in biotechnology and genetic engineering from 1989 to 1994 was 10%. This share has decreased, from 12% in the 1980s to 10.3% in 1989–91, and to 9.7% in 1992–94.[1] As the German share decreased, the US and UK shares increased. Nevertheless, after the USA and Japan, Germany ranks third in the patent statistics.[2]

Although the research base of genetic engineering and biotechnology is quite strong, modern biotechnology does not seem to be very important in economic terms in Germany. According to a study from Ernst & Young, there are less than 100 companies in Germany dealing with modern biotechnology.[3] Estimates of the number of employees in modern biotechnology (R&D, production, services, suppliers) range from 13,000 employees (according to the OECD in 1996) to 58,000–60,000 in 1995.[4] The German government sees genetic engineering as a key technology comparable to microelectronics and information and communication technologies.[5] Funding for genetic engineering from the German Ministry for Research has increased from DM57 million in 1975 to DM309 million in 1997. One example of political support for innovation in the economic use of genetic engineering was the Bio-Regio Contest, where the three best regions received special financial help to develop industrial applications.

Sales of biotechnological products in Germany reached approximately DM2200 million in 1995.[6] More than 90% of this sum was due to pharmaceutical, therapeutic and diagnostic products,[7] but few of these products had been developed and produced in Germany. Production in the food sector and agriculture was negligible.[8]

How can this discrepancy between strong research competitiveness and weak economic performance be explained? With one exception, the large German companies engaged rather late with modern biotechnology.[9] Although there had been hints from the administrative system as well as from science that genetic engineering and biotechnology would offer new fields for innovation, the German chemical and pharmaceutical industries failed to recognise biotechnology as an innovative business sector. According to Dolata, this lack of interest was a symptom of the dominance of synthetic chemistry in these companies – a field in which German companies have been very successful.[10] So the major impulses for development in biotechnology came from the German Ministry for Research and Technology (BMFT), and were funded with public money. The fact that investments by German companies in biotechnology happened mainly in the USA can be seen as a 'catch-up modernisation':[11] German companies tried to buy scientific competence. In the public debate this delay is usually considered to be symptomatic of the resistance of the German population to genetic engineering.

Less important in economic terms than pharmaceutical applications of genetic engineering are applications in agriculture. Although Germany is not an agrarian country, plant breeding is important in economic terms: the market for seeds is approximately US$1500 million[12] – about 3% of the world market. German plant-breeding companies are rather small: the largest is ninth in the world, and the second largest is seventeenth. There were 49 release experiments on genetically modified plants in Germany between 1986 and 1995, compared to 113 in the Netherlands, 133 in the UK, 253 in France and 1952 in the USA. In Europe Germany ranked sixth, after France, the UK, the Netherlands, Belgium and Italy.[13] On the level of their knowledge base in modern biotechnology, German plant-breeding companies are competitive on the world market, but they are concerned that as other countries develop and use genetically modified plants which are, for example, designed for the food industry, the German market share will decline.[14] So in both pharmaceuticals and plant breeding, the use of genetic engineering seems to be driven by the need to fend off the substantial economic disadvantages that will result from competitors' successes in biotechnology.

Implementation of genetic engineering in Germany was not unproblematic.[15] The overwhelming acceptance of technology that dominated the period of the *Wirtschaftswunder*, the rapid growth of the post-war economy in Germany, has disappeared.

Longitudinal studies show that public acceptance of technology declined considerably in the 1970s and 1980s.[16] Accordingly, discussions about new technologies in Germany increasingly took the form of political conflicts.

New political and scientific institutions have developed as a consequence of the controversy over nuclear power. From 1981 onwards, the Green Party, formerly an environmental group, became a widely accepted member of the German political system, participating in the government of several federal states. In the late 1970s, as a counter-reaction to traditional scientific risk expertise allied to the political system, new scientific institutes like the Eco-Institutes in Freiburg and Darmstadt were founded to provide scientific expertise in support of resistance to nuclear energy. Furthermore, strong NGOs like the Association for the Protection of the Environment and Nature (BUND) were established in this period. According to Rudzio, Greenpeace, the BUND, and the German Association for the Protection of Nature have a total of about a million members.[17]

The history of public policy

An overview of the development of legal regulation of genetic engineering is shown in Figure 1, where the phases of the regulation process can easily be detected.

Phase 1, up to 1984: neocorporatist self-regulation. In the first phase, which lasted up to the early 1980s, the political system was more involved in providing incentives than in regulating. Already in 1970, before we can use the term 'genetic engineering', the German government saw opportunities for developing modern biology, and started a programme to focus research efforts on 'biology, medicine and technology'. Regulation was not seen as a task for the political system. In this period, corresponding to the neocorporatist political culture, genetic engineering was regulated, not by the political and administrative system, but by the active institutions themselves.

In the medical field, decisions were to be made only by organisations of the medical profession. The first of these to regulate research were the ethics committees, the first of which was established in 1971. The scientific use of genetic engineering was to be regulated by the scientific community. As a result of that stage in the development of genetic engineering, the main focus of regulation was laboratory safety. The first legal regulation in that area, the guideline on protection from dangers resulting from *in-vitro* recombined nucleic acid, was passed in 1978. All five versions of this guideline regulated only the research of federal research institutes and research funded by the federal government. But this guideline was not legally binding: researchers were left to regulate themselves.

At the same time, in 1978, the Research Ministry, BMFT, presented a draft of a law concerning protection from the risks of genetic engineering. This draft did not lead to formal regulation because of resistance from scientists and industry.[18] They saw it as a threat to the freedom of research, and felt there was no need for formal regulation. The federal government gave up further plans for regulation.

In summary, then, the neocorporatist approach owes more to the strong and successful resistance of social sectors in defending their independence than to the self-restraint of the public regulatory authorities. In this first phase, which lasted until the early 1980s, the German public was not really interested in these questions.[19]

Phase 2, 1984–90: the road to the German Gene Law. The situation changed in the early 1980s. After the birth of the first test-tube baby in Germany, an ethical and moral discussion on biotechnology emerged. The BMFT called for a public discussion on acceptance. To avoid a fundamental discussion about genetic engineering, the scientific community was urged to start the discussion itself, to display its knowledge and authority.[20] This encouraged the channelling of the emerging public debate with the aim of referring discussion back to the scientific experts.[21] At the same time, the German government started several new programmes to support research on biotechnology.

Gill observed that, at that time, the German government tried to concentrate the political debate on technological details.[22] However, discussions on genetic engineering only really started in 1984, when both the working group on *in-vitro* fertilisation, genome analysis and gene therapy (the Benda Commission) and the parliamentary Enquete Commission on benefits and risks of genetic engineering were established. For the first time, genetic engineering was discussed in broader contexts than just the protection of embryos and containment guidelines.[23]

The Benda Commission was established by the ministries of Research and of Justice. It had two tasks: to examine the potential of modern reproductive technologies, and to assess the need for regulation. In November 1985 the Commission presented a number of suggestions for regulation. The consensus in this working group, which represented a wide range of interests (science, churches, employers and employees, humanities and legal sciences, and medical associations) was broad, and its recommendations led to the Law for the Protection of Embryos on 13 December 1990.

In parallel to the establishment of the Benda Commission, the Green Party and the Social Democratic Party demanded a parliamentary Enquete Commission on genetic engineering. The two parties had different interests. While the Green Party wanted to prohibit the use of genetic engineering (it should be allowed only in exceptional and revocable circumstances), the Social Democratic Party wanted to discuss the risks and benefits of genetic engineering in ecological, economic, legal, social and safety-technological contexts. The governing coalition agreed to the Social Democrats' proposal with the precondition that the Enquete Commission should talk equally about benefits and risks. Not included on the Commission were environmental and consumer organisations and farmers' representatives.

The Green Party had to recognise that their formulation of the task of the Enquete Commission would not prevail. So the Party changed its strategy and concentrated on mobilising the public.[24] The first measure was a congress in 1985 entitled 'Women against Genetic Engineering and Reproductive Medicine'.[25] As a consequence, the representative of the Green Party distanced herself from the final report of the Commission and provided her own report.

The Enquete Commission made the distinction between human genetics and industrial applications of genetic engineering. On human genetics, there was a broad consensus that human cloning and the creation of chimeras should be prohibited. The Commission recommended reformulating an existing law on contagious diseases, rather than formulating a separate gene law. This raised a juridical discussion over whether a genuine gene law would be necessary or not. Especially opposed to a formulation of a gene law were representatives of scientific organisations. Despite objections, the government passed, after the final report of the Commission in 1986, standards of a planned gene law. Two decisions increased the pressure to formulate a genuine gene law. First, the Bundesrat decided that commercial applications of genetic engineering in plant production should be integrated into the federal law on protection from emissions. This integration would mean that public consultation would be necessary before the approval of a plant. Industry feared, that, as a consequence, the approval process would be prolonged. Secondly, in November 1989, the Supreme Administrative Court of Hessen prohibited the chemical company Hoechst from building a factory to produce human-insulin-producing plants using genetically modified bacteria.[26] This decision was based on the argument that a biotechnological plant could not be approved by the Court in the absence of a legal basis for approval. This judgement prompted industry to push for the swift formulation and passage of the German Gene Law, a process which was concluded in May 1990. The Hoechst plant was built, in compliance with the terms of the law, six years later.

The two EU guidelines (90/219/EWG, 90/220/EWG) were reflected in the formulation of the German Gene Law, but with several changes.[27]

Phase 3, 1991–96: deregulation. Soon after its implementation, the Gene Law was criticised for being too restrictive. The legal regulation of genetic engineering in Germany was seen as threatening the economic competitiveness of the German biotechnology industry. As a consequence, all the opportunities offered by the EU guidelines for simplifying the legal procedures were implemented in a modification of the Gene Law, which was passed in December 1993.

In 1995 and 1996 came several prescriptions for simplifying and accelerating the regulation of genetic engineering: the Biotechnology Safety Prescription from 21 March 1995; the reformulated Biotechnology Hearing Prescription (the 4 November 1996 version), which restricted public hearings to proceedings concerned with safety levels 3 and 4; and a further five prescriptions.[28] Although the political regulation of modern biotechnology seemed to reflect a relaxation of the situation, the public debate continued.

Phase 4, 1996–onwards: regulation on a European level. There is no longer any special German legislation for any new problems. This can be seen in regard to the discussion on patenting, as well as in the discussion of the Novel Food Prescription (285/97). Although there is a great need in German society for regulations on these topics, a national law or prescription has not been formulated. According to Article 15 of the Novel Food Prescription, the prescription is binding for each state of the EU. But up to now, there have been no concrete implementation regulations.

Media coverage

The print-media sector in Germany is a highly complex and diversified market dominated by four big multinational companies. The daily newspaper appears to be the most popular and significant print-media type. However, about 75% of all daily newspapers have a circulation of less than 50,000. In comparison, *Bild*, the well-known German tabloid newspaper, reaches a circulation of more than 4 million per day. Traditionally neither the typical daily newspapers nor the tabloid press are regarded as opinion-leaders. This role is played by five national quality newspapers (which have a total circulation of about 1.3 million) and a group of weekly news magazines (such as *Focus, Der Spiegel*

and *Die Zeit*), because of the number of news agencies they use, their considerable and strong network of international correspondents, and their skill at using investigative journalism. Finally, there are also a huge number of special-interest magazines in Germany.

Description of the sample

Media coverage of biotechnology was studied in Germany through a sample consisting of two papers: the weekly news magazine *Der Spiegel*, and the daily newspaper *Frankfurter Allgemeine Zeitung. Der Spiegel*, famous for its investigative journalism, is the most important German news magazine – and until the 1990s it was the only one with a circulation over 1 million. Politically, it stands on the moderate left. On the other hand, the *Frankfurter Allgemeine Zeitung (FAZ)* stands on the right of the political spectrum. This national quality newspaper has a circulation of 400,000. Both papers have a science section, and so have specialist editors for science and technology. Both papers are opinion-leaders, and influence the agenda of most other German papers.

Therefore, we consider our sample as representative of German coverage of biotechnology from 1973 to 1996. The unit of analysis is defined as any semantic complex of journalistic origin dealing in any sense with biotechnology. This includes written texts as well as pictorial material. Letters to editors were also included, but not advertising (for example, campaigns by the biotechnology industry).

To collect the articles we constructed two systematic samples: for the *FAZ* we used two artificial weeks, selecting at a particular time two different days of every week. Following this sampling method, one-third of the entire coverage was screened. For *Der Spiegel* we screened every second edition. Every article about biotechnology was then chosen for the sample. The selection was made by hand for the whole period.

In total, the sample consisted of 418 articles from the *FAZ* and 170 articles from *Der Spiegel*, giving a total of 588. Because there is no reasonable theory of how to compare the 'reader effect' of articles from daily and weekly papers, we did not apply any weighting measure to our analysis. With the limitations mentioned above, our results reflect the average image of German coverage of biotechnology over the last 25 years.

Phases of the coverage of biotechnology in Germany

Figure 2 shows the intensity of coverage from 1973 to 1996. The peak year, 1996, in which 174 of an estimated total of 1254 articles were published, is used as an index year for the development of the coverage.

We used the intensity of coverage to identify the different phases because intensity of coverage indicates journalistic attention and, therefore, possible public attention on certain issues. Our interest in media coverage stems from its possible effects on public attitudes and opinions as preconditions for political behaviour. It is therefore reasonable to see media attention as the most important criterion for detecting the politically interesting phases of biotechnology coverage.

Thus we were able to identify three phases of media attention, which we can characterise by frames, themes, actors, risk-and-benefit evaluations, and locations (see Table 1).

Phase 1, 1973–84: the perspective of scientific progress. This first phase contained about one-fifth of all the articles. It was dominated by the 'progress' frame: 60% of the coverage, more than in all other phases, was linked to the idea of scientific progress. However, only in this phase did the frame 'Pandora's Box', denoting uncontrollable catastrophic consequences, play a relatively important role (7%). Nevertheless, 60% of all articles expected only benefits from the new technology, mainly with reference to research and health. When risks and benefits were mentioned in the same article, which happened in a quarter of cases, the main opposition was moral and health risks vs economic and health benefits (see Table 2).

In the early years the coverage was clearly dominated by medical themes and basic research issues. Scientists were the main actors, and most of the biotechnology coverage referred to locations in the USA.

In summary, the media portrayed biotechnology as a scientific breakthrough accompanied by some (inevitable) risk, while at the same time promising huge benefits for society as a whole. From the media's point of view, biotechnology was widely considered as science's own business. One may assume that at this time the German media tended to regard biotechnology as something remote from the public for whom they were reporting.

Phase 2, 1985–91: the debate over regulation and public accountability. The second phase comprised about 35% of all articles. Now the media's focus of attention had shifted to Germany: half of the articles placed the reported event in Germany. This development was accompanied by a significant change of all characteristics constituting the former phase: the progress frame was less dominant than before, and the role of scientists as the major actors in biotechnology was challenged by political, industrial and NGO actors. The proportion of exclusively benefit-oriented articles slid down to a still considerable 43.4%. Finally, the spectrum of biotechnological themes was no longer dominated by medical and basic research issues. Overall, the media

coverage of biotechnology in the second phase was characterised by a distinct increase of those factors that indicate a turn towards the public discussion of the risks, benefits and regulation of biotechnology. More than one-fifth of all articles framed the story as one of public accountability, and 12.7% applied an ethical frame. This observation corresponds to the reported themes: 45% of the articles dealt with safety and risks (13.9%), concrete questions of regulation (12.8%), political matters (11.6%) and ethical aspects (5.5%). The second phase was the phase with the highest percentage of political actors (about one-quarter), and the only phase in which the media themselves were actors in the biotechnology debate (7.1%). It was also the first time that NGOs entered relatively often into public discussions (4.5%).

Furthermore, the years from 1985 to 1991 show the highest rate of articles (37.1%) that mention both risks and benefits. Issues that attract attributions of benefit as well as of risk are of special interest for the media analysis, because they form a reservoir of topics where the journalists' efforts to reduce uncertainty, in one way or another, have failed. This observation can be read as an indicator of public conflict and insecurity about the possible consequences of biotechnology. These conflictual risk/benefit articles placed health, economic and research benefits in opposition to environmental, moral and health risks (see Table 2). Nevertheless, more than 40% of the articles mentioned only benefits, and the number of articles giving exclusively risk estimations did not increase.

In summary, the second phase of biotechnology coverage in Germany can be characterised on the one hand as highly influenced by a strong public and political debate about risks and benefits of the forthcoming key technology, followed by considerations of how to regulate these possible outcomes. On the other hand, there are clear indications that journalism still portrays biotechnology as a benefit-oriented branch of scientific progress that does not seriously endanger society.

Phase 3, 1992–96: the globalisation of biotechnology. The third phase included about 46% of all articles. After a decrease to less than 30% of the index-year level, there was a continuous and steep increase of the coverage intensity until 1996, the last year included in our study.

Whereas the progress frame maintained a score of 48%, the frame of public accountability dropped to its lowest level of all three phases (11.2%). Corresponding to this development, the ethical frame apparently lost its appeal (now 5.9%). The sole increase with regard to story framing concerned globalisation, which was a feature of almost 11% of the articles. This increase was mainly due to the articles that reported economic issues (10%). More than one-third of these were in the globalisation frame. This corresponds to the observation that the globalisation debate as a whole was fostered by economic issues. Like the second phase, the media agenda was constructed from several heterogeneous issues. It comprised medical (19.1%) themes and basic research issues (down to 12%) as well as regulatory (12.3%), political (11.1%) and safety and risk topics (7.4%).

Scientific sources were still the most important actors in the coverage of biotechnology (42.8%), which strikingly contradicts the common assumption that scientists face serious difficulties in gaining enough media attention for their work.[29] In contrast to the monocultural source reference of the early years, the second and third phases were characterised by a kind of normalisation, in the sense that different actors became able to make their different claims in the media. Industry in particular (17.8%) played a more and more important role, but NGOs (5.8%) also established themselves in the public debate. The trend of globalisation is illustrated by the fact that for the first time a transnational actor – the EU (4.5%) – received considerable attention in the media coverage.

If it is reasonable to interpret articles with combined risk and benefit estimations as indicators of public conflict, the German media coverage of biotechnology in the 1990s becomes more and more benefit-oriented,[30] and hence less conflictual. More than the half of the articles mentioned only the benefits of biotechnology (especially for research, but also for health and economics; see Tables 1 and 2). Some applications of biotechnology (for example novel foods), however, may still have provoked public outrage. The conflictual risk/benefit articles decreased in quantity from 1992 onwards, but they still made up more than a quarter of the coverage. These articles – which are perhaps the most interesting for analyses of the style and content of the debate – placed economic (42%) and health (27%) benefits in opposition to health (31%), moral (18%) and environmental risks (14%); and for the first time they discussed risks to the consumer (14%). A considerable proportion (13%) of the articles mentioned neither risks nor benefits, compared to the smaller number (5%) that mentioned only risks.

The globalisation of the debate placed events in biotechnology in a transnational or international environment,[31] especially with respect to other European countries (about 26%). Furthermore, nearly 30% of the stories covered events in the USA. It is striking that two-thirds of the coverage concerned other countries. German opinion-leading papers chose to report on biotechnology even though the majority of events did not take place in Germany.[32]

On one hand the third phase of the German biotechnology coverage showed a decrease of the formerly strong public and political debate about risks and benefits of biotechnology; but on the other hand it did not return to the former dominance of the frame of scientific progress. The reported image of biotechnology seemed to be more pragmatic, weighing up the risks and benefits, but clearly in favour of biotechnology. The proportions of different themes and actors became more equal (referring to an enlarged scope of interests), and there was a distinct tendency to broaden the debate to an international and increasingly complex context. Viewed through the media coverage, the current German debate about biotechnology can be characterised as a pragmatic, problem-oriented discussion of the potential benefits of biotechnology.

Public perceptions

Acceptance

At first glance, Germany is one of the countries most concerned about modern biotechnology. First analyses of the 1996 Eurobarometer survey showed that the German public, together with the Austrians, expressed the lowest support in the EU for genetic engineering.[33] This view of the German situation is supported by the debate on 'acceptance' of genetic engineering: industry and politicians complain that the sceptical views of the German public are a threat to further economic development in the global market.

But at second glance, the image of an emotional and irrational German public resistant to biotechnology cannot be supported. Indeed, it can be shown that Germans have less positive expectations about biotechnology than the average European, and than the populations of most other European countries. Compared with the 45% of Europeans who think that genetic engineering or biotechnology will positively influence their lives within the next 20 years, the rate in Germany (36.2%) is rather low.[34] But from the opposite perspective, in its expectation that biotechnology will make life worse, Germany is no longer an exception in Europe. Indeed, the proportion of Germans who think that genetic engineering will have negative effects on their lives (23.2%) is higher than the European average (19.2%), but people in other northern and central European states are more concerned about the further implications of biotechnology than are the Germans. If we restrict this analysis to the sub-sample that explicitly evaluated genetic engineering, the positive expectations are reduced to 32% in Germany and negative expectations are shared by more than a quarter of the population (27.6%). But in other countries, such as the Netherlands, Denmark, Austria and the UK the scepticism is higher.

Looking at the evaluation of the likelihood that several positive and negative developments will occur – for example, that genetic engineering will create new diseases or that genetic engineering will help to solve world hunger problems – it can be shown that not only regarding positive expectations (which is not surprising),[35] but also regarding negative expectations,[36] the Germans rank below the European average.

That the evaluation of biotechnology is not fundamentalist in nature is also demonstrated by the fact that only a few people use the extreme points of the attitude scales, and by the fact that Germans evaluate the importance of biotechnology lower than do most other Europeans. On a 10-point scale (10 is extremely important), we get an average of 6.1 in Germany, while the European average is 6.5. Even the proportion of people who think that modern biotechnology is extremely important is lower in Germany (9%) than in most of the other European countries.

Nevertheless, it would be misleading to conclude that genetic engineering and biotechnology are non-themes in Germany: they are part of public and private communication. Fewer Germans than other Europeans reported that they had never talked about biotechnology before (33% compared to 52% in the EU). Only in Denmark was this proportion even smaller. On the other hand, the proportion of people who often talk about biotechnology is higher in other countries (Denmark, the UK, the Netherlands and Austria).

Social differentiation

Compared with other central European states, knowledge levels in Germany are rather low. The average number of correct answers to knowledge question in the 1996 Eurobarometer survey was lower in Germany (4.27) compared to Europe in general (4.43) and especially compared to other central and northern European countries. Regarding the knowledge questions, there is one extremely conspicuous result in Germany which is rather singular in Europe: almost half (44%) of Germans think that ordinary tomatoes do not have genes.

As in Europe in general, knowledge in Germany is not responsible for the evaluation of biotechnology.[37] Nor does educational level explain rejection. However, the higher the educational level, the more positive are expectations about genetic engineering. In West Germany, positive expectations of genetic engineering are held by only 26% of those who finished their education before the age of 15, but by more than 46% of people with higher education (in East Germany the rates were 30% and 41%). The higher the educational level, the lower is the proportion of people who say that this technology will have no effect; there are also fewer 'don't know' answers.

But respondents with negative expectations of genetic engineering do not differ according to their educational level. This is, however, only true for West Germany: in East Germany, more highly educated people have more negative expectations of genetic engineering (25%) than people with lower levels of education (17%).

As expected, people with a training in natural science have more positive expectations (58%) than people with a training in the humanities (48%), but this difference is not significant. These results are very similar for West and East Germany.

Men and women show significant differences. More men (about 40%) are optimistic about genetic engineering than women (about 33%), while more women (28% in West Germany, 22% in East Germany) are sceptical about genetic engineering than men (21% in West Germany, 17% in East Germany). With increasing age we find a substantial decline in positive expectations.[38] This does not lead to a general scepticism but to greater indifference. The older people are, the more they think that the implementation of genetic engineering will have no effect. The most supportive group in West Germany are young women (up to age 24), followed by men aged between 40 and 54. In East Germany both men and women between 25 and 39 are the most supportive. The most critical group are West German women aged 55 and over.

Applications of biotechnology

The attitudes of Germans to biotechnology, which should be described as sceptical rather than negative, were endorsed by evaluations of attitudes to different applications of biotechnology.

Respondents were asked, for each application, whether or not it is useful, risky and morally acceptable, and whether it should be encouraged (see Figure 3). While genetic testing and medicines were evaluated as useful, morally acceptable and to be encouraged, food production was seen as risky and applications involving the genetic manipulation of animals were deemed morally unacceptable.

Evaluation of the risks of different applications of biotechnology was more modest in Germany than in many other European countries, but this was also true of the evaluation of benefits. Fewer Germans than other Europeans think that genetic engineering is really useful, which relates directly to the configuration of the positive and negative expectations. These results support the view that Germans are not averse to but are sceptical about biotechnology and genetic engineering. But there is one substantial difference between Germany and other European countries: with some exceptions, Germans are more aware of the moral problems of biotechnology than most other Europeans.

It is known that moral acceptability, and then usefulness, are the most important predictors of support, and that the perception of risk is relatively unimportant.[39] But the question still remains: are there applications where usefulness is more important than moral acceptability? We undertook a regression analysis for each of the applications to look at the relative importance of perceived risk, usefulness and moral acceptability in predicting support of the application.[40] These models explain between two-thirds (food) and three-quarters (applications in animals) of the variance of the support. So moral acceptability and usefulness are the best predictors of support in Germany too. But there were some differences in the relative importance of these factors. For applications related to humans and animals, which include medical and pharmaceutical applications, moral acceptability is much more important ($\beta = 0.53–0.59$) than usefulness ($\beta = 0.33–0.34$). The only exception to this general rule is the application of genetic engineering in food production, where the relative importance of usefulness ($\beta = 0.46$) is slightly stronger than the importance of moral acceptability ($\beta = 0.41$).

Trust and control

Technology consists not only of the technological artefact and the technological method: technology is the artefact, the method and its social context. Control and regulation of genetic engineering are of great importance, according to the estimations of the respondents.

In terms of trust, Germany shows the same pattern as the other European countries: people trust NGOs, but not governments and not science. More than 40% of Germans trust consumer organisations the most. If we add to these 40% the 26% who expressed the highest trust in environmental organisations and the 7.5% who had the highest trust in animal welfare organisations, almost three out of four Germans trust most the critical NGOs. Only 8% say that they trust schools and universities the most, compared to 12.5% in the EU. Political organisations, public authorities and industry are not among the first to be mentioned when we ask people which they would trust the most. If we differentiate between different applications of genetic engineering, we find that only with medical applications do people trust the professionals (i.e. doctors) the most.

There is a strong demand for control and regulation, and not only existing regulations and laws (58%) but also institutional controls are regarded as insufficient. About 60% of respondents believe that scientists do whatever they want, with no respect for the law and regulations. This refers directly to the result that trust in political and administrative institutions – the sources of regulation and control – is very low. Respondents showed a strong need to get involved in decision-making processes. Only 21%, which along with the Netherlands and Finland is the lowest proportion in the EU, think that public consultation is a waste of time. Seventy-five per cent want to make their own decisions: they demand labels for genetically modified food. The strong demand for regulation, together with the low levels of trust in political organisations and institutions, lead to the question of who should be assigned to control genetic engineering adequately and competently. The result that international organisations are very important for regulation and control is surprising because these organisations cannot be controlled by the people. However, it does seem that people think that genetic engineering cannot be regulated on a national level.

Germans have lived in two different states with fundamentally different political perspectives, and there are still some differences in the perception of political processes between West and East Germans. More West Germans (22%) than East Germans (17%) think that public consultation is waste of time, and more West Germans (29%) than East Germans (24%) are willing to take some risks to enhance their economic competitiveness. Far more East Germans (38%) than West Germans (27%) would entrust the regulation of biotechnology to international organisations. But other differences are surprisingly small.

Resistance to genetic engineering in Germany seems to be more of a symbolic act where people declare their concern about the path of technological development and further rationalisation of their 'lifeworld' (*Lebenswelt*) than the result of a technocratic balancing of benefits and risks. This feeling of being at the mercy of a development which is not controlled and not controllable seems to be one of the causes of the irritation that was apparent during the introduction of biotechnology in Germany.

Commentary

The congruence of policy and media phases

There is a striking congruence in the phases of policy-making and of media coverage. Specific phases mentioned here therefore refer to both areas.

Because of the preference of the political system in Phase 1 (1973–84) for neocorporatist self-regulation, the frequencies of themes concerning regulation and political matters were the lowest of the entire media sample. There were few political actors in the media: the agenda was clearly dominated by non-political actors. Within this group, the scientific actors predominated more than they ever have since. These observations from the media study support the result of the policy analysis that the first phase of the discussion was characterised by relatively low political involvement.

This situation clearly changed in the mid 1980s. In the second policy phase, the German government was very active in fostering a public debate about the regulation of biotechnology. This process can be interpreted as the emergence of a broader public discussion, in which the churches, science, industry, NGOs and the political system competed with each other in their claims-making activities. This forced the political system as a whole to rethink its regulative and environmental policy, especially in respect of biotechnology. The strong conflict about all aspects of the regulation of biotechnology can also be found in the media coverage.

The same strong congruence between policy and media coverage can be found in Phase 3: political deregulation corresponds to media coverage whose focus is less on political and conflictual events and more on the pragmatic discussion of the benefits of biotechnology. Coverage also loses the predominantly national perspective of the late 1980s. It is not by chance that this change in media coverage occurs directly after the first German Gene Law had been passed.

This clear congruence of phases should not be seen as a one-sided mirroring of the political events by journalists or as crude political opportunism towards public opinion expressed by media coverage. Rather it should be interpreted as a quite sensitive interactive and collective process for producing social meanings, in which different actors in different arenas compete with and provoke each other.

Public opinion and media coverage

Due to the complexity of modern Western societies, their members cannot take part in every potentially important decision. It is not even guaranteed that they will even know about events that could be of interest to them, and about which they could develop or express an opinion. Journalism operates as a professional observer of this social complexity. The development of a public opinion – but not necessarily its tendency – is highly therefore highly determined by the agenda-setting function of journalism.

It therefore makes sense to compare public opinion as measured by the latest Eurobarometer survey with that indicated by the media coverage. The agenda-setting function of journalism seems to have worked quite well in Germany: in the Eurobarometer survey the percentage of Germans who have never spoken with others about gene technology is much lower than the European average.

Other results of the public opinion survey correspond less well to the results of the media study – or they are, at least, harder to interpret. This is true

especially of the evaluation of positive (one-third of respondents) and negative consequences (23%) for future life – the German media coverage clearly sees the benefits of biotechnology and gene technology (in the last phase 55% of the articles exclusively mention benefits compared to only 5% exclusively mentioning risks). One possible explanation is that the comparison of percentages alone cannot explain the complex process of public opinion formation over time. For an event or technology to be perceived as risky, it seems to be enough for there to be a continuous portion of coverage that deals with risky aspects of biotechnology. The high proportion of articles dealing with benefits and risks (an average of around 30% over period of the study) may support this argument empirically.

Perhaps the Germans also remember the conflicts over benefits and risks during the regulation phase up until the beginning of the 1990s. A possible conclusion is that public opinion is still 'recovering' from this regulation debate, observing the more benefit-oriented coverage of recent years with scepticism or even distrust. Looking at the trust expressed in the biotechnology actors reveals, surprisingly, that the actors with the lowest trust ratings are those who appear most frequently in the media (such as scientists). Consumer and environmental NGOs have the highest trust rates, but receive much less attention in the media.

The story behind the story

From the perspective of decision-making elites, public perception of genetic engineering in Germany is often characterised as a fundamentalistic, highly emotionalised, widely uninformed (if not ignorant) public discourse. This situation is furthermore described as a result of a primarily negative and sensationalist journalistic coverage. Thus the journalists' (political) attitudes and their professional bias are regarded as responsible for the German malaise. Our analyses of the media coverage and of the attitude structure of the German public do not support this point of view. Media reporting is not negative, and Germans are not fundamentally against genetic engineering.

It is true that Germans are not enthusiastic about modern biotechnology, but we could not find any sign of extreme attitudes. What we can observe is a distinct mistrust by the German public of the scientific and political elites. This mistrust is mutual and may have its root in the beginnings of the debate as well as in the issue itself.

In the early 1980s different social movements – such as the anti-nuclear, peace, feminist and environmental movements – became very powerful and influential political actors. Genetic engineering as a social problem stimulated, in part, awareness of all these different single-issue movements. At this time Germany also experienced some fierce conflicts on nuclear energy (at Gorleben and Wackersdorf). It was this period of a great willingness to mobilise that gave birth to the broader discussion on genetic engineering.

Right from the beginning, the elite perspective on genetic engineering was characterised by the overwhelming effort to restrict participation in the debate. In particular the scientific elites in Germany tried to stem the broad public discussion about genetic engineering. The early plans of the German government to pass a law on genetic engineering were thwarted by interest groups from science and industry. Combined with the political mistrust against participatory elements and a traditional overconfidence in scientific self-regulation, the technology assessment of genetic engineering was neither prepared for nor open to a broad public debate. Even in the early 1980s, the German government wanted to channel the debate and to delegate it to the scientific community. At the same time the German public learned from nuclear accidents such as Three Mile Island and Chernobyl that scientific risk expertise may fail, and may get instrumentalised by political decision-makers. Seen in this way, the German conflict on genetic engineering appears to be a prolongation of the conflict on nuclear energy.

Because of the institutionalisation of the nuclear-power conflict, it was easy for the opponents of genetic engineering to oppose the risk-communication strategy of the decision-making elites. Their efforts to gain control of the discussion and to avoid a broad conflict of values may be best represented by the work of the German Enquete Commission. The Commission has restricted its task to the technical analysis of the benefits and risks of genetic engineering.

What we experienced in Germany was a debate without discussion.

Acknowledgement

The German project is associated with the project 'Chancen und Risiken der Gentechnik aus der Sicht der Öffentlichkeit', which has been funded by the German Ministry of Education, Science Research and Technology, no. PLI 1444 .

Figure 1. A regulatory profile for Germany

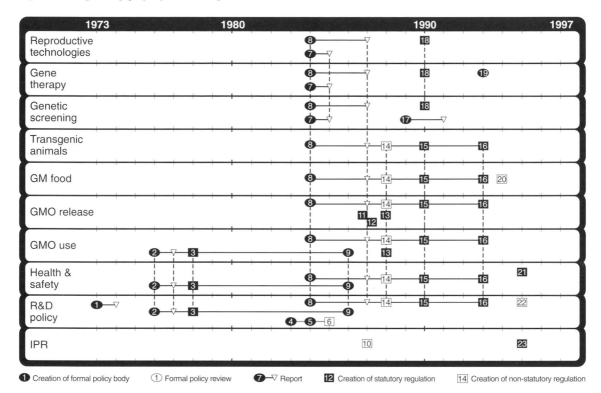

	1973	1980	1990	1997

● Creation of formal policy body ① Formal policy review ❼–▽ Report 🔳12 Creation of statutory regulation 🔲14 Creation of non-statutory regulation

1 *1972–74.* In 1972 the Federal Ministry for Research and Technology (BMFT) appointed the Commission on Biology, Ecology and Medicine, and engaged the German Society for Chemical Apparatus (Dechema) to design a programme for the promotion of biotechnology. In 1974 a study group from Dechema responsible for biotechnology submitted its proposals to BMFT in a report. This report became the basis for decisions on biotechnology research and funding.

2 *1976.* The BMFT appointed an experts committee to tackle questions on the standardisation of regulations. In March 1977 The BMFT produced a draft for discussion on safety guidelines. An interministerial committee under the leadership of the BMFT revised this draft and introduced it to the BMFT, who had overall control of this subject for decision purposes, as 'a guideline on the protection from dangers resulting from *in-vitro* recombined nucleic acid'.

3 *January 1978.* The Cabinet passed the guideline on the protection from dangers resulting from *in-vitro* recombined nucleic acid (so-called gene guidelines).

4 *1983.* The BMFT appointed a public-funded consultant commission for large-scale research in the field of biotechnology.

5 *1984.* The BMFT appointed an expert commission on biotechnology to develop a new biotechnology programme.

6 *1985.* The BMFT established the first definitive programme, on applied biology and biotechnology,

based on the expert commission on biotechnology.

7 *May 1984.* The BMFT and the Federal Ministry of Justice established a common working-group on *in-vitro* fertilisation, genome analysis and gene therapy (Benda Commission). In November 1985 The Benda Commission introduced its report.

8 *August 1984.* The Enquete Commission (on opportunities and risks of genetic engineering), established by the Bundestag, started its work. In 1987 it presented its final report.

9 *June 1986.* Fifth version of the gene guidelines.

10 *February 1987.* The Federal Supreme Court modified its patent-related procedure of jurisdiction.

11 *July 1987.* Modification of the directive on the origins of sewage.

12 *December 1987.* Revision of the directive on prescriptions for dangerous substances (road).

13 *September 1988.* The Bundesrat decided to include plants for commercial applications of genetic engineering in the law on protection from emissions. This amendment of the fourth directive came into force.

14 *November 1988.* The federal government decided to formulate the central points for a genetic engineering law.

15 *May 1990.* The German Parliament and the Bundesrat passed the law on genetic engineering which came into force in July.

16 *December 1993.* The amendment of the law regulating genetic engineering came into force.

17 *1989.* The BMFT established an expert commission on ethical and social aspects on research of the human genome (Bökle Commission). In October 1990 its report was released: 'Ethical and social aspects of researching the human genome'.

18 *October 1990.* The Bundestag passed the Law on the Protection of Embryos which came into force on 1 January 1991.

19 *May 1993.* A federal task group at the Federal Ministry of Health discussed a separate law on gene therapy.

20 *December 1994.* The Bundesrat favoured a comprehensive labelling of genetic engineered food.

21 *January 1995.* Along with a new version of the safety directive on genetic engineering, further safety standards dealing with transgenic organisms were weakened.

22 *March 1995.* The Federal Ministry for Education and Research initiated a programme to share in capital for smaller technology companies, a funding programme on research cooperation, and the Bio Regio competition.

23 *1996.* Change in the patenting law (BGBl. I 1996, S. 1546).

Figure 2. Intensity of articles on biotechnology in the German press

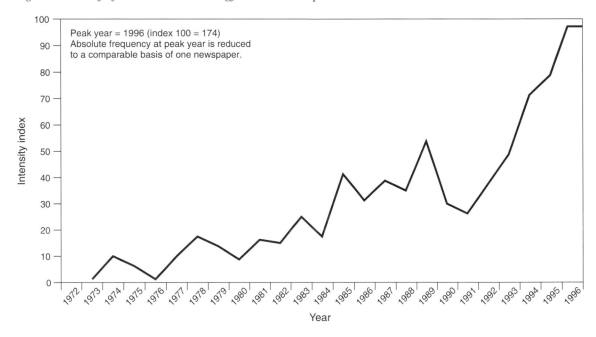

Figure 3. Attitudes to applications of biotechnology in Germany, 1996

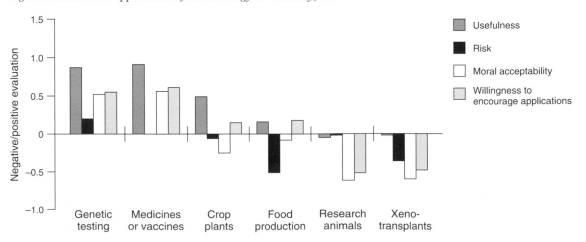

Table 1. German media profile (for an explanation of terminology please see Appendix 5).

Phase	Freq.[a] (%)	Frame (%)		Theme (%)		Actor (%)		Benefit/risk (%)		Location (%)	
1. 1973–84	19	**Progress**[b,c]	57	**Medical**	23	**Scientific**	68	**Benefit only**	59	**USA**	37
		Accountability	13	*Basic research*	23	Political	13	Both	25	Germany	29
		Economic	9	Regualtion	9	Industry	11	*Risk only*	11	*Other Europe*	14
		Pandora's Box	7	Safety & risks	9			Neither	5	UK	8
		Globalisation	5	Political	6					*Other countries*	5
				Economic	6						
2. 1985–91	35	**Progress**	45	**Medical**	18	**Scientific**	41	**Benefit only**	43	**Germany**	50
		Accountability	22	Basic research	15	*Political*	26	*Both*	37	USA	27
		Ethical	13	*Safety & risks*	14	Industry	13	Risk only	11	Other Europe	9
		Economic	7	*Regulation*	13	*Media*	7	Neither	8	World	5
		Pandora's box	5	*Political*	12	NGO	5				
				Economic	7						
				Ethical	6						
				Agriculture	5						
3. 1992–96	46	**Progress**	48	**Medical**	19	**Scientific**	43	**Benefit only**	55	**Germany**	33
		Accountability	11	Regualtion	12	Political	18	Both	28	USA	29
		Globalisation	11	Basic research	12	Industry	18	*Neither*	13	*Other Europe*	18
		Economic	8	Political	11	NGO	6	Risk only	5	UK	8
		Ethical	6	*Economic*	10	EU	5			*World*	7
				Safety & risks	7					*Other countries*	5
				Agriculture	6						

a Percentage of corpus in the period; total $n = 588$.
b Bold indicates highest frequency within phase.
c Italics indicate highest frequency within category.

Table 2. Type of benefit and risk evaluations

	1973–84 (%)		*1985–91 (%)*		*1992–96 (%)*	
Only benefits	Research	57	Research	56	Research	41
	Health	31	Health	19	Health	27
	Economic	17	Economic	21		
Risks & benefits	Moral risks	29	Environmental risks	33	Health risks	31
	Health risks	25	Moral risks	21	Moral risks	18
	Environmental risks	18	Health risks	16	Environmental risks	14
			Consumer risks	14		
	Economic benefits	43	Health benefits	32	Economic benefits	42
	Health bencfits	36	Economic benefits	29	Health benefits	27
			Research benefits	15		
Only risks	Environmental		Environmental		Social (in)equality	
(<5%)	Health		Moral			
			Health			

Notes and references

1 Streck, W R and Pieper, B et al., *Die biotechnische Industrie in Deutschland; Gutachten im Auftrag des Bundesministeriums fuer Wirtschaft* (Munich: ifo-Institut fuer Wirtschaftsforschung, 1997), p128.

2 Streck, W R and Pieper, B ct al., (1997), p135.

3 Lee, K B and Burrill, G S, 'Biotech 97: Alignment' in Ernst & Young, *The Eleventh Industry Annual Report* (Palo Alto: Ernst & Young, 1996).

4 Streck, W R and Pieper, B et al. (1997), p113.

5 BMFT, *Biotechnologie 2000 Programm der Bundesregierung* (Bonn: Bundesministerium fuer Forschung und Technologie, 3rd edn, 1992).

6 Becher, G and Schuppenhauer, M R, *Kommerzielle Biotechnologie; Umsatz und Arbeitsplaetze 1996–2000, Arbeitspapier für das BMBF* (Basel, 1996).

7 The chemical industry has a turnover of DM175,239.5 million. The turnover of pharmaceutical products is DM34,636.1 million. See VCI, *Chemiewirtschaft in*

Zahlen; Ausgabe 1996 (Frankfurt a.M.: Verband der Chemischen Industrie, 1996).

8   Streck, W R and Pieper, B *et al.* (1997), p195; Becher, G and Schuppenhauer, M R (1996), p2ff.

9   Dolata, U, 'Nachholende Modernisierung und internationales Innovationsmanagement; Strategien der deutschen Chemie- und Pharmakonzerne in der neuen Biotechnologie', in Schell, Th von and Mohr, H (eds), *Biotechnologie-Gentechnik, Eine Chance für neue Industrien* (Berlin: Springer, 1995), p462.

10 Dolata, U (1995), p63.

11 Dolata, U (1995), pp456–80; and Ammon, U, 'Arbeits- und industriepolitische Perspektiven der Biotechnologie', in Schell, Th von and Mohr, H (1995), pp489.

12 Streck, W R and Pieper, B *et al.* (1997), p55.

13 Streck, W R and Pieper, B *et al.* (1997), p42.

14 Streck, W R and Pieper, B *et al.* (1997), p56.

15 The first pharmaceutical product to be developed in Germany was human insulin – see Thielemann, H, 'Kommunikation und Konflikt um die gentechnische Insulinherstellung bei der Hoechst AG', in Renn, O and Hampel, J (eds), *Kommunikation und Konflikt; Fallbeispiele aus der Chemie* (Würzburg: Koenigshausen & Neumann, 1998).

16 Kliment, T, Renn, O and Hampel, J, 'Die Chancen und Risiken der Gentechnik aus der Sicht der Bevoelkerung', in Schell, Th von and Mohr, H (1995), pp558–83; and Renn, O and Zwick, M, *Risiko- und Technikakzeptanz* (Berlin, Heidelberg: Springer, 1997).

17 Rudzio, W, *Das politische System der Bundesrepublik Deutschland* (Opladen: Leske & Budrich, 4th edn, 1996).

18 Brocks, D, Pohlmann, A and Senft, M, *Das neue Gentechnikgesetz; Eine praxisgerechte Einfuehrung mit Gesetzestext und Verordnungen* (Munich: Beck, 1991).

19 Gill, B, *Gentechnik ohne Politik; Wie die Brisanz der Synthetischen Biologie von wissenschaftlichen Institutionen, Ethik- und anderen Kommissionen systematisch verdraengt wird* (Frankfurt: Campus, 1991).

20 BMFT, *Ethische und rechtliche Probleme der Anwendung zellbiologischer und gentechnischer Methoden am Menschen*; Dokumentation eines Fachgespraechs im BMFT am 13./15.9.1983 (Munich, 1984).

21 Gill, B (1991), p107.

22 Gill, B (1991), p198.

23 Gill, B (1991), p109.

24 Gill, B (1991), p172.

25 In 1985 the Oeko-Institute in Freiburg started to develop a working group on genetic engineering.

26 Thielemann, H (1998).

27 Eberbach, W and Herdegen, M, 'Gentechnikgesetz; Einleitung, Teil B, Europarecht', in Eberbach, W,

Herdegen, M and Ronellenfitsch, M (eds), *Recht der Gentechnik und Biomedizin*; GENTR/BioMEDR (Heidelberg: Müller) (loose-leaf).

28 BMBF, *Rat für Forschung, Technologie und Innovation: Biotechnologie, Gentechnik und wirtschaftliche Innovation; Rechtliche Grundlagen im Ueberblick-Bestandsaufnahme, Vollzugsprobleme, Vergleich; Ergaenzung zum Ratsbericht* (Bonn: Bundesministerium für Bildung, Wissenschaft und Forschung, 1997).

29 Kepplinger, H M, Ehmig, S C and Ahlheim, C, *Gentechnik im Widerstreit; Zum Verhaeltnis von Wissenschaft und Journalismus* (Frankfurt: Campus, 1991).

30 Ruhrmann, G, Goerke, A and Kohring, M, *Berichterstattung über Gentechnologie in deutschen Tageszeitungen; Ergebnisse einer systematischen Inhaltsanalyse*; unpublished final report for the Deutsche Forschungsgemeinschaft (Muenster and Osnabrueck, 1992). Ruhrmann, G, 'Genetic engineering in the press: a review of research and results of a content analysis', in Durant, J (ed), *Biotechnology in Public: a Review of Research* (London: Science Museum, 1992), pp169–203.

31 Kohring, M, Goerke, A and Ruhrmann, G, 'Konflikte, Kriege, Katastrophen; Zur Funktion internationaler Krisenkommunikation', in Meckel, M and Kriener, M (eds), *Internationale Kommunikation; Eine Einfuehrung* (Opladen: Westdeutscher Verlag, 1996), pp283–298.

32 Ruhrmann, G, Goerke, A and Kohring, M, *Gentechnik in den internationalen Medien; Eine Inhaltsanalyse meinungsführender Zeitschriften*; unpublished final report (University of Jena and Akademie für Technikfolgenabschätzung Stuttgart, 1998).

33 There is one West German and one East German sample. Because there are only small differences between the 'two Germanies', we combined both samples. Results for East and West Germany are only reported separately if there are differences.

34 Only in Austria (27.9%) and Greece (30.1%) is this rate lower. In Greece this is due to the extremely high proportion of people with no opinion.

35 In Germany, on average 2.3 of 5 positive developments are evaluated as likely, compared to 2.6 for Europe.

36 In Germany 2.05 of 5 negative developments are evaluated as likely, vs 2.2 for Europe.

37 Biotechnology and the European Public Concerted Action Group, 'Europe ambivalent on biotechnology', in *Nature*, 387 (1997), pp845–48.

38 In West Germany 28% of age 55 and older; in East Germany 32% compared to 37% on average.

39 Biotechnology and the European Public Concerted Action Group, 'Europe ambivalent on biotechnology', in *Nature*, 387 (1997), pp845–48.

40 For this analysis we used the West German data only.

Address for correspondence

Dr Jürgen Hampel, Akademie für Technikfolgenabschätzung in Baden-Württemberg, Industriestr. 5, D-70565 Stuttgart, Germany. Fax +49 711 9063 175.

Greece

Athena Marouda-Chatjoulis, Angeliki Stathopoulou and George Sakellaris

Demographics and political history

Greece is composed of the mainland, covering the southernmost tip of the Balkan peninsula, and some 9000 islands of which only 200 are inhabited. Its entire area covers 132,000 km², and the overall population is 10.4 million.[1] Two-thirds of the population are located in the two largest cities: the capital city Athens (4.5 million) and the northern port city of Thessaloniki (2 million). Greece has been a member of the EU since 1981.

Greece's recent political history has been characterised by change. There have been three different systems of government in less than 35 years: a constitutional monarchy prior to 1967, a military dictatorship from 1967 to 1974, and a parliamentary democracy since 1974. The Greek Constitution of 1975 established a one-chamber parliament of 300 deputies elected every four years. The Prime Minister is also elected every four years, and the President is elected every five years by the deputies. Elections may be called at any time and there are frequent modifications to electoral law, with the current law requiring that the ruling party hold more than half of the seats in Parliament. All citizens aged 18 and older are required by law to vote. There are two main political parties in Greece: PASOK (traditionally, the socialist or social democratic party), and New Democracy (traditionally, the conservative party). Three other parties are represented minimally in Parliament: the Communist Party; Synapsismos, a spin-off of the Communist Party; and DIKI, a spin-off of PASOK. Both New Democracy and PASOK were founded in 1974 after the collapse of the military dictatorship. New Democracy held power between 1974 and 1981 but since then, with only one exception, PASOK has held power.

The Greek economy is largely based on services and agriculture. Specifically, in 1993 services contributed 66% to the GDP, industry 22.2%, and agriculture 11.8%.[2] With a total labour force of 4,193,000 in 1994, the service sector accounted for 55.6% of all civilian employment, with 34.4% of the labour force self-employed.[3] The agricultural sector of the economy is one in which biotechnology has been particularly applicable in Greece. However, the so-called 'novel foods' have been controversial – they are the only application to be opposed since their introduction.

In 1998, gross domestic expenditure on R&D was 0.6% of the GDP, compared to an OECD median of 1.5%. Researchers (defined as those with at least a university degree working in research) accounted for 0.2% of the labour force and numbered 8050 when adjusted for full-time hours, compared to 240,802 in Germany.[4] However, that said, the situation in Greece has been improving. For example, the percentage of researchers working in the public sector who hold doctorates increased from 24.6% in 1987 to 41.7% in 1991. Funding of biotechnology at seven different research centres tripled between 1986 and 1992. In 1992, biotechnology received 4.2% of state R&D funding.

European collaboration

Particularly with membership of the EU, Greece has seen an increase in R&D, stimulated by access to EU funding. Foreign investment accounted for only 2.4% of national expenditure on research and technology in 1986, compared with 19.3% in 1991.[5] Under the 1998 political structure, the General Secretariat of Research and Technology (GGET), an office of the Ministry of Development, has played a major role in coordinating Greece's participation in EU programmes. For example, under the Third Framework Programme of the EU, Greece was awarded 3.6% of the total funding available to EU countries, whereas Greece's population is 1.3% of the EU, and its number of researchers 0.63%.[6] One of the largest EU/Greek cooperative ventures is the Enterprise Programme for Research and Technology (EPET). A listing of the projects funded under this programme and on-going in December 1997 demonstrates that funding in the biosciences (including health and agriculture) accounted for approximately 26% of total funding (8.2 out of 31 billion drachmas).[7] In addition, based on the keywords 'biotechnology', 'genetic', 'proteins', and their derivatives, biotechnology accounted for approximately 44%, or almost half, of the funding in the biosciences (10 of 26 projects), representing 12% of total funding in this programme. The figures are likely to be an underestimate, as the exact nature of certain projects is not clearly evident from the project titles.

In summary, Greece has lagged behind other European countries in biotechnology. The economy is largely service-oriented, without a recent strong tradition in scientific research. However, increased funding available largely through EU cooperative ventures has begun to stimulate both academic and industrial research, including numerous projects in the field of biotechnology.

Public policy

The biotechnology debate in Greece has been following rather than leading that in other European countries. In fact, unlike most other European countries, there was no policy activity in the 1970s in reaction to the US debate following Asilomar. Instead, the Greek government became involved in the debate during the early 1980s (see Figure 1). The debate, per se, has taken place primarily in informal forums such as scientific meetings, conferences and congresses. The public has not been actively involved. As such, the key constituencies have been mainly organisations of scientists, such as the Greek Association of Chemists, the Greek Association of Biologists, the Greek Biotechnology Society, the National Hellenic Research Foundation, and so on. Thus, the more vocal constituencies have traditionally taken a 'pro-technology' stance. More recently, Greece has witnessed active protest via such constituencies as Greenpeace.

Numerous government ministries have played a role in the biotechnology debate, by far the most frequent player being the GGET, as part of the Ministry of Development. Other ministries include those of Economy, Health, Education and Agriculture. Aside from ministries, the Agricultural Bank of Greece and two companies, Biohellas SA and, more recently, Zeneca, have participated in the debate.

National policy regarding the biosciences has been largely voluntary, as witnessed by enactment of only four statutory regulations since 1980. Other regulations have been non-statutory and primarily formulated in line with EU directives rather than under direct pressure from constituencies. Such is also the case with modern biotechnology, herein defined as biotechnologies based on recombinant DNA techniques and the use of GMOs.

Phase structure of the debate (Figure 1)

Greek biotechnology policy-making has gone through three phases: Phase 1 (1981–88), promoting and regulating biotechnology in Greece; Phase 2 (1989–91), leaving science to the scientists; and Phase 3 (1992–97), coping with the reality of biotechnology in agriculture. Phase 1 is characterised by a series of discussions and initiatives promoting biotechnology in Greece, particularly in the fields of agriculture and genetically modified foods, in combination with the formulation of statutory and primarily non-statutory regulation largely in line with EU policies. Phase 2 appears to have been a period of quiet in terms of public debate and policy-making, perhaps because the real effects of modern

biotechnology had not yet been felt by the Greek population, and thus biotechnology was not at the forefront of the minds of politicians and the public. However, Phase 3 was characterised by intense activity regarding biotechnology in the food sector, as GMOs and novel foods become a reality and the Greek public was brought face-to-face with issues of ethics and health, and effects on society and the environment.

Phase 1, 1981–88: promoting and regulating biotechnology in Greece. The first phase of Greek biotechnology policy-making began in 1981, with a series of discussions by scientists regarding the creation of a 'common body of action' to monitor and govern the progress, use and applications of biotechnology. The following year, the company Biohellas SA was founded with investments from the GGET, the Agricultural Bank of Greece and the Greek Bank of Industrial Development, with a focus on agricultural applications of biotechnology. This was followed closely in 1983 by the start of a collaboration between EL.KE.PA, the Greek Productivity Centre, and the University of Crete to provide training and education and to conduct applied research in the field of agricultural biotechnology and environmental protection. At this time, both the Ministry of Economy and the GGET supported the development of a biotechnology R&D policy programme with a controlling committee. In 1984, Biohellas SA was accepted as a state association acting as a formal policy body regarding biotechnology, with the goal of promoting and validating biotechnology research in Greece.

In 1983–84, the government held two conferences on biotechnology, and three more conferences were organised by scientific associations. Specifically, in 1983 the GGET held a congress promoting biotechnology R&D and addressing environmental policy and consumer protection as per the EU directives of 1981 (971.81 and 650.79, respectively), which were adopted. Also in 1983, the local Athens government organised a conference on environmental problems and biotechnological solutions. The conferences organised by scientific associations covered a range of issues from genetic engineering and traditional techniques to ethics, economics, patents and the role of the media.

The years 1984–85 were also relatively prolific in terms of policy-making. Three non-statutory regulations were passed (including one on gene therapy), as well as two statutory regulations (including one on the protection of workers). In addition, a formal policy body on biomedical technology was formed, reporting to the Ministry of Health.

The end of Phase 1 was characterised by enthusiasm for biotechnological developments in the field of agriculture, as well as evidence of institutional learning in light of increasing applicability of modern biotechnology. The Agricultural Bank of Greece, through its Centre for Training and Development, participated in the EU project COMETT (administered by the GGET) by developing an educational and training programme in English on agricultural biotechnology, and organising training seminars. Regarding institutional learning, with the advent of biotechnological applications the Greek Association of Biotechnology was founded, as was the Institute of Technical Applications. Further, the GGET established postgraduate studies (master's degrees and doctorates) in the field of biotechnology. Finally, two patent laws for technological developments were passed as per EU directives.

Phase 2, 1989–91: leaving science to the scientists. As stated previously, Phase 2 stands out clearly as it is characterised by an absence of scientific meetings and congresses, national policy-making and national policy debate. It appears to have been a period in which the field of biotechnology was allowed to develop on its own without the adoption of new regulations, and it was paralleled by a dearth of media coverage.

Phase 3, 1992–97: coping with the reality of biotechnology in agriculture. This third phase of policy-making was characterised by a focus on genetically modified foods and organisms, particularly with respect to regulation and public protest. Following EU directives, non-statutory regulations were adopted governing genetic screening, novel foods, the use of GMOs, and the deliberate release of GMOs. The Ministry of Agriculture established the Organisation for the Control and Certification of Biological Products, with the goal of promoting biological agriculture and providing for the control and certification of biological products. In addition, the Association of Agricultural Research organised a meeting that resulted in the development of a network to facilitate national information dissemination regarding problems and research goals in agricultural biotechnology.

The years 1996 and 1997 saw heightened public awareness of and concern about the risks of biotechnology. The National Hellenic Research Foundation organised a conference on biotechnology in society and the environment, and a panel discussion on social and ethical aspects of biotechnology. Following the announcement by Greenpeace of a large shipment of genetically modified soyabeans from the USA, a national debate arose among the following constituencies: Zeneca, seeking permission to develop genetically modified tomatoes; the General Secretariat of Industry, which granted permission to Zeneca via a non-statutory regulation; and Greenpeace, which staged a protest march at the Secretariat, forcing it to revoke its permission.

In response, a formal policy review was initiated by the Ministry of Development, of which the GGET is a part.

Thus Phase 3 saw a renewal of activity, focusing again on agriculture in the form of genetically modified foods. Specifically, several non-statutory regulations were adopted. Greenpeace entered the debate, and the impact of biotechnology appeared to have come to the attention of the Greek public.

Media coverage

The sample

We investigated the coverage of biotechnology in the Greek media through a time-series analysis of two written sources of information, namely two daily newspapers: *Kathimerini* (which means 'Daily') and *Eleftherotypia* ('Free Press'). These newspapers, particularly *Kathimerini*, are considered to serve as opinion-leaders for the Greek community. All the information given here about the character of these newspapers and their readership profiles refers to the current state of affairs in 1998, and thus may not apply to the entire period of study.[8]

Kathimerini is a national morning daily newspaper. It was first published in 1919, and is considered to be conservative; but it has been openly critical of the right-wing political parties. The average daily circulation of *Kathimerini* is 24,119, making it the leader among the morning dailies.[9] *Eleftherotypia* is an afternoon daily newspaper first published in 1975. It is considered to be oriented towards the socialists and the left. Its average daily circulation is 52,076 ranking it third among the afternoon dailies.[10] Table 1 shows the readership profile as determined from a 1996 survey.

The media analysis was designed to cover the period 1973–96. However, these two papers were not published in 1973 during the last year of the dictatorship. For the period 1974–85, the newspapers were sampled by hand. In order to facilitate the process, the newspapers were sampled every other year, including years important in terms of biotechnology news. For each year and each newspaper, a random sample of 100 days was generated, and the years studied were 1975, 1977, 1979, 1980, 1981, 1983 and 1985. Thus, seven sample years were evaluated for 100 days each, providing a sample of 700 issues per newspaper over the period 1974–85. Articles were evaluated and included in the study if: **1** they covered 'modern biotechnology' defined as 'the intervention, handling and/or analysis at the level of the gene(s)', regardless of whether or not the word 'biotechnology' appeared in the article; or **2** the main theme was not biotechnology, but the word 'biotechnology' appeared in the article. For the period 1986–96, a database of 'almost all articles' of all Greek newspapers, coded with keywords by theme according to journalists, was searched using selected keywords. Those articles whose themes were identified as modern biotechnology were included for analysis.

Phase structure of media coverage

Figure 2 illustrates the intensity of media coverage in the two opinion-leading newspapers from 1977 to 1996, with a total sample size of 65. Media coverage of biotechnology in Greece has been minimal, in fact it is by far the lowest among European countries studied. Given the small sample size, the phases of media coverage have been assigned based strictly on intensity, as the other variables such as frame, theme, etc. do not provide additional insights into alternative phase assignments. However, each phase is characterised with respect to each of the variables.

Using intensity as an indicator, three phases of media coverage have been identified (Table 2): Phase 1 (1977–85), medical progress at the risk of opening Pandora's Box; Phase 2 (1986–92), more progress, more benefits; and Phase 3 (1993–96), progress at the expense of morality? The first phase, spanning nine years, accounts for 32% of the articles with an average of 3.7 articles per year. The second phase, spanning seven years, accounts for 16% of the articles with an average of 1.8 articles/year. The third phase, spanning four years, accounts for 52% of the articles with an average of 8.8 articles/year. Although the intensity increased significantly during Phase 3, it is important to note that the majority of the articles did not focus on biotechnology, but rather just mentioned it in context. Progress was the main frame throughout, and articles focused on the benefits rather than the risks of biotechnology.

Although the media analysis extended only to the end of 1996, it is worth noting that Dolly the cloned sheep, and cloning in general, were covered extensively in the Greek media in 1997, particularly in the press. In contrast, the aforementioned soya case received very little coverage, in both 1996 and 1997. This finding parallels those from the previous media phases in which medical themes dominated and agricultural ones were rare.

Phase 1, 1977–85: medical progress at the risk of 'opening Pandora's box'. The first phase of media coverage was dominated by the progress frame (56% of articles), although progress gained further significance in the subsequent phases. There was also concern about 'opening Pandora's Box', which framed 18% of the articles, but which was not a substantial frame in subsequent phases. In parallel, almost half of the articles discussed only the benefits of biotechnology, the most common of which was with respect to health. One third of the articles

mentioned both benefits and risks – benefits to health, and moral risks. Medicine was the main theme (42%), and the main actors were scientific (68%), as is true for the subsequent phases. Approximately three in 10 articles (31%) discussed events in the USA, three in 10 (31%) discussed those in European countries other than Greece, and slightly fewer (27%) discussed events in other non-European countries. This suggests that the media portrayed an image of these events as occurring far from Greece, and thus provided some feeling of immediate security. Thus, the media coverage of the first phase characterised biotechnology as rapidly progressing and promising welcome health benefits, but involving certain unknown risks of which society must be wary.

Phase 2, 1986–92: more progress, more benefits. As already mentioned, the years 1988–91 were very quiet in terms of public debate and biotechnology policy-making. Thus, it is not surprising that Phase 2 of media coverage contains the fewest articles of all three phases: only 16% of the total. An overwhelming 82% of articles were framed by progress, and only one of the 11 articles in this phase mentioned any risks of biotechnology. Medicine was still the main theme (63%) and scientific actors dominated (50%), but industry emerged as a key actor as well (25%). As in Phase 1, benefits were almost exclusively health benefits. The media coverage again suggested that events were taking place at a safe distance from Greece, with more than one-third of the articles focused on the USA (36%) and an equal number set against a European backdrop. Thus, progress gained ground, Pandora's Box faded from media frames, risks were rarely mentioned, and industry emerged as a key player.

Phase 3, 1993–96: progress at the expense of morality? This phase witnessed the highest number of articles in the shortest period of time. However, as mentioned previously, biotechnology was more often a peripheral issue rather than the focus of the articles. The positive attitude that dominated Phase 2 continued into Phase 3. Although Pandora's Box did not re-emerge as a significant frame, the ethics of biotechnology framed 14% of articles. It appears that the fear of doom that countered the progress frame of Phase 1 was replaced by more pragmatic issues resulting in questions about morality and ethics. In parallel with the positivism of progress, and similar to Phase 2, 69% of articles mentioned only benefits of biotechnology, and 23% mentioned both benefits and risks. As in Phase 2, medical benefits and moral risks dominated the press. However, in Phase 3, research benefits were mentioned far more often than previously, although they still came second to medical benefits. Basic research became a main theme (35%), together with of medicine (36%);

while the key actors remained scientific (67%) and industrial (23%). The primary backdrop of the articles became the rest of Europe (40%), which gained prominence, while the USA remained almost the same (33%), and Greece finally emerged in the media (15%). Thus, Phase 3 characterised biotechnology as beneficial, making excellent progress and advancing research, but also opening new issues of ethics and morality; and events were finally brought closer to home.

Public perceptions

The following analyses of Greek public perceptions of biotechnology are based on the results of the Eurobarometer survey of 1996. The Greek profile is presented and, where differences exist with respect to the EU average, the differences are highlighted. Thus, an absolute measure of Greek data is presented, as well as its relative valuation in comparison with other European countries.

Knowledge and personal importance of biotechnology

Greece's biotechnology debate has largely been confined to scientific groups and government ministries. Press coverage has been sparse. It is therefore perhaps not surprising that, on an objective knowledge scale based on correct answers to true/false questions about biotechnology, the Greek sample participating in the Eurobarometer 1996 achieved the lowest score of all European countries (3.76, against an EU average of 4.95, with the Netherlands leading at 6.27). Greece also scored lowest in objective knowledge in the previous two Eurobarometers in 1991 and 1993.

In response to the question 'What comes to mind when you think of modern biotechnology?', 61% of the sample replied that they did not know. The responses of another 4.2% were mistakenly not about 'biotechnologia' but the similar-sounding though unrelated 'biotechnia', which means a small industry or craftsmanship. Thus, almost two-thirds of the population did not associate anything with biotechnology.

The most common associations, which were generally mentioned in a positive tone, were related to medical issues (4.9%), progress (4.8%) and fertilisation (4.8%). Of the remaining negative associations, the most common were related to issues of nature (30%), such as 'it is a disaster for nature', to moral issues (22.5%), such as 'it is against tradition', and to fears about the future (17%).

Not surprisingly given the lack of knowledge about biotechnology, which was reflected both by lack of objective knowledge and by lack of word association, an overwhelming majority of Greeks (72%, the highest in the EU, and compared to an EU average of 51%) had never discussed biotechnology before participating in the survey, and 69% (again the highest compared to an EU average of 47%) had not heard about biotechnology in the three months preceding the survey.

Of those who had heard of it, 62% had heard from television, followed by 31% from newspapers. Regarding official information, Greeks are like other Europeans: they prefer to receive information from environmental organisations (24%), schools or universities (20%), and consumer organisations (20%). On a par with the EU average, 42% of Greeks surveyed considered biotechnology to be important to them personally. In fact, 20% considered it extremely important, as compared to an EU average of only 12%.

An interesting contrast emerges from these results. Greeks have far less objective knowledge of biotechnology and speak about and hear of it less than other European citizens, yet just as many as the EU average consider biotechnology to be important to them personally.

Regulation and use of biotechnology

Greeks are relatively pessimistic and distrustful when it comes to the regulation and use of biotechnology. They tend to believe that regulation is insufficient to protect them from the adverse effects of biotechnology, and that risk is unacceptable even if it increases economic competitiveness in Europe (44%). In particular, only 18% of Greeks (the second lowest in Europe) believe that current regulations are sufficient. They would like to be informed and involved in the debate, as almost two-thirds favour public consultation; but 41% feel that regulations are irrelevant because biotechnologists disregard them (compared to the EU mean of 54%). Thus, it appears they would at least like to know of the risks even if they are not protected against them. There appears to be a distrust of industry, as only 14% (again the second lowest in Europe) want regulation left to industry. Instead, like other Europeans, they favour regulation by international organisations such as the UN or the WHO (32%), followed by scientific organisations (27%).

Greece was second only to Austria in terms of the percentage of people who believe that only traditional breeding methods should be used (69% vs 56% for Europe). Only one-third believe that traditional breeding is not as effective in changing hereditary characteristics as is biotechnology. Greeks, like people in the rest of Europe, are very wary of genetically modified foods: eight out of 10 want to see labels on these foods in the marketplace. Approximately two-thirds of Greeks surveyed would tend not to buy genetically altered fruits even if they tasted better than unmodified ones (67% vs 57% for Europe).

Evaluation of specific biotechnological applications

Greeks are particularly interested in health care. Protection of social benefits and health care was the primary issue that would influence their vote at the next general election, and was selected by approximately half the Greeks surveyed. Similarly, they consider genetic testing and the application of biotechnology in medicine to be very useful (see Figure 3). In fact, they are more positive about genetic testing than the citizens of any other European country. Specifically, 71% strongly agree that genetic testing is useful (Europe, 53%), 48% strongly disagree that it is risky (Europe, 22%), 62% strongly agree that it is moral (Europe, 35%), and 66% definitely agree that it should be encouraged (Europe, 40%). However, while they consider the application of biotechnology to research animals and xenotransplants to be more useful than do most other Europeans, they are neutral to negative about risk and morality.

Although Greeks are not very positive about the risks associated with the application of biotechnology to crop plants, they consider it useful and basically moral, and tend to encourage it. However, they are much less positive about its application to food production, which they consider risky, immoral, not useful and to be discouraged (Figure 3). In fact, Greece is among the countries with the largest difference between attitudes towards medical applications and those towards food applications.

Multiple regression analyses were carried out on the dependence of encouragement on estimation of usefulness, risk and morality. The degree to which Greeks encourage certain biotechnology applications is strongly related to the applications' perceived moral acceptability, followed closely by perceived usefulness. In contrast, perception of risk does not appear to influence encouragement to any significant extent. These results are similar to those for the rest of the EU, for all six applications.[11] For example, the average ß (an index of strength of association) for medical applications was 0.537 for morality, 0.373 for usefulness and –0.74 for risk, whereas for food applications the values were 0.458, 0.418 and –0.106, respectively.

Future horizons of technology

Although Greeks are optimistic about improvements in their lives in the next 20 years as a result of developments in areas such as telecommunications, computers and information technology, and solar energy, they are much less optimistic about new materials and substances than Europeans on average, and they are pessimistic about biotechnology (see Figure 4). Their focus on genetic testing and fear of genetically modified substances is mirrored in their evaluation of future applications of biotechnology. Although more people in Greece than in any other country believe that most genetic diseases will be cured (76% vs 56% for Europe), they are also first in the belief that designer babies will be produced (58% vs 36% for Europe). Furthermore, more Greeks than any other citizens believe it likely that biotechnology will create dangerous new diseases (87% vs Europe 69%), that it will decrease the range of fruits and vegetables available (45% vs 27% for Europe), and that existing foods will be replaced with new varieties (71% vs 43% for Europe).

In Greece, as in the rest of the EU,[12] increasing knowledge as measured by correct responses to objective knowledge questions does not imply increasing optimism about the future effects of technology. Rather, with increasing number of correct responses to objective knowledge questions, there are fewer 'don't know' responses about the future of technological development. In other words, Greeks who know more about biotechnology are more likely to form an opinion, be it positive or negative.

Summary

In general, almost half of Greeks surveyed consider biotechnology important to them personally, with one-fifth considering it to be extremely important. However, Greeks are not well informed about biotechnology, nor do they discuss it among themselves. They are distrustful of regulations and of the biotechnologists who must adhere to them; and they are pessimistic about biotechnological developments in the future, and about the effects of these developments on their lives. As in other EU countries, pessimism was not correlated with lack of knowledge; knowledge simply increased the expression of a specific opinion, whether negative or positive.

Although there is overall pessimism regarding the future as related to biotechnology, there is a sharp discrepancy between attitudes toward medical and agricultural applications. Specifically, Greeks more strongly favour genetic testing than any other citizens, but they are very negative about biotechnology applications to food production. As with Europeans in general, the Greeks' level of encouragement is most strongly correlated with perceived moral acceptability, and is almost unrelated to perception of risk.

Commentary

Greece was not well positioned to pick up the 'biotechnology ball' and run with it. The Asilomar conferences in the 1970s took place against a backdrop of dramatic political reform in Greece, namely the overthrow of an eight-year military dictatorship and the establishment of a parliamentary democracy. The political, institutional and social changes occurring during this time clearly overshadowed the significance of the biotechnology debate that was occurring in other countries. Greece continues to lag behind other European countries in the biotechnology race, but is making great strides scientifically via support from government ministries in cooperation with the EU. The public, however, has largely been absent from the debate.

Based on measures used in this study, biotechnology has not been a particularly controversial issue in Greece, nor has it been such an active area of research or law-making as in other European countries. In addition, until recently, biotechnology in Greece has been almost exclusively the realm of scientists and to a lesser extent law-makers, with little public participation or perhaps even awareness. However, although the vast majority of Greeks are poorly informed about scientific concepts related to biotechnology and they do not discuss the topic with others, many have quite definite opinions regarding its regulation, use and value.

Comparing policy-making and media coverage

Many parallels exist between the biotechnology debate and policy-making, on the one hand, and media coverage on the other. However, although the media coverage appears to reflect the frames of the policy debate, it does not reflect the agenda for it, as the actors and themes are generally different. Three phases have been identified for each, and those for the policy debate closely overlap those for the media coverage. In particular, the second phase shows a dearth of policy-making and of media coverage. Here, overall comparisons are presented initially, followed by a more in-depth review of similarities and differences in each phase.

In all phases of policy debate, organisations of scientists were the key constituencies, representing a pro-technology stance. Similarly, in all phases of media coverage, articles were predominantly framed by progress, scientists were the main actors, and the majority of articles focused only on benefits of biotechnology. However, all phases of media coverage focused almost solely on medical benefits, whereas the policy debate was much more often linked to agricultural applications. This can be explained by the fact that the media portrayal focused almost exclusively on foreign countries, whereas the policy debate obviously was a national one.

In the first phase, the media's Pandora's Box frame and mention of health and moral risks of biotechnology are reflected in conferences and regulations covering environmental and worker protection and gene therapy. However, industrial

and political actors were insignificant in media coverage, whereas they were central to activity in Greece (e.g. Biohellas SA and the GGET). Again, the lack of media coverage of politicians is in part related to the fact that almost one-third of articles were set in the USA and another third in other non-European countries.

By the third phase, some controversy had arisen in the policy debate involving a negative response to industrial production of genetically modified foods. Similarly, an ethical frame emerged in the media coverage as well as a renewal of coverage of moral risk, and industry became an important actor, second to scientists. Whereas research became an important theme in the media, and research benefits were frequently mentioned along with medical ones, the key players in the policy debate were involved with applied research, specifically in the field of agriculture.

Thus, the policy debate has been dominated by scientists, government ministries and industry with a particular bias toward agricultural issues. In general, the debate has been pro-biotechnology until quite recently. The media coverage has focused on the progress of medical science and its benefits, thus again presenting a generally positive image with some precautionary moral and ethical concerns. However, the optimism of the scientists, government and journalists is largely lost on the general public, as reflected by the 1996 Eurobarometer survey.

Elaboration of the debate – national sensitivities

Paradoxically, while policy-making and media coverage of biotechnology have been characterised as positive, the Greek public is pessimistic. The simplest explanation, that anxiety is due to lack of knowledge or information, has been shown not to be true. Although Greeks scored very low on objective knowledge of biotechnology, there was no correlation between knowledge and attitude. It seems more likely that their pessimism has its roots not in a lack of knowledge, but in a lack of trust, particularly of national players. This lack of trust has been demonstrated: specifically, Greeks prefer regulation to be handled by international organisations; they believe that current regulation is insufficient; and they believe that biotechnologists ignore regulations anyway. Thus, the Greek public distrusts: **1** the government representatives whose job it is to make legislation and to inform the public of laws and regulations; **2** the biotechnologists and their obedience to legislation; and **3** perhaps even the media who have not extensively covered biotechnological issues arising in Greece. The public was largely left out of the policy debate, and media coverage of biotechnology was not intense during the period of study. The majority of those who had heard about biotechnology in the three months prior to the survey had received their information from television, not from newspapers.

The fact that Greeks favoured medical applications of biotechnology much more than agricultural and food applications, and in fact were negative in most evaluations of genetically modified foods, can be explained by an issue of trust. Almost all media coverage concerned medical themes and medical benefits, whereas agriculture was not included, perhaps suggesting to the public that they are not being well informed about agricultural biotechnology developments in Greece. In addition, agricultural applications may represent a clearer reality in a largely agricultural country, whereas the medical ones, such as cloning, seem a bit remote to most people. It would not be unreasonable for the public to feel more in control of personal 'exposure' to medical biotechnology such as genetic testing (so-called informed consent) or to feel able to benefit without directly participating (e.g. curing genetic diseases), whereas they may feel more vulnerable to unknowingly ingesting genetically modified foods. Thus, the ability to monitor their own 'exposure' to biotechnology rather than relying on those in authority would again enable stronger support of medical rather than agricultural applications. It appears that the public's discouragement of food applications is also related to an inability to see their

usefulness, as shown by the Eurobarometer, and can be interpreted as a local perspective based on the wide variety and high quality of fresh foods available in Greece.

What does the future hold?

Increased funding and thus activity in the field of biotechnological research is likely to increase public awareness and knowledge, but whether this will enhance optimism or increase pessimism remains to be seen. In general, knowledge was not strongly correlated with support in the Eurobarometer surveys of 1996. The Greek public have eagerly embraced medical applications of biotechnology; however, the debate over foods appears to be growing and is a particularly sensitive issue in Greece, where fresh fruits and vegetables are available all year round. An interesting scenario would be one in which farmers, who may benefit tremendously from heartier crops with higher yields and easier maintenance, and who have often held strikes due to dissatisfaction over income and benefits, are recruited as a lobby group for genetically modified foods. The increasing influence of the EU on Greece through statutes and funding, as well as the continued modernisation of Greece, will surely play an important role in the formation of attitudes in the future.

Figure 1. A regulatory profile for Greece

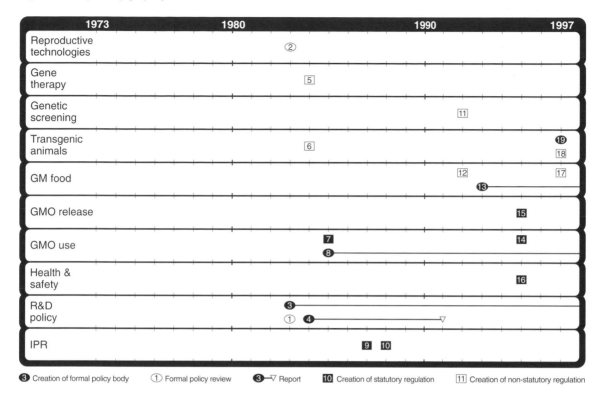

Figure 2. Intensity of articles on biotechnology in the Greek press

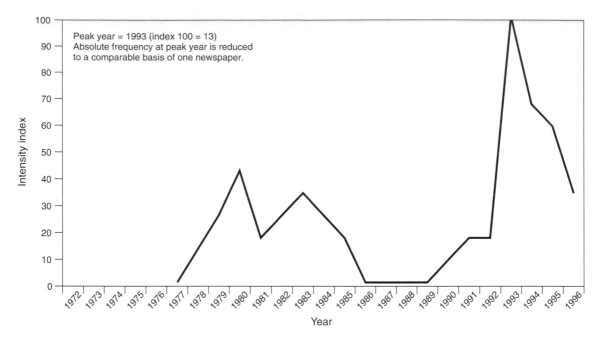

Figure 3. Attitudes to applications of biotechnology in Greece, 1996

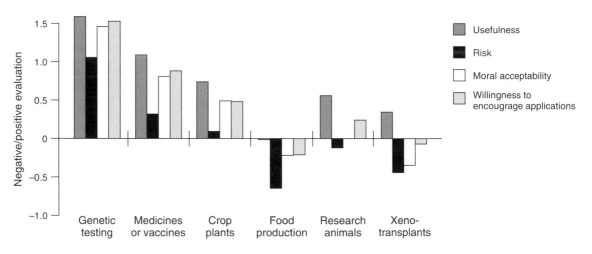

Figure 4. Attitudes to future applications of science and technology, 1996

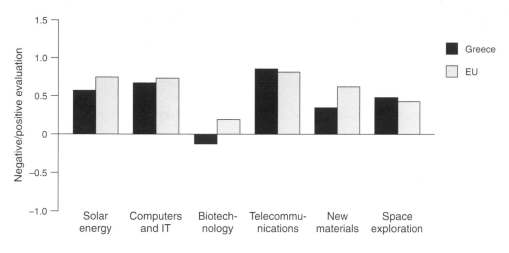

Table 1. Profile of readers of Kathimerini *and* Eleftherotypia, *1996*

	Kathimerini (%)	**Eleftherotypia (%)**
Gender		
Male	63.4	56.2
Female	36.6	43.8
Age		
13–17	3.6	4.1
18–24	9.8	1.0
25–34	17.0	24.6
35–44	22.0	25.8
45–54	25.6	20.0
55–70	22.0	15.5
Social class		
ABC1	55.0	46.8
C2	34.0	39.2
DE	11.0	14.0
Educational level		
Academic qualification	50.0	46.0
Secondary/High School	42.7	43.0
Primary School	7.3	11.0

Table 2. Greek media profile (for an explanation of terminology please see Appendix 5).

Phase	Freq.[a] (%)	Frame (%)		Theme (%)		Actor (%)		Benefit/risk (%)		Location (%)	
1. 1977–85	32	**Progress**[b]	56	**Medical**	42	*Scientific*	68	**Benefit only**	46	USA	31
		Pandora's Box[c]	18			Political	20	Risk only	32	Other Europe	31
						Industry	10	Both	10	*Other countries*	27
2. 1986–92	16	*Progress*	82	*Medical*	63	*Scientific*	50	*Benefit only*	73	*USA*	36
						Industry	25	Both	18	Other Europe	36
3. 1993–96	52	**Progress**	80	**Medical**	36	**Scientific**	67	**Benefit only**	69	USA	33
		Ethical	14	Basic research	35	Industry	23	Both	23	*Other Europe*	40
										Greece	15

a Percentage of corpus in the period; *n* = 65.
b Bold indicates highest frequency within phase.
c Italics indicates highest frequency within category.

Notes and references

1 OECD, *Labour Force Statistics: 1974–1994* (Paris: OECD, 1996).
2 OECD, *National Accounts* (Paris: OECD, 1996).
3 OECD, *Labour Force Statistics: 1974–1994.*
4 STI Data, February 1996.
5 Geniki Grammateia Erevnas kai Technologias, *Deiktes Erevnas kai Technologias, period 1980–1992* (Athens: Publications Dromeas, EPE, 1995), p13.
6 Ministry of Development, General Secretariat for Research and Technology, *New Approaches Towards the Development of Research and Technology*, Vol. 2 (Athens: Ministry of Development, 1996), p11.
7 Geniki Grammateia Erevnas kai Technologias, *Katalogos Ergon*, EPET2 (internal publication, December 1996).
8 Zaousis, A and Stratos, A, *The Newspapers 1974–92* (Athens: Gnosi Publications, 1993).
9 National Athenian Agency, week 16–22 May 1996.
10 Bari Research Company, 8 January – 31 March 1996.
11 The 1996 Eurobarometer survey.
12 Biotechnology and the European Public Concerted Action Group, 'Europe ambivalent on biotechnology', *Nature*, 387 (1997), pp845–7.

Address for correspondence

Athena Marouda-Chatjoulis, Director of Operations & Scientific Development, Centre for Counselling and Development (CCD), 24 Fleming St 151 23, N. Filothei, Athens, Greece. Tel. +301 6816850, fax +301 6891460.
E-mail ccd@athens.mbn.gr

Italy

Agnes Allansdottir, Fabio Pammolli and Sebastiano Bagnara

The political system

The Italian Republic was founded on a constitution in 1948 after a national referendum in 1946. MPs and the Senate are elected in general elections and sit for four years, but elections can be, and are, called at any time. Parliament and the Senate elect the President of the Republic, who sits in office for seven years. In the Cold War era the Italian political landscape was best described as polarised pluralism: the extreme left and right of the political spectrum were formally excluded from power in a permanent, although frequently changing, coalition of centrist political forces, usually led by the Christian Democrats, formerly the largest of the Italian political parties.

In the first half of the 1990s Italy underwent a period of profound political change and corresponding political instability. A national referendum in 1991 abolished the former voting system of multiple preference; a majoritarian (first-past-the-post) electoral system was approved by a national referendum in April 1993 and first applied in the parliamentary elections in March 1994. By that time the Christian Democratic Party had been abolished and its members were divided into smaller factions, and the formerly strong Socialist Party had been wiped out. The election of 1996 was won by a coalition of centre-left parties, and this government is still in office in the spring of 1998. In the 1990s public funding has been reduced, and in some cases suspended or seriously delayed, as part of the government's effort to reduce public expenditure in order to meet the requirements of the Maastricht Treaty.

Research infrastructure

Italy has a relatively strong basic science base and a good tradition in biology, but lags somewhat behind other European countries in the development of modern biotechnology. The Italian science base is for the most part centrally managed and organised by the National Research Council (CNR), the most important Italian research body. The CNR allocates research funds, and organises domestic research, training, and international collaborations. The majority of the research centres dedicated to biotechnology are university departments or laboratories attached to the CNR, which have not yet reached a suitable size for performing research within projects structured at the different levels of 'biotechnology filiere'.

After a period of intense industrial development up until the second half of the 1960s, the Italian chemical industry experienced a progressive decline; but in the late 1970s Italian firms began to show signs of interest in new developments in biotechnology. At the time, the major national companies had already acquired the capability of absorbing or incorporating new technological developments. The state-owned ENI (*Ente Nazionale Idrocarburi*), and in particular Montedison (a private company), had good connections with the emerging biotechnology industry in the USA. However, industry was going through a phase in which public policies focused more on restructuring than on supporting innovative activities. The industrial base for the development of biotechnology – pharmaceutical, chemicals, food and agriculture – was structurally weak and fragmentary.[1] In all, the reaction of Italian industry to new technological opportunities was rather slow and timid.

The early 1980s saw the beginning of industrial research, mostly in the form of small contracts with university laboratories. By the mid 1980s, the large companies, ENI and Montedison, had started to establish research departments and to arrange external contracts with university laboratories and new biotechnology firms. Today the Italian pharmaceutical industry shows a clear divergence between the evolution of demand and the trend of supply: Italy is the fifth largest pharmaceutical market in the world, but lags behind on the supply side. The healthcare sector is most prominent on the Italian biotechnology scene, while plant and animal applications, as well as food processing, have been slower to take off. The chemical sector is growing, while the areas of energy and environmental protection are still marginal.[2]

The role of the government. An important factor which has so far hindered the emergence of consistent innovative activities in biotechnology in Italy is the absence of any focused and conscious government intervention. The government was not able to produce either a coherent 'dirigiste' policy framework geared to stimulating and orientating research (as in the case of France) or a 'softer' set of measures directed towards the general support of bridging activities and collaborative, precompetitive research (as in Britain). On the contrary, government intervention was weak, patchy and belated, and not sufficient to overcome the structural weakness of the industrial system. The profound political changes and instability of the 1990s further weakened the effects of government intervention. In short, public policies intended to enhance the development of modern biotechnology are generally considered to be successful in terms of the quality of basic research, whereas the stimulation of industrial developments has been less successful. This state of affairs can be attributed to the lack of structures facilitating technology transfer, and the absence of necessary financing instruments such as venture capital.[3]

Public participation in Italian policy-making is generally slight, and most issues related to science and technology are dealt with by government experts. The constitution allows for a public petition calling for a national referendum to abolish or reform existing laws. A national referendum in November 1987 confirmed a general public hostility towards nuclear power plants. No such broad consensus was ever reached in relation to modern biotechnology or in the debate about the ethical aspects of genetic engineering. One plausible explanation is that the issue of nuclear power did not touch upon any fundamental conception of or intervention in human life, as did the debate around genetic engineering where it was closely associated with the politically sensitive issue of abortion and the use or creation of human embryos for research purposes. But it has to be emphasised that compared to the issue of abortion, or even nuclear power, modern biotechnology has not been the subject of heated public debate in Italy.

Public policy

The first public policy-making initiatives in Italy were instigated by the activities of the Federation of Technical and Scientific Associations (FAST), which in 1984 published the first Italian report on modern biotechnology. The report emphasised the need for support and coordination of basic scientific research, and for links between industry and universities, as well as the lack of finance for new innovative ventures. However, it did not propose specific policy measures. In 1985 Assobiotech, an industrial association for the enhancement of biotechnology in Italy, was founded within the ranks of Federchimica, the Chemical Industrial Association. This group published a new report on modern biotechnology, and a policy proposal, in 1986. The initial policies called for by industry were modelled mostly on the British experience of the industrial development of modern biotechnology.

Initiatives to introduce an integrated programme of interventions began in 1985, when a National Committee for Biotechnology was established at the CNR. The Committee was composed of the scientists and industrialists of the original biotechnology association, and in 1986 it issued its proposal for a National Programme for Biotechnology. This was approved in 1987, and it funded research into industrial innovation in biotechnology from 1988 to 1993. In June 1987 a Targeted Project was launched at the CNR to fund basic research from 1988 to 1993. Between 1987 and 1993, several Targeted Projects and other National Programmes were launched that contained an important component dedicated to the support of research activities related to modern biotechnology. Another government intervention in 1987 was the establishment of a new firm, Tecnogen, dedicated to biotechnology research. Research is now mainly funded through public funds allocated within national research programmes.

The influence of ethics

While in the first half of the 1980s, industrial and scientific experts were actively promoting biotechnology as an industrial enterprise, another political discourse, within which modern biotechnology (or, more accurately, genetic engineering) would later be discussed, was gathering strength. The origins of this discourse lie in the 1970s, in the reaction to the development of reproductive technologies that became a sensitive social and political issue after the birth in England of the first test-tube baby in 1978, and in particular after the controversial Italian Abortion Act that was passed by Parliament in 1978 and later confirmed in a national referendum in 1981. Although there is a formal separation between the Italian state and the Catholic Church, the Catholic discourse has traditionally had a voice in Italian politics and policy-making, particularly in terms of human health. In 1982 the Pope, John Paul II, pronounced against experiments on embryos, and at the same time encouraged the application of genetic engineering and gene transfer to improve human health, for example to cure genetic diseases. In the spring of 1998 reproductive technologies were still regulated nationally only by a 1984 memorandum from the Ministry of Health to the Regions, whereby artificial insemination, apart from in its simplest form, was prohibited in all public institutions. This is notwithstanding numerous public debates and parliamentary discussion of bills and proposals that were never to become regulations. This absence of formal regulation reflects the importance of ethics in Italian politics, and in particular the difficulties of reaching a political consensus on all matters ethical, especially those relating to human life.[4]

The early 1980s saw two distinct areas of policy activity regarding modern biotechnology. On the one hand, policy activities got under way to enhance the biotechnology industry and basic scientific research from 1984, culminating in governmental intervention prepared in 1986 and approved in 1987. On the other hand, discussion about the ethical and moral aspects of modern biotechnology, mostly in relation to reproductive technologies and abortion, was shaped at the beginning of the decade, was further established on the agenda by the Minister of Health in 1984, and became an on-going political concern in subsequent years.

The year 1986 was something of a watershed in the public debate about genetic engineering (to use the same term as did the debate). It should be emphasised that genetic engineering was not the central issue either in the public debate or in the concerns about intervention in and the technical creation of human life. The anti-abortion campaign, the Movement for Life, gathered over 2.5 million signatures on a public petition that was presented to Parliament in September 1987, and which gave rise to a parliamentary motion in July 1988, where calls were made for the government to impose a moratorium on all research and experiments related to the manipulation of embryos. Although the context was a revision of the Abortion Act from 1978, there was at this point a clear convergence between the Catholic-inspired and environmental concerns opposing generic engineering, but these concerns were carried forward by separate discourses each with its own vocabulary. Communist parliamentary groups, along with the Green movement, proposed a committee for the scientific evaluation of the risks associated with new biotechnological developments, while the discourse of the Catholic opposition to biotechnology focused on the moral acceptability of intervention in human life at the level of the gene, and in particular on the creation and use of human embryos in research. The parliamentary debate was settled by an agreement of a proposal, put forward by the Christian Democrats, to set up a governmental expert National Bioethics Committee under the Cabinet, with the aim of shaping future legislation involving ethical issues. This committee was formally set up in 1990 and was later enlarged; it became a permanent governmental consulting body in 1992.

Regulation and intellectual property issues

The Ministry of Health and its Higher Institute of Health (ISS) set up a policy working group in 1987 to draft a proposal for future national regulation of the production, use and diffusion of modern biotechnological products. This working group was composed of industrial and academic experts, and it published its report in 1990. By that time the European directives on the contained use and deliberate release of genetically modified organisms were already in preparation, and the activities of the Ministry of Health turned towards the implementation of those two directives.

The years from 1986 to 1991 mark a distinct phase in policy activities related to modern biotechnology. The agenda was divided between the technical task of enhancing basic research and industrial developments on the one hand, and the more politically sensitive ethical issues on the other hand. The issue of intellectual property – the possibility of patenting living organisms – was where those issues or discourses intersected. This was particularly the case in 1988, with the establishment of the Human Genome Project in which Italy initially took part. Since then, the issue of patenting has been a component of the left-wing secular and environmental cases against modern biotechnology, and also of the Catholic opposition since the papal pronouncements in 1988 against the patenting of living organisms.[5] Thus the general agenda had a national focus in this phase of Italian policy-making in biotechnology.

Government initiatives

The year 1992 marks the beginning of a new phase in the story. The government and the Ministry of Health set up the Scientific Committee on the Risks Deriving from the Use of Biological Agents, a committee of government and academic experts to formulate the implementation of the EC directives and national regulations on genetically modified organisms. There was little public attention to the issue of GMO laws, which were dealt with in the traditional manner as a technical issue in an exclusively expert domain. The EC directive 90/219/CEE on the contained use of genetically modified organisms was implemented as Act 91/1993 in March 1993. The directive 90/220/CEE on the deliberate release of genetically modified organisms was implemented as Act 92/1993 in March 1993, instituting a new governmental expert committee, the Interministerial Commission for the Investigation of Biotechnology.

The year 1992 was also a turning point in Italian politics. Investigations into the relationship between bribery scandals, political parties and national industries led to a major political upheaval, and a period of intense political instability with frequent changes of government. It is therefore not surprising that modern biotechnology was not a major concern for politicians and policy-makers in general.

Figure 1 is an attempt to chart the development of the policy debate around modern biotechnology. The first thing to note is that the approach to policy-making is more general, rather than being geared towards specific issues. A few governmental expert committees are the main policy-making bodies in several areas of modern biotechnology. Further, the profile draws attention to the separation of ethical and technical issues in the policy arena. Policy-making in the area of research and development started earlier than policy activity in other areas. The first activities marked date back to 1982 and consisted of an Act that was not specifically geared towards biotechnology but was more general in scope. It nevertheless enabled the development of the first public-policy initiative in 1987. Typically, then, government intervention is more general, and lacks specificity in the case of modern biotechnology. A particular feature of the Italian profile is the relatively high levels of activity that did not lead to a specific policy outcome.

The profile does however capture the different phases in the story of policy-making. There are no initiatives specifically related to modern biotechnology before 1984. It is particularly interesting to note the low level of activity from 1986 to 1991, since what the profile does not show is the intensity of public debate in that period. The profile clearly indicates a shift or a change in the debate in 1992: preparations were under way to implement the EC directives, and at the same time the National Bioethics Committee became a permanent policy body. Further, the relative absence of activity after 1993 might be due to national political instability and change in which modern biotechnology was not a major issue.

The profile draws further attention to the absence of regulation in those areas that are socially defined as ethical, most notably issues concerning the regulation of reproductive technologies. A profile of the intensity of public discussions and debates around genetic engineering would show exactly the opposite: high levels of public attention to and debate about what is understood as ethical and potentially applicable to humans; and a corresponding scant public attention to and low levels of debate about what is defined as exclusively technical and industrial. For example, up to the end of 1996, agricultural applications of modern biotechnology were not a salient issue in the public debate, and have traditionally been considered as an exclusively technical issue to be settled by government expert committees outside the boundaries of public debate.

Media coverage

The Italian public is served by a media system largely dominated by television. In general, newspaper readership is rather low in comparison with other European countries, but there is a marked preference for regional or local press.[6] Elite opinion is served by a number of national newspapers. *Corriere della Sera* has the highest circulation figures, followed by *La Repubblica* and *Il Sole 24 ore*, the daily financial paper.

Because it is mostly the national press that serves elite opinion, press coverage of biotechnology is an indicator of the salience and intensity of elite public attention. Our longitudinal analysis of press coverage considered, for the period 1973–96, the daily newspaper *Corriere della Sera*. This paper has long aspired to be the key national newspaper, although the circulation figures remain strongest in the north. *Corriere della Sera* has always been close to the ruling elite, and its political stance is broadly centrist with slight variation over time according to variations in Parliament.[7] In this sense, the paper can be regarded both as an opinion-leader, serving the country's elite, and as an index of the relative salience of issues in the public arena. Like most Italian newspapers, *Corriere della Sera* embraces journalistic styles from the traditions of both quality reporting and the more sensationalist coverage that elsewhere would be confined to the popular daily press.

Intensity of press attention

Figure 2 shows a relative index of press attention to modern biotechnology from the beginning of 1973 to the end of 1996. The articles were found by manual keyword searching in indices published by the paper from 1973 through 1984. From 1984 to 1991 an electronic archive was used for a keyword search in headlines and abstracts. From 1992 to 1996 the search was performed on-line of the full text of articles using the same keywords. Out of this corpus of material (in all, 936 articles), 340 articles were subject to content analysis, including all published articles from 1973 to 1985 and a fixed quota of 20 randomly drawn articles for each year from 1986 to 1996. The graph can be interpreted as an indication of intensity of media attention to modern biotechnology from 1973 to 1996. It is a relative index, where the peak year, 1992 (110 articles published), is set at 100 and the intensity of the debate at other time points is relative to this peak.

The index shows low but constant levels of media attention to modern biotechnology from 1973 to 1979. The intensity of coverage slowly peaks in the early 1980s, marking the second phase in the press debate from 1980 to 1985. This second phase is followed by a sharp increase in coverage: the first real wave of press attention to biotechnology, which takes off in 1986, peaks in 1987 and then diminishes in intensity over the next few years, reaching a low in 1991. In 1992 a second wave of attention begins to decrease steadily to a low in 1996. This second wave of press attention constitutes the fourth phase in the media debate. Continued monitoring of biotechnology coverage in *Corriere della Sera* beyond the period of the current study shows that 1997 marks another substantial rise in media attention to biotechnology, indicating the beginning of a new wave of attention.

Content analysis

A separate analysis of the content of coverage, the framing of the debate, the main actors mentioned and the most common themes shows that the cycles of intensity of press attention correspond to shifts in the main focus of content. Modern biotechnology was discussed in the Italian press within an interpretative frame of progress that was strongly associated with medical applications, and the protagonists were mainly scientific actors. One of the most interesting findings is how this dominant media representation of biotechnology in terms of medicine and progress changed throughout the period in question. The framing of the debate in terms of progress steadily decreased as a proportion of articles, while the medical theme maintained its relative importance. Scientists were identified as the principal actors in the story, although their presence in the press debate or coverage halved from 81% in the first phases of the debate to stabilise at just over 40% in the mid 1980s, when other actors entered the scene. As a general trend, negative evaluation got stronger over the whole period and positive evaluation weakened, apart from a rise in the early 1980s. Most of the articles sampled mentioned both positive and negative consequences, with the notable exception again of those published in the early 1980s. The different phases of the media debate around modern biotechnology, which are summarised in Table 1, can be characterised as follows.

Phase 1, 1973–79: concern. The first period was one of concern and bewilderment. The coverage in this period brought together medical themes and basic research; and the scientific community was by far the most important actor in the stories, with the only other actor being references to public opinion and other media (i.e. public discourse). Half of the articles mentioned controversy of some kind, associated mostly with ethical and medical themes.

Phase 2, 1980–85: optimism. In the second phase the concern of the previous phase appeared to have been placated, and the storyline turned towards the actual or potential industrial applications of discoveries and new techniques. What was notable about this period was how anomalous it was within the overall media debate in Italy from 1973 to 1996. In this phase, economic framing themes were conspicuous in the debate. Further, this phase was marked by the relative absence of political and ethical actors, and the appearance of industry as an important actor. There was a sharp drop in mentions of moral risks or hazards. The dominant framing of the discussion was mainly in terms of progress but seemingly without the concerns of previous years, as indicated by the drop in the proportion of ambivalent stories, the strength of positive evaluations, and

the fact that 60% of the articles mentioned only benefits. The media discussion moved away from the predominance of basic research and scientific agency towards economic prospects and industry. Only 23% of articles mentioned controversy, mostly associated with the less salient issue of ethics and basic research. Although this is clearly a phase of optimism and progress, the ethical discourse of concern was present, though at a low level.

Phase 3, 1986–91: ethics and health. The year 1986 marked the beginning of the first big wave of media attention, which peaked in 1987. This period was quite clearly marked by the surge of the ethics discourse and the rise of public concern. The framing of discussion about biotechnology was less in terms of progress, and an ethical framing of the debate appeared. Medical themes became more prominent, but they were no longer in conjunction with basic scientific research or economic prospects; instead, they were increasingly related to ethical themes. The optimism associated with benefits brought about by the industrial development of biotechnology was challenged by the ethical and political discussion. Half of the articles mentioned controversy, mostly associated with medical and ethical themes. The content of articles in this period was quite diverse. One of the main topics was the Human Genome Project and Italy's involvement in it, which was discussed in terms of great promise and sometimes great concern; and as always there was attention to the issue of the national regulation of reproductive technologies.

Phase 4, 1992–96: the gene, risk and safety. The year 1992 marks the beginning of a new cycle of media attention where, after a sharp rise, intensity of coverage steadily decreased to the end of 1996 (Figure 2). This period still maintained the characteristic features of the debate set in the years before, even if there were important changes. Framing of the discussion in terms of progress continued to lose ground, with just over half of the stories being framed in this way (Table 1). The ethical frame remained salient, but most importantly there was a return to the early media debate on unknown consequences, and the public accountability frame resurfaced.

The importance of medical themes declined somewhat, while new themes of safety and risk entered the discussion for the first time since the 1970s. The coverage became more ambivalent in the sense that half of the stories mentioned both positive and negative consequences. Some 60% of all the articles sampled from this phase mentioned controversy, mostly in terms of ethics, safety and risk, political themes, and regulation. Ethical actors were absent with respect to previous phases, perhaps because national or local events were more likely to be discussed and commented upon by ethics experts than events happening abroad.

The most striking difference from the previous phase was in terms of location, which had moved away from Italy to the USA. This was when the gene discourse – as in articles about the genetic basis of various diseases and behaviour – really started. In the beginning of this fourth phase many articles dwelt upon the regulation of reproductive technologies. Further, in the second half of this phase, the theme of safety and risk became the second most important theme under discussion, although previously it had been absent. There are two possible reasons for this: on the one hand, the rise of the risk discourse as articulated in Ulrich Beck's theory of the 'risk society',[8] and on the other hand simply that press reporting was mostly about events taking place elsewhere, and thus was taking up themes that might have been central to the debate in other cultures.

The story of the media coverage, or rather of cycles of media attention to issues related to biotechnology, can be summed up as a narrative where at the outset something new, unfamiliar, and threatening appeared on the horizon, and was clearly perceived to be within the boundaries of discourses of progress where, however, the negative side was more salient through potential human applications. This perception characterised the first phase. After the initial reaction, the threatening possibility implied in progress was overtaken by the potential promise and benefits of such new technology, mostly in terms of improvement of human health. This perception marked the second phase of the debate. The third phase (health and ethics) and the fourth phase (the gene, safety and risk) shared some important features. Both opened with a sharp increase in the amount of coverage that decreased rapidly over the following years until the next wave came along. In fact, 1997 saw a virtual explosion in the intensity of coverage after a low in 1996.

Modern biotechnology in the Italian press has clearly been a story about progress and health, but it is highly relevant in the national context that ethics as a proportion of themes remained constant throughout the whole period (at 12% of all themes within a given phase), with the notable exception of the second most optimistic phase. While benefits were mostly in terms of health, the costs and risks associated with modern biotechnology were primarily perceived as moral hazards (see Table 2).

Public perceptions

When the results of the Eurobarometer survey from November 1996 are compared with those of a previous Eurobarometer survey conducted in 1993, it is evident that levels of knowledge of basic biology have greatly increased in Italy, more so than in any other European country. This increase in knowledge does not correspond to more positive attitudes towards science and technology in general. Between 1993 and 1996, there was virtually no change in the percentage of the population believing that biotechnology or genetic engineering will improve our way of life, while there is a substantial increase in those who think it will make things worse. This increase is accompanied by a corresponding drop in the proportion of the undecided (Table 3).

For some questions, half the 1996 sample were asked about genetic engineering and half were asked about biotechnology. Table 3 shows the joint percentages for those two groups. 'Genetic engineering' seems to have more negative connotations than the term 'modern biotechnology': 22% of respondents believed that genetic engineering will make life worse, while 11% believed that modern biotechnology will make life worse.

Comparing the 1996 survey with the 1993 Eurobarometer survey, more Italians now believe that modern biotechnology is more effective than traditional breeding methods; but compared with the 1993 data, the public is now more in favour of using only traditional breeding methods, rather than changing the hereditary characteristics of plants and animals through modern biotechnology: 63.1% of respondents agreed in 1996, compared to 57% in 1993.

Survey results: risk and regulation

Italy is among the European countries most anxious about risk and regulation of biotechnology. The Italians appear on average rather risk-averse, with 60.8% disagreeing with the statement 'We have to accept some degree of risk from modern biotechnology if it enhances economic competitiveness in Europe'; 22.9% agree with that statement and 16.4% do not know. Further, the Italians appear to have little faith in regulations: only 17.9% of the sample agree with the statement 'Current regulations are sufficient to protect people from any risks linked to modern biotechnology', while 46.1% disagree and 36% don't know. Those who either have talked about biotechnology or have heard of it are significantly more likely to find current regulations insufficient. However, this might not imply that the public believes regulation to be insufficient in itself, since 52.5% of the whole sample believe that 'Irrespective of the regulations, biotechnologists will do whatever they like' (36.5% disagree); it may instead be an indication of a lack of trust in institutions to enforce the regulation of biotechnology. In the Italian case, this might be related more to a general lack of trust in public authorities than to the issue of modern biotechnology. Qualitative research reveals that in general people do not know about regulation, but they are concerned about control and about the ability of national public bodies to regulate and enforce the regulation of biotechnology. When survey respondents were asked who should regulate modern biotechnology, 39% chose international organisations and a further 30% scientific organisations; while national public bodies were at the bottom of the league with 2.5%. The national Parliament fared slightly better at 4.7%

Survey results: the image of biotechnology

In November 1996 modern biotechnology was not a burning issue for the Italian public. When asked what came to mind in association with the term modern biotechnology, 30% of respondents said they did not know. However, in general the overall proportion of 'don't know' answers to questions in the Italian survey is rather high. Nevertheless, 62% of respondents claimed to have either received some information about modern biotechnology or to have discussed the issue with others, thus indicating a familiarity with the subject. The images associated with biotechnology are mostly about progress (28% of all respondents), which is more common for men (33%) than for women (23%); while the second most common association is medicine (13% of all respondents), which is more common among women (16%) than among men (10%). Medical applications are most frequently mentioned by 19% of respondents. In other words, one out of three Italians does not have a precise idea of modern biotechnology, while those who do tend to hold an image of progress and medical applications.

Survey results: attitudes to applications

Italy is among the countries in Europe most supportive of medical applications of biotechnology and least supportive of the creation of transgenic animals for medical research purposes. This goes some way to characterising public perceptions of biotechnology in Italy. In general, modern biotechnology tends to be linked to medical progress and as such to be regarded rather favourably. Figure 3 shows the high levels of support for medical applications, such as vaccines and genetic tests, which are perceived as useful, morally acceptable and safe; such applications should be encouraged. For medical applications, some familiarity with the subject seems to lead to greater acceptance (t = 5.3, sign. = 0.000).

Not all applications that could potentially bring benefits to human health are regarded favourably by the Italian public. There is generally low support for applications involving transgenic animals: the Italian public is concerned about their use and creation. Transgenic animals are perceived to be risky for

society and morally unacceptable, although they may also be rather useful; encouragement of those applications is low. Qualitative research and interviews with members of the public revealed that modern biotechnology is often discussed in terms of a dilemma between benefits for health and general human well-being on the one hand, and an image of progress out of control and transgressing natural boundaries on the other. An example neatly captures this: in a focus-group interview with middle-class women, the question of xeno-transplants fuelled a heated discussion. According to one respondent: 'It is against nature to inject human genes into pigs, but if it would be a matter life or death for my child I think it is a good thing and I would like that possibility to exist.' In the 1996 survey, it was only in the case of transgenic animals that there were no significant differences in the level of support or acceptance between those who had some familiarity with the argument and those who did not. When further asked whom they trust to tell the truth about xenotransplants, 53.3% chose the medical profession.

Moving down the hierarchy of living things, transgenic plants were not a common cause for concern for the Italian public in November 1996 (see Figure 3). The results indicate general support for the use of biotechnology in agricultural production, but at the consumption end of the chain the Italian public become sceptical. Using biotechnology in food production is perceived by the public as being risky for society and less useful than the other applications; and although it is also perceived as morally acceptable, support for this application is quite low. In this context it is relevant that 59% of all respondents said that they would not buy genetically modified fruit even if it tasted better than naturally bred fruit. The concern about or aversion to genetically modified foodstuffs might be more closely associated with scepticism towards modern mass production of food. In focus-group interviews it became evident that genetic modification of foodstuffs was generally perceived as merely yet another example of manipulation and intervention in modern food production, and not as a categorically different form of intervention. Many of those interviewed were uncertain about the actual existence of such products, but in general people felt that it was not a big issue. By the end of 1996 the Italian public did not seem concerned about the potential environmental impact of the cultivation of genetically modified crop plants, which was perceived as another example of sophistication in agriculture. When respondents were asked whom they trust to tell the truth about genetically modified food, 25% chose environmental organisations, 20.8% consumer organisations and 18.2% farmers' associations.

Commentary

To trace the development of the public debate about modern biotechnology in Italy is to follow the course of several separate topics. The slow and sometimes faltering advancement of the Italian biotechnology industry has been accompanied by the discourse of bioethics, where genetic engineering has at times been associated with the profoundly sensitive issues of abortion and human-embryo research. Modern biotechnology has never been an important political issue in its own right in Italy; industry was ill-equipped for the exploitation of this new technology, and governmental interventions were for the most part slow and ineffective. Political instability in the 1990s has not favoured recovery. Modern biotechnology has, on occasion, become a transient political issue, primarily through the potential ethical aspects and consequences.

The notion of bioethics needs further clarification in the Italian context, where it inhabits a domain of contrasting positions in secular and religious bioethics. The religious bioethics perspective is promulgated by the Catholic Church, and religion has traditionally been an important subtext in Italian politics. The secular bioethics perspective is traditionally linked to more progressive centre-left politics, and has some affiliations with environmental movements. It would be wrong to set up a contrast between bioethics and technical progress: both perspectives are pro-science, and actively support the scientific and technological effort to improve the human lot; but there is a clear divergence between the two perspectives in the political question of how to govern, regulate and monitor the progress of modern biotechnology. Intellectual property rights is an area where the bioethics discourse comes into conflict with the discourse on industrial and scientific progress, and has often been a point of contention in Italian politics.

The relationship between the media and policy activity

This profile has attempted to identify phases in both policy-making and media coverage. There are clear temporal links between the two, but what is interesting is the relationship between the contents of the phases. The evidence indicates that media coverage of modern biotechnology actually preceded policy activity. Press coverage in the 1970s dwelt upon great hopes of improving health, eradicating diseases and resolving problems faced by a famished world, but at the same time it was uneasy about the imminent 'end of nature'. Press coverage in the second phase was more optimistic, and is neatly summed up in a headline stating that now science helps nature, in a story about the quest for a cure for cancer finally coming to an end. The optimistic

second phase in media coverage actually preceded the period of governmental intervention intended to advance both basic research and a budding modern biotechnology industry.

The relationship between media and policy activity becomes more interesting as the story develops. The public debate ('debate' may be overstating the case) started in 1986, opening up a new phase in both media coverage and policy-making when Italy was enjoying a phase of prosperity. These were the glorious years of 'made in Italy'. In this phase, the intensity of media coverage clearly precedes policy-making activities. These are crude measures, but they indicate that as the attention and activities of policy-makers increased, so the intensity of media attention declined. This can be interpreted as the sensitivity of Italian politics to public concerns, as articulated by the media, and in particular the sensitivity towards ethical questions. Another factor that might account for the relationship between media coverage and policy-making is that as the new decade began, the locus of activity became less and less national. The next wave of media attention began when the activities in the policy arena were generating an outcome in 1992. There was a clear shift in the content of the media debate: as the ethical concerns of the previous phase were taken over by policy machinery, the attention of the media turned toward other issues.

The peak in media coverage in 1992 corresponded in time to events that were to leave traces in the collective memory. In the absence of any regulation of reproductive technology, Italian doctors gained worldwide fame for assisting formerly impossible pregnancies, such as that of a woman in her early sixties, and that of a woman who gave birth to her daughter's child.

The latest phase in the media debate, from 1992 to 1996, corresponded to a period of great political upheaval and reforms in Italy, when the regulation of modern biotechnology was simply not high on the political agenda. Policy-making was focused on ensuring the implementation of EC directives.

Public opinion and the public agenda

The Italian public was rather unfamiliar with modern biotechnology in late 1996, but public opinion appeared to be largely structuring itself around the main lines of the policy debate and the main features of the content of newspaper coverage. That is, public images of biotechnology are predominantly those of medicine and progress, but at the same time there is a notable moral unease about the consequences of transgressing natural boundaries. While the Italian public is strongly supportive of medical applications, transgenic animals provoke moral concerns even when their purpose is to improve human health. Knowledge and familiarity do not necessarily lead to higher acceptance (in the case of xenotransplants, knowledge makes no significant difference). The Italian public have little faith in the abilities of authorities and national public bodies to regulate modern biotechnology. The Eurobarometer survey was carried out when the intensity of media attention was at its lowest (after 1986), and therefore it is not surprising that modern biotechnology is not a major issue for the Italian public.

This profile spans the years from 1973 to 1996, and comes to a close just as a new phase, the debate about modern biotechnology, is really getting started in Italy. The Italian media is paying more attention to modern biotechnology than ever, and there have been a number of developments in the policy arena. There were two separate issues that put modern biotechnology back on the public agenda: the dispute over genetically modified maize; and, more importantly, the public reaction to Dolly, the cloned sheep. In the first week of March 1997, following intense media coverage as well as questions to Parliament, the Minister of Health, Rosy Bindi, issued a ministerial decree imposing a 90-day ban on all experimentation that could be related to human cloning. During that time a new framework for the regulation of reproductive technologies was to be prepared. In the spring of 1998 the parliamentary Commission for Social Affairs was finalising a new proposal to put before Parliament. Dolly arrived on the Italian media scene when the human embryo was a major political issue; in the context of biotechnology, the conflict centred on the possibility of carrying out experiments on human embryos within the first 14 days. In the media coverage of cloning, the threshold of attention to modern biotechnology was radically lowered and coverage both increased and became more diverse than before. Dolly was clearly something of a catalyst in the Italian debate.

In the weeks before the cloning story broke, the question of genetically modified maize became an issue in Italy. First of all, there were negative media reactions to the decision of the European Commis-

sion to allow the import and cultivation of genetically modified maize. More importantly, a new Parliament had been elected in the spring of 1996, and the resulting coalition of centre-left forces included the Green Party. Traditionally, the direct impact of biotechnology applications on human health has been given more weight than its environmental impact, but in 1997 there was some evidence that this dominant framing was being broadened to give more attention to the environment. This trend was evident in the story of the debate around genetically modified maize. The Minister for Agriculture and Environmental Resources (Green Party) had actively opposed it, and had raised concerns with the Ministry of Health, the relevant national authority. After the Dolly case, the issue of the environmental impact of genetically modified organisms became more salient. When a ban was imposed on cloning, another ministerial decree was issued which banned, first for 90 days, the cultivation in Italy of the maize in question. Recently there has been a shift in the focus of the Italian debate: the general framing of progress, medicine and morals has given up space to the issues of safety and risk not only in terms of health, but also of the environment. These changes are not merely attributable to the advancement of modern biotechnology in itself, but have to be interpreted in the context of increased political stability: the same government has been in office for over two years, and it is a government that is both more sensitive to environmental causes and supportive of scientific and technological progress.

The fact that there has been a shift in the national debate about biotechnology does not necessarily imply that the Italian public are becoming more sceptical. Survey results indicate that more familiarity with modern biotechnology or genetic engineering leads to more polarised attitudes – Italians become either more supportive or more sceptical. Now that media coverage of those issues has risen, the public are becoming more aware of problems and possibilities of modern biotechnology than they were in the autumn of 1996.

Figure 1. A regulatory profile for Italy

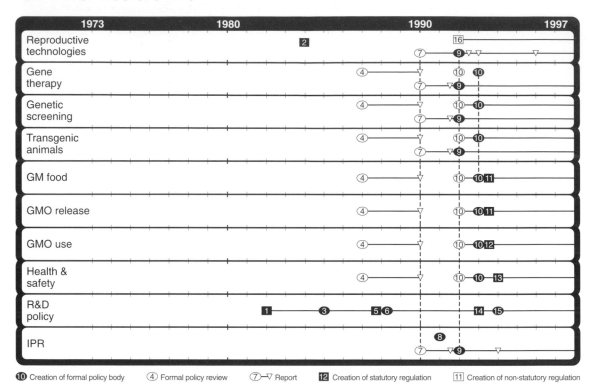

1 *1982.* Act (n.46/82) on Industrial and Technological Research and Development that provides for several ways of supporting research and other innovative activities undertaken by companies. Most importantly national research programmes (see notes 6 and 14). The content of these programmes is defined by public administration and funds research and training projects but the results become property of the state.

2 *1984.* The Minister of Health bans all forms of reproductive technologies from public structures, apart from within married couples. As a consequence reproductive technology falls outside the regulatory mechanisms of the Ministry of Health.
By 1998 no regulation has yet been approved.

3 *1985.* The national research Council (CNR) sets up *a national committee for biotechnology*. Its role is to identify priority activities and means to support modern biotechnology which for the first time is identified as a strategic area. Soon it proposes two CNR (*Progetti Finalizzati*) targeted projects (see note 6).

4 *1987.* National Health Institute and the Ministry of Health set up a working group to drafts a proposal for regulation of research, production, use and diffusion of biotechnology products in Italy. Publishes a report in 1990 when the EU directives are already in sight.

5 *1988.* National Committe for Biotechnology (see note 3) becomes the National Committee for Biotechnology and Molecular Biology.

6 *1988.* Four-year National Research Project for the

advancement of biotechnology in Italy. Two *Progetto Finalizzato* (CNR) targeted the enhancement of basic research, both of which are considered to be successful. On the other hand a national programme (see note 1) under the control of the Ministry of Universities and Scientific and Technological Research (MURST) that was considered somewhat less successful.

7 *1990.* Cabinet sets up National Committe on Bioethics, (CNB) as a consulting body including representatives of both secular and religious bioethics. The choice of a Catholic Senator as the first president was criticised by the secular part and most of the first 40 members were Catholic and male. In 1992 CNB published guidelines for regulation of gene therapy.

8 *1991.* The Minister for the Adaptation of EU Policies, sets up a Committee for the Study of Legal Protection of Inventions in Biotechnology, composed of scientific and industrial experts.

9 *1992.* CNB (see note 7) becomes a permanent governmental advisory body. After that, in 1992, the CNB published a report on prenatal diagnosis and another on the patenting of living organisms, and in 1994 repeated its opposition to patents. The same year CNB announced a study on reproductive technologies, published in 1994, but in the meantime published a report on foreign legislation of reproductive technology. The report published in 1994 recommended a permanent monitoring body and the regulation of private clinics. The Commission failed to reach an

agreement on the status of the embryo. In late 1994 the then prime minister Silvio Berlusconi changed the composition of the commission resulting in the resignation of some members of the secular wing and the CNB becomes predominantly Catholic. In the summer of 1996 it publishes a controversial document on the legal status of the embryo.

10   *1992.* As specified in Act n.142/92, art. 40 the Cabinet sets up the Scientific Committee on the Risks Deriving from the Use of Biological Agents) to give technical advice to the government on the implementation of EC directives GMOs (219/90) and (220/90) on as well as on workers' safety (679/90). In 1993 this body was transformed into a permanent advisory body, the Interministerial Commission on Biotechnology.

11   *1993.* Act n.92/93 implementation of EC directive n.220/90 (Gazzetta Ufficiale n. 78, 3 April) on deliberate release of genetically modified organisms. The relevant national authority is the Ministry of Health which organises the Interministerial Commission on Biotechnology permanent advisory body. The role of this committee is to advise the Ministry of Health to evaluate the risk analysis presented by whoever asks for authorisation of deliberate release of genetically modified organisms (plants) and the possible negative effect of the use in human or animal food of products containing genetically modified organism. The coordination is in the hands of Ministry of Health but the committee is composed of representatives from the ministries of: Agricultural and Environmental Resources; Industry; Labour; MURST and the National Institute of Health. 1997 Ministry of Agricultural and Environmental Resources raises concern over the control and monitoring of field trials.

12   *1993.* Act n.91/93 (Gazzetta Ufficiale n: 78, 3 April). Implementation of EC directive n.219/90 on contained use of GMOs. The relevant national authority is the Ministry of Health, along with, when necessary, the ministries of Agricultural and Environmental Resources, Labour, Industry, Commerce and MURST. The Ministry of Health set up an interministerial advisory body (see note 11) whose role is to examine the notices of contained use of genetically modified organisms, and if necessary consult the Health Council and *i*Scientific Committee on the Risks Deriving from the Use of Biological Agents.(see note 10). In general the Italian biotechnology industry regards this Act favourable (see however note 13) but the Scientific Committee on the Risks Deriving from the Use of Biological Agents along with the Ministry of Health raised concerns in 1996 over the lack of notification of use of GMO, in particular on the behalf of universities.

13   *1994.* Act n.626/94. The implementation of EC directive (n.679/90) on workers' safety turns out to be notoriously difficult to apply and is still only partly in force. There are some discrepancies between this and the law on contained use of genetically modified organisms (n91/93, see note 12) and industry maintains that the law is too restrictive and difficult for the biotechnology industry to fulfil.

14   *1993.* R&D, Procedural modification of the law n.46/82 (see note 1) on industrial and technological research, to favour small and medium sized companies. Formerly the procedures had favoured big industry and had carried unacceptable delays.

15   *1994.* MURST sets up National Committee for Advanced Biotechnology. Its role is to update the National Research Programme (see notes 1 and 6) and to elaborate series of new proposals.

16   *1992.* Deontological code of practice concerning reproductive technologies by the Italian Medical Association.

Figure 2. Intensity of articles on biotechnology in the Italian press

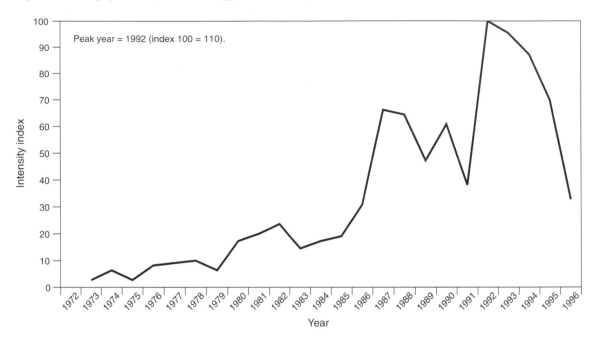

Figure 3. Attitudes to applications of biotechnology in Italy, 1996

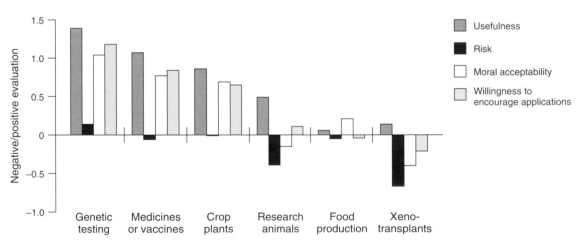

Table 1. Italian media profile (for an explanation of terminology please see Appendix 5).

Phase	Freq.[a] (%)	Frame (%)		Theme (%)		Actor (%)		Benefit/risk (%)		Location (%)	
1. 1973–79	12	**Progress**[b,c]	**81**	**Medical**	**34**	*Scientific*	*81*	**Both**	**49**	**Italy**	**27**
		Pandora's Box	10	*Basic research*	*32*	Public opinion	13	Benefit only	32	USA	20
		Accountability	5	Ethical	12	Political	7	Risk only	12	*UK*	*20*
				Safety & risk	8	Ethical	5	Neither	7	World	12
				Agriculture	6						
2. 1980–85	23	**Progress**	**74**	**Medical**	**34**	Scientific	60	**Benefit only**	**59**	**Italy**	**38**
		Economic	6	Basic research	19	*Industry*	*25*	Both	27	USA	27
		Ethical	5	*Economic*	*11*	Public opinion	10	Risk only	7	*World*	*19*
		Pandora's box	5	Animal	7	NGO	6	Neither	7		
3. 1986–91	35	**Progress**	**59**	*Medical*	*44*	Scientific	44	**Both**	**44**	*Italy*	*40*
		Ethical	*17*	Ethical	12	Industry	19	Benefit only	37	USA	32
		Nature/nurture	7	Animal	9	Political	11	Risk only	10		
		Economic	5	Basic research	6	*NGO*	*9*	*Neither*	*9*		
				Regulation	6	Public opinion	7				
						Ethical	5				
4. 1992–96	30	**Progress**	**54**	**Medical**	**39**	Scientific	43	**Both**	**44**	*USA*	*41*
		Ethical	12	*Safety & risk*	*13*	Industry	17	Benefit only	36	Italy	25
		Nature/nurture	11	Ethical	12	Political	12	Risk only	11	UK	11
		Pandora's box	8	Animal	8	Public opinion	11	Neither	3		
		Economic	*8*	Regulation	8	NGO	6				
		Accountability	*5*	Economic	5						
				Political	5						

a Percentage of corpus in the period; total $n = 340$.
b Bold indicates highest frequency within phase.
c Italics indicates highest frequency within category.

Table 2. Type of benefit and risk evaluations

	1973–79 (%)		1980–85 (%)		1986–91 (%)		1992–96 (%)	
Benefit	Health	60	Health	59	Health	52	Health	43
	Research	21	Economic	19	Economic	15	Economic	16
	Economic	15	Research	14	Research	13	Research	16
							Social equality	7
Risk	Moral	40	Moral	28	Moral	55	Moral	41
	Health	29	Health	28	Health	12	Environment	16
	Research	16	Social equality	15	Environment	10	Social equality	16
			Research	10	Social inequality	9	Health	12
			Economic	10				

Table 3. Attitudes to biotechnology in Italy

Effect on quality of life	1993 (%)	1996 (%)
Will improve	54.5	55.8
No effect	5.0	5.3
Make worse	9.0	16.2
Don't know	30.0	23.8

Notes and references

1 Gambardella, A and Orsenigo, L, 'Le biotecnologie in Italia: una opportunità perduta?', in Bussolati, C, Malerba, F and Torrisi, S (eds), *L'evoluzione delle industrie ad alta tecnologia in Italia* (Bologna: Il Mulino, 1996).

2 Spalla, C, *Il progresso delle biotecnologie in Italia* (Milan: Federchimica-Assobiotec, 1990); Spalla, C, *Le biotecnologie in Italia e nel mondo* (Milan: Federchimica-Assobiotec, 1996).

3 Gambardella and Orsenigo (1996), Spalla (1996).

4 Mazzoni, C M, 'La bioetica ha bisogno di norme giuridiche', in Mazzoni, C M (ed.), *Una norma giuridca per la bioetica* (Bologna: Il Mulino, 1998).

5 Bonsignore, A, 'Il problema della liceità delle invenzioni relative alla genetica', in Menesini, V (ed.), *Il vivente brevettabile* (Perugia: University of Perugia, 1996); Ricolfi, M, 'Bioetica, valori e mercato: il caso del brevetto biotecnologico', in Mazzoni, C M (1998).

6 Lumley, R, 'Peculiarities of the Italian newspaper', in Forgacs, D, and Lumley, R (eds), *Italian Cultural Studies* (Oxford: Oxford University Press, 1996); Sorrentino, C, *I percorsi della notizia* (Bologna: Baskerville, 1995).

7 Sorrentino, C (1995).

8 Beck, U, *Risk Society: Towards a New Modernity* (London: Sage, 1992).

Address for correspondence

Agnes Allansdottir, University of Siena, Scienze della Comunicazione, Via del Giglio 14, I-53100 Siena, Italy. Tel. +39 (0)577 298 461, fax +39 (0)577 298 460. E-mail allans@lettere.media.unisi.it

The Netherlands

Cees Midden, Anneke Hamstra, Jan Gutteling and Carla Smink

Political system and the economy

The Netherlands is a small, highly industrialised nation. This industrialisation extends to its agricultural sector: the lack of space and high population density have pressured agriculture to intensify, and a strong scientific foundation and economies of scale have enabled the Netherlands to become an important food-exporting country. It occupies a relatively central position in Europe from a trade point of view, and so the Netherlands' main harbour (Rotterdam) and the main airport (Schiphol) are both significant distribution nodes.

Of the sectors for which biotechnology is or may become important, the 'food and feed' sector is the largest (its turnover was Dfl75 billion in 1995), followed by the chemical sector (Dfl40 billion) and the agricultural sector (Dfl36 billion). The other relevant sectors are the environmental sector (Dfl8 billion) and pharmaceuticals (Dfl4.9 billion).[1]

Formally, the Netherlands is a constitutional, hereditary monarchy. Constitutionally, the country is a parliamentary democracy with a multi-party political system. The Head of State is not responsible politically for government policy; this responsibility rests with the cabinet ministers. The legislative branch is composed jointly of the government and the two-chambered States General, the latter serving the citizenry through proportional representation.

Executive power, in principle, is vested in the King; in practice, government ministers exercise this power. The judicial branch is composed of (politically) independent judges, who are appointed for life by the Head of State.

Dutch politics changed markedly after the Second World War, as traditional concepts of societal segmentation and religious divisions were abandoned in favour of more democracy and public involvement in the policy process. Indeed, the old politics of reconciliation and depoliticised action were steadily supplanted by a trend toward politicised and polarised activity. The principle of openness and transparency to popular scrutiny stimulated political activity, and the public was able to closely monitor government action and to respond quickly. These developments gave impetus to an increasing number of extraordinary parliamentary actions that have occurred during the last 25 years. Both centre-right and centre-left coalitions of political parties have served in government.

Public policy

Since the 1980s, the Netherlands has invested strongly in the development of biotechnology. After the breakthrough of recombinant DNA, the focus was on fundamental research. At the beginning of the 1980s, the stimulation of innovation became part of government programmes, and was complemented in the late 1980s by the stimulation of industrial biotechnological R&D. From the beginning of the 1990s, fundamental research has been attuned to industrial priorities on the one hand, with attention to market pull (specifying market priorities) for biotechnology on the other.

The percentages of the turnover in the various sectors related to biotechnology in 1996 are estimated (by experts) to be of the order of 5–10%, highest in pharmaceuticals and in chemicals, closely followed by the agricultural sector, food and feed, and environment. For all sectors, significant growth in the contribution of biotechnology is expected, reaching 20–40% of turnover by the year 2010.[2]

The sectors differ widely in their use of and intentions for biotechnology. The pharmaceutical sector is dependent on product innovations (e.g. new pharmaceuticals, diagnostics). Within the agro-food sector, the innovations are small steps, often directed at production processes rather than leading to new products. Especially in the food market, producers fear consumer reactions to major changes.

Biotechnology policy in the Netherlands falls into the following phases.

Phase 1, 1981–89: start of the debate. Perhaps because of the signals coming from the self-imposed moratorium on recombinant DNA research by scientists in Asilomar (Dutch scientists participated and reported back), Dutch policy concerning biotechnology has always been two-sided: to stimulate technology on the one hand (to harness the potential benefits); and on the other hand to consider potential risks. In 1981 a broad committee, the Brede Committee, was established to advise the government on the benefits and risks of recombinant DNA research (see Figure 1).

One of the first discussions in Parliament on recombinant DNA research concerned the approval for building a high-risk (CIII) laboratory (permission was granted in 1981). The most important social actor then was the Royal Academy of Science (KNAW), which was promoting recombinant DNA research.

Recombinant DNA pharmaceuticals (e.g. insulin, 1982) were introduced without any public debate or new regulations. The government used the existing law on pharmaceuticals, and followed the FDA directions. The early government attention to social considerations was reflected in the commissioning in 1985 by the Ministry of Education of a series of studies on social aspects (including public percept-ions) of biotechnology.

One of the NGOs that started considering biotechnology at an early stage was the animal welfare movement (Dierenbescherming). In 1985, they initiated a study on the subject of biotechnology and animals. Also, they were actively pushing for the development of a new law on animal welfare, which was finalised and passed in 1993, and which included a chapter on transgenic animals. A specific directive on animal welfare was passed in 1997.

The first product related to food to reach the Netherlands was the growth hormone bovine somatotropin (BST), which was developed in the USA for increasing milk production in cows. In 1987, the Ministry of Agriculture issued a consumer attitude study on BST, and in 1988 the issue was discussed in Parliament. BST was never allowed on to the Dutch market; and an EU moratorium came into effect at the beginning of 1990.

The first approval for field tests (of a potato with virus X resistance) was given in 1988. It was contested by an environmental organisation, the Foundation for Nature and the Environment, but was approved in court by the highest authority (the Council of State) in 1989. Action groups calling themselves 'Angry Potatoes' destroyed the field tests for the first time in August 1989, and several times after that. These activities, however, did not succeed in mobilising the public to any great extent.

In Parliament, questions continued to be asked about both the usefulness and the risks of biotech-nology. The issues were mostly related to animal biotechnology (animal welfare and ethics) and agricultural biotechnology (safety for health and the environment, patenting, market demands and consumer acceptance).

Phase 2, 1989–91: broadening public debate. Around 1989, the debate also increased outside Parliament. Broader discussions developed, and were reflected in press coverage in newspapers, NGO magazines and professional magazines. The two main issues were transgenic animals (because of the development of transgenic cows, announced in 1989, and the first calf born in 1990) and food (the first case was the recombinant DNA cheese rennet chymosin, which was known publicly to be a reality in 1988 and put before the government for approval in 1989, and about which discussions went on until 1990).

In this period, attention was given to potential public acceptance problems, and subsequently demands for public information were made. A government-initiated study in 1988–89 showed that public awareness and knowledge of biotechnology (traditional or modern) was still rather low. The government supported various initiatives for multi-actor workshops and debates on biotechnological subjects, and installed both an ethical committee on animal biotechnology (Schroten Committee, 1989) and a committee on environmental safety (VCOGEM, 1990, replacing the ad hoc committee on recombinant DNA that had been established in 1979). The government also subsidised the Con-sumer and Biotechnology Foundation (1991), which was to support knowledge-based opinion-forming about biotechnology by NGOs, especially the consumer organisations, and to take part in the debate. A government-funded, broad public-information campaign was started by the Found-ation for Public Information on Science, Technology and the Humanities (Foundation PWT – now known as Foundation WeTeN). This campaign ended in 1998.

Phase 3, 1991–97: realisations. Although chymosin was approved by the government, and consumer organisations and food producers agreed in the Advisory Committee of the Food Law that it would not be necessary to label it (1991), the dairy industry decided not to use chymosin because of the potential reactions of the important German export market. The birth of the first transgenic bull, Herman, in 1990 raised a lot of public debate, both in Parlia-ment and in the newspapers. Although the new law on animal welfare had been under discussion since 1987, the law was not yet ready for this develop-ment, and no adequate legislation was available until the law was in place in 1993. After this, Parliament asked for a revision of the decisions on research with transgenic cows, which had not been approved by the government. In 1992, Parliament decided that under stringent restrictions the breeding programme involving Herman could continue, in order to produce a certain amount of milk containing lactoferrin.

In 1993, just after the new law on animal welfare was in place, a consensus conference was held on transgenic animals. In 1995, another consensus conference considered human genetic screening research. The Health Council published their advice on genetic therapies in June 1997. One of their recommendations was that in the near future the use of genetically modified organisms in humans should be regulated by law. Also, genetic therapy research involving humans should be regulated by protocols.

Finally, in 1996, genetically modified soyabeans came on to the European market. The Ministry of Health approved the use of this soya in the Netherlands. In April 1997, this Ministry issued, within the framework of the Food Law, a labelling directive on the use of genetically modified soya and maize in products. However, in October, the courts decided that the Food Law could not be used to differentiate between products, and the directive was rejected. This did not imply any changes because from November 1997 onwards the labelling of genetically modified soya was directed by European regulations. In practice, however, in 1998 only a few such products that are labelled can be found in the supermarkets.

Media coverage

Six major national newspapers and at least one newspaper in each region serve readers throughout the country. All of the major national newspapers have their headquarters in the western region of the Netherlands. Circulation statistics – collected and published by CEBUCO – indicate that in the years leading up to 1998 the total annual circulation of Dutch newspapers is 4.6 million copies (the Netherlands has about 6 million households). The current trend is that the circulation of national dailies is increasing and that of the regional newspapers is decreasing. Some 89% of all newspaper sales are subscriptions. For the average person in the Netherlands, newspaper-reading is a daily ritual, with most people spending on average half an hour a day reading at least one newspaper.

The history of opinion-leading Dutch press

During the study period, the country witnessed a process of newspaper concentration. Responding to growing pressure on their revenues from television and radio advertising, several newspaper companies merged their operations. From 1946 to 1989, the number of privately owned newspaper companies plummeted from 81 to 20, while the total number of newspapers declined from 124 to 82. Of these 82 newspapers, only 45 have their own staff; the other 37 newspapers serve as regional editions of the national newspapers. In 1988, the top four newspaper companies owned 60% of the market. By 1992 most regions had only one newspaper, the result of both the trend towards press concentration and other social and economic considerations.

Prior to 1960, the Dutch press was either religiously or politically oriented, with most social movements arising from these sources. Indeed, since the end of the nineteenth century, so-called 'persuasive newspapers' had prevailed, their goal being to foster the organisation and commitment of the rank and file around their specific ideological group. Besides ordinary news, these newspapers published articles on political, social and cultural events within their own movement. The 1960s saw a gradual but steady decline in newspapers' commitments to particular social movements, such as trade unions or specific political parties. Similar strides toward autonomy took place within newspaper companies, as editorial staff became more independent and proactive in editorial matters. In 1977, newspaper journalists and editors contractually assumed full responsibility for their newspapers' content, while the newspaper companies were entrusted with commercial activities.

Although all Dutch newspapers offer a variety of reports, newspapers are distinguishable by the type of information they publish. One group consists of the popular, general interest newspapers (*Telegraaf, Algemeen Dagblad*) which emphasise human interest stories, entertainment (e.g. cinema, cartoons), sporting events, and practical information (e.g. weather reports). The second group, the so-called 'quality' newspapers (*De Volkskrant, Trouw, NRC-Handelsblad*), offer readers more policy-oriented information on, say, social and political issues and organisations, and their key actors. The popular national newspapers have a higher circulation than the quality national dailies.

Science coverage in De Volkskrant

In our biotechnology study we analysed the coverage in *De Volkskrant*. *De Volkskrant* ranks third in the circulation statistics and has the largest circulation of the quality papers. More than two-thirds of *De Volkskrant*'s readers can be categorised as middle or upper class.[3] Overall, the *De Volkskrant* readership is better educated than the Dutch public in general. Additional analyses indicate its readership is also different from the popular newspapers' readership in this respect.

De Volkskrant was founded as a newspaper serving the explicit interests of the Roman Catholic population in the Netherlands. After 1960, Dutch newspapers were no longer divided along political and religious lines and could address broader public issues. In 1965 *De Volkskrant*, which was then owned by a Catholic labour movement, abolished its subtitle 'Catholic Newspaper for the Netherlands'. Dramatic changes followed, and *De Volkskrant* emerged as a liberal, left-of-centre newspaper. As the first national Dutch newspaper, on 2 October 1975 *De Volkskrant* adopted editorial bylaws, which included a declaration of the newspaper's identity:

De Volkskrant is a national daily newspaper, which has the objective of informing readers in as fair and as versatile a manner as possible. *De Volkskrant* arose from the Catholic labour movement. For that reason, among others, *De Volkskrant* wants to be progressive and to plead for the oppressed and those whose rights are violated. *De Volkskrant* is independent in its opinions. In particular *De Volkskrant* aims to stimulate developments which promise a more humane society.[4]

From 1966 to 1971, *De Volkskrant*'s editor of foreign affairs also wrote scientific articles; from 1971 he concentrated on scientific topics. In 1977, *De Volkskrant* launched a half-page weekly column on scientific issues, expanding this in 1981 to a full section entitled 'Science and Society' that appeared in the Saturday edition. (In 1988, this section was renamed 'Science'.) Since then, two editors and a team of regular science reporters have written articles covering up to four pages per edition.[5] Basically, De Volkskrant's editorial policy on science is to treat all issues competitively and objectively. The introduction of modern techniques has changed the appearance of the newspaper over the last 25 years. The amount of news presented to the readers has increased: in April 1973, the average daily number of pages with news was 14.5 (out of a total of 24.1 pages), but by 1996 this had increased to 22.3 pages (out of a total of 40.6 pages). The increase in overall size can be attributed mainly to the introduction of a number of specialist sections that include large advertising spaces.

The results of content analysis of De Volkskrant, *1973–96*

Two methods of selection were applied. For issues published prior to 1993 we used microfiches containing exact copies of the 'paper' version of the newspaper. After 1993 we used the CD-ROM version of *De Volkskrant*, containing all relevant information. Because of the laborious nature of scanning microfiches and our limited resources, we scanned the following years: 1973, 1975, 1977, 1979, 1981, 1983, 1985 and 1986–92. For the years 1973–85 we scanned every other day, starting on the first day in the year that the newspaper was published. From 1986 complete years were scanned.

Microfiches and CD-ROMs were screened for articles containing references to (applications of) biotechnology, DNA, genetic manipulation or modification, the genome project or transgenic techniques, etc. Using the microfiches all headlines were scanned as well as the lead paragraph of the article. The CD-ROMs were scanned electronically with search keys. All selected articles were copied or printed, and coded according to the international project's coding frame.

In total, 1119 articles were selected and coded according to the procedures described above. Figure 2 presents the longitudinal data. Making tentative estimates of the frequencies of articles in the non-scanned years and doubling the frequencies for the 50% sample years would bring the number of articles to a total of 1201. So, we assume our database of 1119 articles contains approximately 93% of all biotechnology articles over the period 1973–96. We observed a particularly sharp increase

in biotechnology coverage from 1993 onwards, probably due to the CD-ROM scanning technique.

Overall, almost 58% of the articles were published on Saturdays. Tuesdays to Fridays accounted for an average of 10%, and Mondays for only 3% of biotechnology articles. Almost 38% of the articles were published in the (Saturday) Science section, and almost 10% referred to articles in other media (mostly scientific magazines such as *Science, Nature*, etc.). National news accounted for 19% of the articles, and 8% were coded as international news. Approximately 7% took the form of 'letter to the editor', and 8% were found on the business pages. About 57% of the articles were shorter than 500 words. About 21% had over 1000 words.

Based on the empirical criteria observed – highs and lows in the frequencies over the years, as well as the different modes of scanning – we divided the time period into five phases: 1973–81 (lowest), 1982–87 (relatively high), 1988–90 (relatively high), 1991–92 (relatively low) and 1993–96 (CD-ROM period, highest). Table 1 details these phases. Looking at these 'biotechnology phases' as they are broken down by frames, themes, actors, risk and benefit information, and location, we observe a rather remarkable stable presentation of biotechnology.

In 1973–81, progress was by far the dominant frame, and it remained so in all other phases. Public accountability and economic opportunities were other frequently observed frames. The medical application of biotechnology was, in all phases, the theme mentioned most frequently, followed by basic research. Safety and risk, economic, animal and agricultural themes also appeared regularly in all phases. Scientists were by far the most dominant actors in *De Volkskrant*'s accounts of biotechnology. Political actors followed from 1982, and NGOs played significant roles before 1982 and between 1988 and 1992. From 1993 onwards, media and public opinion seem to have taken over their position in the coverage. Throughout the entire period, too, industrial actors were important. Articles containing information about only the benefits of biotechnology were very dominant, and outnumbered articles with only risk information and articles mentioning both benefits and risks. Increasingly over the phases, articles were coded as describing no risks or benefits at all. In most phases, most articles described biotechnology stories from the Netherlands. Stories about the USA were also found very frequently in all phases. Stories from other European countries gained prominence over the years.

Public perceptions

Biotechnology compared to other technologies

People in the Netherlands are mainly optimistic

about the impact of technologies. For solar energy, computers, telecommunications and new materials, over 80% think that these technologies will improve lifestyles. No more than 10% think they will make things worse. The results show, however, that biotechnology evokes more negative expectations: some 29% think it will make things worse. Both biotechnology and space technology have low expected positive effects (59% and 56% respectively). Yet the number of people who expect biotechnology to have positive effects is double the number of those who expect negative effects. In other words, in a general sense, attitudes towards these technologies seem rather positive.

Looking at the relationships among the different technologies, it appears that biotechnology has the highest correlation with space technology ($r = 0.21$). It shares with this technology a lack of expected benefits. With the other technologies, correlations were lower.

Attitudes towards specific applications

Figure 3 shows the respondents' judgements for six applications on their usefulness, risk and moral acceptability, and whether they 'should be encouraged'. It can be concluded that the most favourable applications concern the production of medicines and medical diagnostics. The least popular are the production of human transplants and the development of animals for research. Genetic testing and the production of insulin are considered the most useful applications; using genetically modified animals for research and for human transplants are the least useful. Applications for human transplants and for food production are seen as most risky. Producing animals for research and for human transplants is considered least morally acceptable, while the opposite is true for genetic testing and the production of medicines like insulin.

In Table 2 are aggregated the characteristics that best explain the more general 'encourage' variable. It can be concluded that whether a respondent feels that an application should be encouraged depends on the level of perceived usefulness and moral acceptability. The level of perceived risks plays a minor role.

The Pearson correlation between the independent variables 'use' and 'moral acceptability' is $r = 0.80$, which indicates a strong relationship between perceived use and moral acceptability. This seems hard to understand within the frame of rational attitude formation. This finding raises the question of whether these attributes are really predictive of general encouragement, or whether the process might be the reverse: these factors justify the general feeling that is expressed in the 'encourage' answers.

Relationship between application judgements

To what extent can the general attitude towards biotechnology be interpreted as a compilation of the six specific applications? Table 3 shows the results of a regression analysis. It can be concluded that the general attitude can hardly be seen in this way: it appears from the regression model that the general attitude can be explained by only one of the applications, namely food production. If we inspect the correlations then it appears that most of the specific applications have very modest relations to the general concept. The analysis suggests that the general attitude cannot be seen as based on a diversity of evaluative judgements on specific applications. Rather, this attitude seems based on very general notions of biotechnology.

Knowledge

Table 4 gives an overview of the knowledge scores on five indicators. The objective knowledge scales are based on items which can be classified as true or false. The 'images scale' refers to notions that may be held among the public and might be powerful determinants of people's mental representations of applications of biotechnology. The general knowledge scale is based on a combination of the objective knowledge scale and the image scale. It can be applied because the objective and 'image' knowledge indicators are not independent, as Table 5 shows.

The objective knowledge scores in the Netherlands appear to be rather high compared to the average scores in Europe. Also, on their acceptance of 'menacing' images, the Dutch score below average.

Predicting attitudes from knowledge

Table 6 depicts the relations between knowledge and attitudes on the general and the specific levels, and shows that knowledge and attitudes do not have any linear relationship. Neither the level of knowledge nor the presence of menacing images is predictive of the attitudes towards the six specific applications, nor are they predictive of the attitude towards biotechnology in a general sense. It might be hypothesised that the relationship is not linear because attitudes have diverged among the more knowledgeable, and that these people have developed more crystallised attitudes. That hypothesis would be supported by a strong relationship between the underlying attributes and knowledge, as knowledge effects on attitudes will be mediated by the attributes which might be less likely to diverge on the basis of more knowledge. The last three rows in Table 6 depict the relationships between knowledge and the three attributes 'risk', 'use' and 'moral acceptability', all aggregated over six applications. The correlations are very low, and not higher than the attitude measures.

Attitudes towards regulation

Table 7 gives an overview of the distribution of attitudes towards the three main aspects of regulation: attitude towards risk management; a technocratic regulation style; and a regulation style oriented towards protecting traditional methods. It can be concluded that the level of risk concern is low to moderate. The majority of the Dutch sample does not support a technocratic regulation style, i.e. leaving decisions to industry and excluding public participation. An approach protecting traditional breeding methods is not very popular either. As is the case in many other European countries, people in the Netherlands think that international bodies are best suited to regulate biotechnology. There is surprisingly little preference for national bodies and Parliament, and for bodies of the EU.

Group differences

Analysing specific attitudes on the sum of scores over six applications of perceived use, moral acceptability and encouragement reveals no substantial differences between groups varying in education, sex, support for green issues, and religiosity. This finding is consistent with other national survey data. It may also be interpreted as an indication that attitudes have not crystallised strongly.

Commentary

In Table 8, we have gathered information from all three modules (policy, media and public perception) regarding five issues that have been important in the Dutch debate on applications of biotechnology. These issues are: GMO field tests; developments in the area of gene therapy; Herman the Bull as an example of a transgenic animal; and finally BST and chymosin as food-related issues. Table 8 refers to the major policy-related events regarding these issues.

The issues have happened in different phases in the history of policy making. The field tests are an example of an issue during the start of the debate. BST became an issue during the initial phase, but was still an issue in the 'broadening public debate' phase, along with gene therapy and chymosin. Herman the Bull is a clear example of an issue during the 'realisations' phase.

Table 8 also presents the first account of each issue in the opinion-leading Dutch newspaper we examined, and the media follow-up and content. It also gives a perspective on the reactions of the Dutch public to these issues in 1993 and 1996.[6]

Regulation of biotechnology

In the Netherlands, unlike in other countries, the debate over biotechnology has not taken on a fundamental character. There have been two occasions when government and Parliament considered devising a general law on GMOs (1990–91 and 1995–96), but both times they thought this unnecessary and undesirable. The general policy has been to see what is needed, and to see which laws are already in place that could deal with the issue, and then to adapt them if necessary. Reflecting on the laws and regulations on GMOs in the Netherlands (the first issue in Table 8), one can say that the Dutch government has taken a pragmatic approach to biotechnology.

The first products of recombinant DNA, in the pharmaceutical sector, were not regulated specifically; the existing pharmaceuticals law was applied, following the FDA advice. The first applications of biotechnology in animals (see the issue of Herman the Bull in Table 8) were accompanied by the preparation of adaptations of a specific regulation (laws on animal testing and on animal welfare). When the first food products made with GMOs approached the market (e.g. the chymosin issue), a regulation was made for novel foods, followed by a ministerial decision on labelling of the transgenic soya that was introduced from the USA in 1996. This decision on labelling was later also adopted by the EU.

The general impression from the policy domain is also reflected in the media coverage in the Dutch opinion-leading press. Through the years its major focus has been on the progress being made in the area of biotechnology, with a strong interest in medical applications and basic research efforts. This interest in medical developments is reflected in Table 8 by the issue of gene therapy, which since 1985 has repeatedly triggered media attention. Other framing of biotechnology news can be characterised as public accountability, referring to the call for public participation in regulatory mechanisms. Over the years, the economic prospects and ethical implications of biotechnology appear to have had similar, but definitely smaller, amounts of media attention. Other themes in the media coverage are safety and risk issues, animal and agricultural issues, and, from the late 1980s, identification issues. As can be seen in Table 8, a typically Dutch biotechnology issue such as Herman has been in the news almost constantly since the first announcement of the effort to exploit transgenic animals commercially for the production of pharmaceuticals. Of the issues in Table 8, Herman the Bull has gathered by far the most media interest.

Actors and public attitudes

The emphasis on scientific endeavours is also

The London School of Economics and Political Science

Department of Social Psychology
St. Clements Building
Houghton Street
London WC2A 2AE
Telephone: 0171-955 7702
Fax: 0171-955 7005
Email: g.gaskell@lse.ac.uk

Professor George Gaskell

25 November 1999

Dear Susanna, Paul and Larry

As background to our meeting on 14th December I thought I should send you a copy "Biotechnology in the Public Sphere". This was designed as a source book based on the first three years of our multi-national project. While this is not one of the most captivating of reads, if you have time to skim Part 1 you will get an feel for the approach we took and some of the concepts. For those interested in survey research, Part 111 outlines the findings of the Eurobarometer and in the appendices you will find details of the methodology for the media analysis. Any of the national profiles shows how we documented policy developments.

... the next project our ambition is to bring a little more to the research, in particular the case studies are planned to address a range of hypotheses on the links between public perceptions, media coverage and policy developments. The development of new concepts and of research questions will be one of the challenges for the project group, and we look forward to your contributions to this collective endeavour.

Martin Bauer and I are currently completing an introduction to a second book to be published by Cambridge University Press. This contains a series of more reflective comparative analyses and case studies. These are first authored by members of the project group other than the three of us in London. Edna Einsiedel, for example, is the first author of a chapter on the reception of "Dolly the Sheep".

I look forward to my first visit to Texas, to meeting you all and to planning a productive collaboration.

George Gaskell

The London School of Economics is a School of the University of London. It is a charity and is incorporated in England as a company limited by guarantee under the Companies Acts (Reg. No. 70527).

reflected in the very dominant role of the scientific community as actors or spokespersons in the newspapers. Perhaps it is illustrative of the Dutch coverage of biotechnology that it is mostly framed as scientific news. Four of the five issues made their first appearance in the science section of the newspaper we studied. Other actors, from industry, politics or NGOs, appear less frequently in the media. Because of this, it is not remarkable that most newspaper articles cover the potential beneficial aspects of applications of biotechnology. Potential risks are not reported as heavily, leaving the general impression that through the years the Dutch coverage of biotechnology has been slanted toward its positive aspects.

A majority of the population is positive about biotechnology in a general sense, although the area is perceived more negatively than other areas of technology. Attitudes towards medical applications are relatively positive; attitudes towards genetic research and human transplants, however, are less positive. These applications are seen as least morally acceptable. All attitude means vary between 2.2 and 3.1, so on average none of the applications is evaluated as very negative or positive. Attitude differences are not related substantially to group differences in religion, support for green issues, sex or education. Compared to other European countries, attitudes in the Netherlands can be characterised as moderate with relatively little polarisation. These conclusions seem to be valid also for attitudes toward the specific issues in Table 8. Biotechnological solutions to improve agricultural products, and

gene therapy, are seen by most people as positive developments. Mixed reactions can be observed toward transgenic animals, BST and chymosin.

Perceived use and moral acceptability are the best predictors of the general judgement 'should be encouraged'. Perceived risk plays no role here. This finding is not consistent with other survey data in which health risk appears to be a significant factor. The high correlation between use and moral acceptability suggests that these attributes might be interpreted as retrospective justifications of attitudes, instead of influential factors. If so, people would justify rather superficial attitudes with these opinions on use and moral acceptability. This explanation might be further explored. The general judgement on biotechnology can hardly be seen as a compilation of specific judgements, which supports the interpretation of superficial general attitudes as the basis for specific judgements.

The level of knowledge is low, although relatively high compared to the European average. 'Menacing' images are not present to a large extent. No linear relationship exists between knowledge indicators and attitudes.

Concerning attitudes on regulation, it appears that the level of risk concern is low to moderate; a technocratic regulation style does not encounter much support, and the perceived need for protection of traditional breeding methods is not very high. Public participation is valued to a large extent. Surprisingly, international bodies, like the UN, are seen as best for regulation, not national bodies, and not the EU.

Acknowledgement

The authors wish to thank the Ministry of Economic Affairs, the Ministry of Agriculture, Nature Management and Fisheries, and the Foundation WeTeN for their support enabling the Dutch part of this research project. We also wish to thank the following people their comments on an earlier version of this chapter: Mrs Anja van der Neut (Dutch Ministry of Agriculture), Frans van Dam (Consumer and Biotechnology Foundation) and Lucien Hanssen (Foundation WeTeN).

Figure 1. A regulatory profile for Netherlands

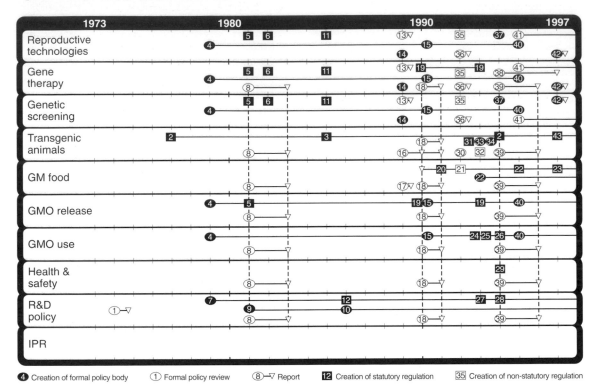

1 *1974.* Advice on recombinant DNA research given (exact date of establishment of advisory group is unknown).

2 *1977.* National law on animal testing designed to weigh up the suffering of animals against the usefulness of the test. This law was modified in 1994 to implement EC directive 86/609/EEC.

3 *1985.* An Order in Council (AMvB) under the general 1977 law on animal testing.

4 *1979.* Ad hoc committee established on recombinant DNA use and release. This committee was replaced in 1990 by the Committee for Environmental Safety (VCOGEM) and then later replaced by the Committee on Genetic Modification in 1995 (COGEM).

5 *1981.* Law passed which covers GMO use. This law is primarily an environmental law established to prevent unhealthy/unsafe substances or organisms from spreading to the environment (contamination).

6 *1982.* Law on hospital facilities which regulates chromosome research and prenatal DNA research.

7 *1979–81.* Government programmes for stimulating R&D in biotechnology.

8 *1981–83.* Far-reaching DNA committee is established to examine the social and ethical aspects of recombinant DNA technologies.

9 *1981.* A formal policy body was established in the form of the Steering Committee on R&D in biotechnology.

10 *1986.* Advisory Committee on Biotechnology established. In 1989 the Advisory Committee submitted

advice on biotechnology research in the Netherlands.

11 *1985.* Hereditary diagnostics are included in the services covered by medical insurance. It is also included in the law on access to medical insurance.

12 *1986.* Government programme for stimulating R&D in biotechnology.

13 *1989.* Report *(Heredity: Science and Society)* on the advice of the Health Council.

14 *1989.* Committee on Ethics in Medical Research (KEMO) founded to examine protocols for gene therapy research. This role will later be taken by the Central Committee after the introduction of a new law on medical scientific research in humans.

15 *1990.* VCOGEM replaced the 1979 ad hoc committee. VCOGEM evolved to become COGEM in 1995.

16 *1990.* The government's viewpoint on the work on the Advisory Committee on the Ethics and Biotechnology on Animals (Committee Schroten – established 1989 – report 1990).

17 *1989.* Formal policy review by the Food Council. Report produced, detailing advice on biotechnology.

18 *1990–91.* Extensive government inventory of existing laws and regulations on biotechnology. Conclusion of this review was that no general GMO law is needed.

19 *1990, 1993.* The Dutch deliberate release regulations formulated in response to the EC directive 90/220 and were later modified in 1993 following the EC modifications. These regulations covered GMO release and gene therapy.

20 *1991.* An AMvB under the frame law on foods. This AMvB was modified in 1995 to become the Warenwetregeling allowing new foodstuffs.

21 *1992.* Non-statutory regulation on food labelling. Partial agreement on consumer information and labelling in the Informal Consultation group on Biotechnology. Informal agreement displaced by the government administered AMvB in 1995.

22 *1993.* The official committee for the implementation of advice on novel foods. Committee still in existence (1997).

23 *1997.* AMvB Government decision on the labelling of GMO foods.

24 *1993.* Law passed to implement EC directive 90/219. This was not an adaptation to the 1981 law but a new extension of the existing law on environmentally dangerous substances. This law regulates activities to do with GMOs in the laboratory and other contained areas.

25 *1993.* Law passed on contained use of GMOs which regulates demands for laboratory installations and permit procedures.

26 *1994.* Law passed on contained use of GMOs.

27 *1993.* Government programme for stimulating R&D in biotechnology.

28 *1994–97.* Government programme for stimulating R&D in biotechnology.

29 *1994.* Law passed on protection of workers following the EC directive 90/679/EEC which regulates the protection of workers against the risk of exposure to biological agents.

30 *1992.* Temporary committee on ethical testing of genetic modification of animals.

31 *1993.* The frame law for animal health and welfare (known as the GWW Frame Law). Discussions in preparation for the frame law began in 1987 and in 1993 the law was accepted by Parliament. Under this frame law a number of AMvB national directives are being developed on concrete issues. An AMvB is made by government formally without putting it before Parliament.

32 *1993 (ongoing).* Statutory decision on transgenic animals (within the frame law in note 1).

33 *1993.* The Council for Animal Affairs was installed as a consequence of the 1993 GWW Frame Law. It works as a consultation and advice body concerning health and biotechnology of animals.

34 *1993.* A working group was established to develop a number of AMvBs for the GWW Frame Law.

35 *1992.* Plan for a law on handling of human embryos and reproductive cells. The result of this plan was a temporary prohibition of research with embryos obtained by IVF. In 1995, this ban was ended by the ministry, because it was thought that temporary prohibitions were unnecessary. In a new proposal, research is only allowed in cases where both parents of the embryo give their written agreement.

36 *1992.* Report (*Genes and Limitations*) on genetic therapy at the scientific institute of the CDA (a political party).

37 *1994.* Formal decision buy the Minister of Agriculture that cloning of animals is subject to permission.

38 *1994–97.* Formal policy review conducted by the Health Council. Report produced on gene therapy.

39 *1994–96.* Government evaluation of all GMO laws and regulations. Conclusion is that no general GMO law is needed.

40 *1995.* COGEM established replacing the earlier VCOGEM to look at release and use of GMOs.

41 *1994–95.* Plans for the future development of a law on medical scientific research on humans. It will provide rules for research with humans (research which is not related in the first place to individual interests of patients).

42 *1997.* Report on the advice of the Health Council on gene therapy.

43 *1997.* Resolution on biotechnology on animals. Ethical tests relating to transgenic animals becomes compulsory.

Figure 2. Intensity of articles on biotechnology in the Dutch press

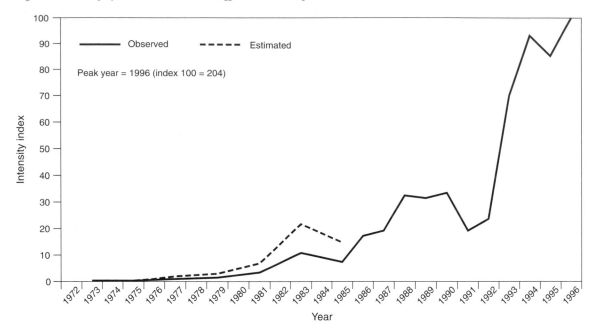

Figure 3. Attitudes to applications of biotechnology in the Netherlands, 1996

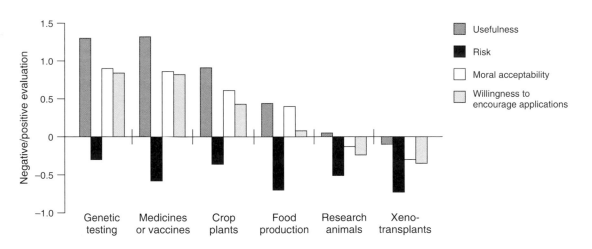

Table 1. Dutch media profile (for an explanation of terminology please see Appendix 5)

Phase	Freq.[a] (%)	Frame (%)		Theme (%)		Actor (%)		Benefit/risk (%)		Location (%)	
1. 1973–81	1	**Progress**[b,c]	**58**	**Medical**	**22**	*Scientific*	*61*	**Benefit only**	**67**	**Netherlands**	**54**
		Accountability	25	Basic research	17	NGO	11	Risk only	17	USA	31
		Economic	8	*Safety & risk*	*17*	Industry	11	Both	17	Other Europe	8
		Runaway	*8*	Other	13					*Other countries*	*8*
				Animal	13						
				Political	9						
2. 1982–87	10	**Progress**	**41**	**Medical**	**19**	Scientific	45	Benefit only	54	*USA*	*39*
		Accountability	*28*	*Basic research*	*18*	Industry	26	Both	28	Netherlands	38
		Economic	*20*	Economic	12	Political	19	Neither	11	Other Europe	15
		Ethical	8	Safety & risk	11			Risk only	7	*Other countries*	*8*
				Agriculture	*9*						
				Regulations	8						
3. 1988–90	18	**Progress**	**33**	**Medical**	**21**	Scientific	44	**Benefit only**	**36**	Netherlands	52
		Accountability	27	Basic research	14	Political	21	Both	23	USA	29
		Economic	14	Safety & risk	9	Industry	14	Neither	22	Other Europe	13
		Ethical	8	*Regulations*	*9*	NGO	10	*Risk only*	*18*		
				Agriculture	8						
				Ethical	*9*						
				Identification	9						
4. 1991–92	8	**Progress**	**36**	*Medical*	*31*	Scientific	43	**Benefit only**	**49**	*Netherlands*	*56*
		Accountability	22	Basic research	11	Industry	19	Both	23	USA	22
		Economic	12	Safety & risk	12	Political	12	Risk only	15	Other Europe	18
		Ethical	*12*	Identification	8	NGO	9	Neither	13		
		No frame	10	Agriculture	8						
				Regulations	7						
				Animal	9						
5. 1993–96	64	*No frame*	*56*	**Medical**	**28**	Scientific	36	*Neither*	*64*	**Netherlands**	**53**
		Progress	**14**	Basic research	13	Industry	20	Benefit only	19	*Other Europe*	*20*
		Ethical	11	*Identification*	*11*	Political	17	Both	11	USA	19
		Accountability	10	Animal	10	*Media/public*	*9*	Risk only	6	*Other countries*	*8*
		Economic	8	*Economic*	*9*						

a Percentage of corpus in the period; total $n = 1119$.
b Bold indicates highest frequency within phase.
c Italics indicates highest frequency within category.

Table 2. Relationship between 'should be encouraged' (mean of six applications) and three characteristics (mean scores of six applications)

Characteristics	β	t	p	r
Risk	−0.05	−30.068	0.002	−0.14
Use	0.42	150.876	0.000	0.80
Moral acceptable	0.46	170.401	0.000	0.80

Dependent variable: 'Encourage'

$R^2 = 0.72$

Table 3. Relationship between general attitude towards biotechnology and specific applications

Specific applications	β	r
Biotechnology for production of foods	ns	−0.39
Using genes from plants to make plants more resistant to insect pests	ns	−0.28
Use human to produces medicines like insulin	ns	−0.23
Developing GM animals for research ('oncomouse')	ns	−0.16
Introducing human genes into animals to produce human transplants	ns	−0.21
Using genetic testing to detect inherited diseases	−0.31	−0.13

Table 4. Mean scores and standard deviations on four knowledge indicators

Knowledge indicators	Mean	SD	N
Objective knowledge (five-item trend scale)	3.43	1.08	1069
Objective knowledge (six-item scale)	4.07	1.24	1069
General knowledge scale (nine-items scale)	5.52	1.80	1069
Images (three items)	1.90	0.98	1015

Table 5. Correlations between knowledge indicators

Knowledge indicators	1	2	3	4
Objective knowledge (five-item trend scale)	1.00	0.92	0.81	0.33
Objective knowledge (six-item scale)		1.00	0.86	0.31
General knowledge scale (nine-item scale)			1.00	0.64
Images (three items)				1.00

Table 6. Correlations between knowledge indicators and attitudes

Attitudes	Knowledge (5)	Knowledge (6)	Knowledge (9)	Images (3)
General (Q2)[a]	−0.02	−0.02	−0.01	−0.03
(Q10a[b])	0.00	0.03	0.05	0.10
(Q10b[b])	0.02	0.06	0.06	0.04
(Q10c[b])	0.06	0.05	0.07	0.08
(Q10d[b])	−0.06	−0.06	−0.05	−0.01
(Q10e[b])	−0.01	0.02	0.01	−0.00
(Q10f[b])	0.03	0.04	0.06	0.07
M risk[c]	0.11	0.09	0.10	0.04
M use[d]	0.01	0.05	0.05	0.05
M mor.acc[e]	0.02	0.05	0.04	0.03

a Attitude to biotechnology in general: optimism or pessimism about the future contributions of biotechnology.
b Attitudes to six specific applications (see Table 3): whether or not each application should be encouraged.
c Mean score for the degree of perceived risk over the six applications.
d Mean score for the degree of perceived use over the six applications.
e Mean score for the degree of moral acceptance over the six applications.

Table 7. Frequency distribution of attitude towards regulation

Score	Risk concern [a]	Technocratic [b]	Traditional [c]
1	6.6	0.8	10.7
2	29.1	0.9	33.1
3	31.5	4.0	28.4
4	32.7	12.0	27.8
5		21.9	
6		60.4	
	100.0%	100.0%	100.0%

a 1 = high risk concern,   4 = low
b 1 = highly technocratic, 6 = low
c 1 = highly traditional,   4 = low

Note refers to distribution of respondents for three factors of regulation of biotechnology (attitude to risk management, to a technocratic regulation style and to the protection of traditional breeding methods).

Table 8. Five issues important to the Dutch biotechnology debate: Data from the policy and media analysis, and information about public perception

Issue	Policy events	First media coverage & follow up
Field tests, GMO release *Triggering event(s):* Developments in other countries.	*1987.* Questions in Parliament to the Minister of Agriculture. Development of concept decision on GMOs. Until this decision is operational, field tests are considered under the Hinder wet, which is a responsibility of local government. *1989.* First application for field test by ITAL Concerned action group destroys test field. *1990.* Decision GMOs by Ministry of the Environment. *1990–92.* Several other fields destroyed by actions groups.	22/2/1986 *Title:* 'Genetically manipulated organisms come outside' *Location:* science section In total 31 articles found, with relative peaks in 1986 (6), 1988 (6) and 1992 (4)
Gene therapy *Triggering event(s):* Medical scientific developments in other countries.	*1989.* Special ethical committee installed for medical research. *1991.* First research proposal for gene therapy. *1997.* Report of Health Council on gene therapy.	15/6/1985 *Title:* 'Gene therapy: new promises, new risks' *Location:* science section In total 41 articles found, with relative peaks in 1988 (6), 1991 (6) and 1994 (9)
Herman the Bull *Triggering event(s):* developments in Dutch agricultural science.	*1989.* Parliament expresses opinions against introducing species-unrelated genes in agricultural animals. Special advisory committee installed *1990.* No support in Parliament for motion asking to abandon all experiments until after introducing new law regulating animal welfare. *1990.* Report of the ethical committee on animal biotechnology is presented to Parliament. *1992.* Parliament decided that under severe restrictions the breeding of Herman may continue in order to gain the desired milk amount for testing. However, the calves of Herman's daughters may not be used for further breeding.	14/7/89 *Title:* In future cows will produce mothers' milk *Location:* front page In total 68 articles found, with relative peaks in 1993 (8), 1994 (23) and 1995 (13), 1996 (7)
BST	*1986.* Start of agricultural research project *1988.* Questions in Parliament regarding allowing BST. Answers relate to EU regulation *1989.* EU proposes moratorium on BST *1990.* Dutch minister says no to BST (for the time being) *1991.* Approval of BST in USA *1994.* Questions in Parliament about banning dairy products from countries allowing BST	25/10/1986 *Title:* Super cow: a development nobody is waiting for *Location:* Science section In total four articles found
Chymosin	*1991.* Producers and consumer organisation agree that chymosin will not appear on food labels	1/10/1988 *Title:* Manipulated yeast cell produces cheese *Location:* Science section No other articles found

Notes and references

1 Deegenaars, G H and Janszen, F H A, *Biotechnologie op Weg Naar het Jaar 2000. Een (Toekomst) Perspectief voor de Nederlandse Industrie* (Erasmus Universiteit Rotterdam, Faculteit Bedrijfskunde, 1996), in opdracht van het Ministerie van Economische Zaken.

2 Deegenaars, G H and Janszen, F H A (1996).

3 Kaiser, A, 'Een wereld van papier. Werkwijze en uiterlijk van de Nederlandse krant', in Bardoel, J and Bierhoff, J (eds), *Media, Feiten en Structuren* (Groningen: Wolters-Noordhoff, 1990).

4 Hemels, J, *De Emancipatie van Een Dagblad. Geschiedenis van* De Volkskrant (Baarn: Ambo, 1981).

5 Heuvelman, A and V.d. Staak, J, 'Wetenschap en technologie in de journalistiek', in Heuvelman, A and V.d. Staak, J (eds), *Communicatie over Wetenschap en Techniek* (Houten: Bohn, Stafleu en Van Loghum, 1992).

6 Based on Heijs, W J M, Midden, C J H and Drabbe, R A J, *Biotechnologie: Houdingen en Achtergronden* (Eindhoven: TUE, 1993); Heijs, W J M and Midden, C J H, *Biotechnologie* (Eindhoven: TUE, 1996).

Media content	*Public reaction*
1986. First US based developments. *1988.* Economic opportunities for Dutch companies (Mogen/Gist-Brocades). *1991/92.* actions of concerned action groups ('Angry Potatoes').	In 1993 the majority of the Dutch public reacted positively to biotechnological solutions to improve agricultural products (>70% supportive). In 1996 this support has decreased somewhat (>65% supportive).
1988. Basic research with many promises for medical applications. *1991.* Preparation of test with humans (TNO). *1994.* Promising developments and proposal to regulate genetic research with embryos.	No specific information about public perception of gene therapy available, but in 1993 and 1996 a majority of the population approve of human medical curative applications of biotechnology involving genetically modified rats (> 60%).
1993. First results of Herman as breeding bull and consequent reactions. *1994.* Results with Herman not according to expectations, government interference, discussion about alternative methods to produce lactoferrine, first articles about economic link between Herman and Nutricia (food-producer). *1995.* Much discussion about genetically engineered animals, 'creator' Herman de Boer shows 'remorse'. *1996.* Should Herman (the bull) be butchered or not?	No specific information about public perception of Herman the Bull available, but in 1993 and 1996 a majority of the population approves of medical curative applications of biotechnology involving genetically modifying cows to produce pharmaceuticals (approve approx. 50%, disapprove approx. 30%).
The first articles announces several major changes in future dairy farming, among other things instigated by modern biotechnology. Other articles mainly refer to policy status quo in various arenas.	The majority of the Dutch population is clearly against the application of BST. In 1993 disapprove >55%, approve 20%. In 1996 disapprove >60%, approve 23%.
Article describes development with Gist-Brocades and the expected difficulty in getting a permit to produce.	In 1993 the Dutch population had mixed feelings about chymosin: 43% approved, 30% were neutral, and 27% disapproved. In 1996 these figures were 42%, 35% and 21%, respectively.

Address for correspondence

Prof. C J H Midden, Technical University of Eindhoven, Faculty of Technology Management, PO Box 513, 5600 MB Eindhoven, the Netherlands. E-mail c.j.h.midden@tm.tue.nl

Poland

Andrzej Przestalski, Bolesław Suchocki and Tomasz Twardowski

The 1970s and 1980s were witness to the period of the so-called 'real socialism' in Poland. Among the centralised, state-owned institutions were all research and development centres and all industries making practical use of biotechnologies. Although the economic crisis between the end of the 1970s and the late 1980s resulted in a substantial reduction of funds for the development of natural sciences, they were still supported by the government. In 1980, investment in science amounted to 1.5% of the Polish GNP.

In mid 1989, as a result of a free election negotiated between the Communist Party and the opposition, the political leadership of Poland was assumed by the pro-capitalist and pro-Western opposition. In 1990, the new leaders started a radical economic transformation, entailing among other things privatisation, the deregulation of prices, the strengthening of the domestic currency and the introduction of market forces. This process, which also included restructuring other social areas (social insurance, medical care, education, culture, sport, etc.) is still far from complete. However, the first results have been manifesting themselves during the late 1990s in the rapid growth of GNP (6–7% annually), slower inflation, and a steady decrease in unemployment.

In 1998, the major political parties represented in Parliament included:

- an election parliamentary coalition called the Solidarity Election Campaign, led by the political representatives of the Solidarity trade union and also containing larger or smaller satellite centre and centre-right parties, the majority of which subscribe to Christian religious values.

- the Freedom Union, a post-Solidarity centre-liberal party.

- an election parliamentary coalition called the Democratic Left-Wing Alliance, led by the post-communist Social Democrats and also containing small satellite left-wing parties.

- the centrist Polish Peasant Party.

Since September 1997, a coalition of the Solidarity Election Campaign and the Freedom Union has been in power.

Research into and applications of biotechnology

In Poland, research is well developed in molecular biology and classical biotechnology – for example in the application of microbes to the manufacture of bioproducts, the production of vaccines and diagnostic tools for *in-vivo* and *in-vitro* applications, and the biological treatment of sewage. Research on modern biotechnology is primarily concerned with health, the food industry, agriculture and protecting the natural environment. In 1994–95, the first research was published on experiments with transgenic plants (wheat, triticale, potato and lupin) and animals (cows and goats). These experiments exclusively concerned contained use under strictly restricted conditions.

Classical biotechnology has been used on a large scale in the fermentation industry, and to a lesser extent for seedling production using *in-vitro* cultures. Currently, there are over 40 laboratories in Poland that produce these seedlings. Classical biotechnology is also commonly used in the pharmaceutical industry. As practically all industrial applications of biotechnology take place in state-owned institutions, the scope and extension of these applications largely depends on the top decision-making bodies.

So far, modern biotechnology has not been used outside the laboratory. In 1996, the first genetic screening for cancer was carried out on a limited number of volunteers. In 1997, the first releases into the environment of genetically modified potato, maize and beet, as well as of transgenic cannola (rapeseed) under a strictly restricted conditions, were carried out under the close supervision of experts from the Ministry of Agriculture.

Main actors

The basic groups of actors in the biotechnological arena are scientists and their organisations. Research into biotechnology (both classical and modern) is carried out in some 40 centres; and the number of researchers who consider themselves biotechnologists is close to 1000. They are not organised in a separate association, but they are members of the Biotechnology Section of the Polish Biotechnology Committee, which is a member of the European Federation of Biotechnology. Biotechnologists as a community are also represented by the Biotechnology Committee affiliated to the Polish Academy of Sciences, which is an advisory body that gives opinions on biotechnological research, informs particular centres about current research projects and their results, and represents the biotechnology community to other agencies. The Biotechnology Committee is also the publisher of the only Polish periodical on biotechnology, the quarterly *Biotechnologia.*

The government remains the main (and so far, practically the only) distributor of funds designated for scientific research. There are a number of governmental bodies involved in the development of biotechnology, including: the Committee for Scientific Research (which, among other responsibilities, is the distributor of funds for particular research projects); and the Ministries for Agriculture and the Food Industry, Environmental Protection and Water Economy, Health and Social Welfare, and the Economy.

In terms of social and economic programmes, a clear majority of the Polish parties are centrist: there is considerable consensus on issues such as the welfare state, control of the market and the availability of basic social benefits to everybody. One characteristic difference between parties concerns views on Poland's role in the international market: alongside the tendency to integrate with the European economy, there is a protectionist approach towards the national economy that expresses a fear of being dominated by stronger competitors. The Social Democrats, the Freedom Union and some centre-right parties believe that opening up to European competition will eventually strengthen the Polish economy. A considerable part of the Solidarity Election Campaign, the pro-Church centre-right parties and the Polish Peasant Party represent the protectionist approach.

This difference is also reflected in the attitude to cultural and ideological values: the 'liberals' are open to the influence of the West European culture and ideology, while the 'protectionists' do not approve of those values, perceiving them to be secularised and cosmopolitan, and to pose a threat to national culture and the Christian values embedded in it. This difference would affect the policy on biotechnology only in regard to issues that stir up moral controversies (such as xenotransplants using human genes, genetic aspects of *in-vitro* fertilisation, and cloning). So far, however, political differences have not impinged upon biotechnology policy, because these potentially controversial processes have stayed strictly within the laboratory in Poland. However, with the practical application of these biotechnologies, the 'nationalist' parties may become important in supporting their control and restriction.

The Catholic Church will probably play a role in the development of modern biotechnology in regard to all of its morally controversial elements. The Polish Church, particularly the middle and lower clergy, is rather traditional as far as customs, culture and moral values are concerned; it is ill-disposed towards Western influences in those areas,

considering them a threat to the preservation of values. Nowadays, the Church has a significant, though indirect, political influence; but it no longer aims at active participation in political life, and instead concentrates on the control of cultural life and ideological and ethical values. So far, biotechnology has not been the subject of any direct statements or actions by the Church which follows the official line of the Holy See. But the Church may be considered a potential actor: its wait-and-see attitude is not due to a perception of biotechnology as uncontroversial, but rather to the fact that biotechnology has not yet been applied on a large scale.

Non-governmental organisations do not play a significant role in Poland. Before 1989 they did not exist, and they are currently in the early stages of development. The consumer movement is weak, and the environmental movement is divided. The latter, however, has its own low-circulation press. Recently (since autumn 1996), environmentalists have included modern biotechnology issues such as GMOs in the scope of their interests by announcing, in November 1996, with financial support from the Scandinavian 'Greens', a report entitled *Playing God: Transgenic Food in Central and Eastern Europe.* The report received a lot of publicity through the press, and started a more comprehensive discussion in the media on biotechnology.

Public policy

Biotechnology in Poland is still in an early stage of development, which is characterised by the institutionalisation of scientific research, the first experimental field trials, and the drafting of the basic legal acts regulating practical applications of biotechnology. This period is marked by the establishment of the Biotechnology Committee of the Polish Academy of Sciences (PAN) in 1986. Its three phases show an acceleration in the course of events culminating in 1996.

Phase 1, 1978–90. Until 1990, the actors were exclusively representatives of the scientific community of biotechnologists. In 1978, the Polish Biotechnology Working Group affiliated to the Technical University of Łódź became a member and co-founder of the European Federation of Biotechnology. In 1984, the first report on the state of biotechnology was published.[1] In 1986, the Biotechnology Committee of PAN was established. Despite its strictly scientific role, the Committee was also a voice through which the biotechnology community could express its opinions and suggestions to the authorities. The establishment of this committee marked the beginning of biotechnology policy in Poland. It believes that the development of modern biotechnology in Poland is important not only in order to keep up with developments in the international scientific arena, but also because biotechnology could play an important role in Polish industry. In 1990, Poland participated in the Rio de Janeiro summit which informed the preparation of legislation on both modern biotechnology and environmental protection.

Phase 2, 1991–96. This period, between 1991 and autumn 1996, was characterised by relatively active involvement in R&D activities with simultaneous attempts to draw the authorities' attention to the benefits of practical applications of biotechnology. The second report on the state of biotechnology served that purpose among others.[2] Under the auspices of different ministries, advisory teams of experts were established to develop recommendations, guidelines and regulations concerning different applications of GMOs. In 1995, the Ministry of Environmental Protection participated in Cairo in the Global Consultation of the United Nations Environment Program on International Guidelines for Biosafety in Biotechnology.

However, attempts to persuade the authorities to allocate substantial funds for development have not been successful, apart from generating a few governmental programmes supporting innovative technologies. Successive governments, concerned with issues vital for the process of political and economic transformation as well as with the ideological issues pervading public opinion, have put aside the problem of scientific development. Total investment in science in the mid 1990s in Poland amounted to only 0.5% of its GNP.

Nevertheless, between 1991 and 1996, the first legal regulations on biotechnology were established. In the 1970s and 1980s, Polish biotechnologists supported the American model of biotechnology regulation: guidelines instead of laws. Only in the mid 1990s, when the prospect of Poland joining the EU became real and Poland started to approximate its legal regulations to EU standards, did the community of biotechnologists accept the inevitability of European-style regulations.

In 1991, the Environmental Protection Law was passed. In 1993, the Budapest Treaty concerning the deposition of microorganisms was signed, and a law was passed on patent protection of drugs, chemical compounds, food and food additives, techniques of isolation and identification of natural compounds, and gene technology. In 1994, Poland signed the 'Australian Group' agreement, obliging governments to control the distribution of chemical and biological preparations that might be harmful. In 1995, the Biodiversity Convention was implemented. These first legal regulations show the influence of international organisations on Polish modern biotechnology, through their conventions

and treaties as well as agreements concluded between them and Poland: the OECD in the case of patent protection and biosafety laws, and UNEP in the case of the Biodiversity Convention. A governmental draft of a new law on the protection of personal data and information which was proposed in August 1996 was an important initiative in intellectual property rights.

At the end of August 1996, a commercial company first applied for approval of a release of GMOs: it planned field trials of genetically modified rape and maize. The request was approved before spring 1997.

Phase 3, 1996–97. In this period, which commenced in the autumn of 1996, the biotechnological issues clearly gained in importance in the public arena. In January 1997, commercial companies continued to apply to the Ministry of Agriculture for permission to introduce GMOs into the environment. In March 1997, a meeting of the Ministries of the Environment and Agriculture was held with several high-level officers of administration and scientific advisors present, at which the preliminary approval was given for GMO releases in Poland, for the purpose of scientific experiments, to be held under the strict supervision of competent authorities. The community of biotechnologists together with the Biotechnology Committee prepared its third report on the state of biotechnology, which was published at the end of 1997.[3]

In 1996 a new and unexpected actor appeared on the biotechnological scene: the 'Green' movement. It was critical of the genetic manipulation of plants and, in particular, protested against the involvement of Polish ships in the import of transgenic food products from the USA to Europe. The Greens took part in public debates on the same subject in November 1996, and in December 1996 took up the issue of the breeding of transgenic carp in Poland and their rumoured uncontrolled release into the environment. In the spring of 1997 the Greens were concerned with the cloning of Dolly the sheep and its implications for humans. Those events had a significant impact on the biotechnological awareness of the general public; and biotechnology became one of the issues raised frequently by the press.

That short period was also significant from the legislative point of view. In November 1996, a new parliamentary law concerning environmental protection was submitted for debate; a separate section of that law was devoted to biotechnology. It provided for a competent authority to grant permits for GMO releases – an authority that would cooperate closely with the Ministries of the Environment and Agriculture. In September 1997, the President signed the law. At the end of 1997, Polish experts prepared a draft of a law on GMOs, based on EU Directives 90/219 and 90/220. In this way, via its legislative power, the EU would also have a strong influence on the shape of modern biotechnology in Poland. So far this influence has extended only as far as the formulation of regulations, but as soon as the law is passed the EU's influence will be felt in real events. The passing of the law on GMOs, which regulates and facilitates practical applications of modern biotechnology, will conclude the preparatory stage of biotechnology policy in Poland.

The specific features of the development of Polish biotechnology are shown in Figure 1. First, biotechnology policy-making processes started in Poland much later than in other European countries. Second, until 1997, R&D and the related publicity had been the only policy issues. Thirdly, 1996–97 saw a high concentration of active policy-making.

Media coverage

Until the political breakthrough in late 1989, the press in Poland was either directly published or controlled by the Communist Party and state authorities. The opinion-leader among the newspapers was the official organ of the Communist Party, *Trybuna Ludu (The People's Tribune)*, a national daily presenting not only the political line of the Party and government, but – unlike any other daily – also providing the reader with relatively comprehensive information and comments on science and culture. In 1982, a new daily *Rzeczpospolita (The Republic)* appeared, which was the official organ of the government. It was much less politically oriented than *Trybuna Ludu*, and it included a large spectrum of issues, including a generous science section. It may be regarded as another opinion-leader. The most popular papers in the 1980s were the local morning and, especially, evening papers, which tended to be sensationalist rather than political, and were published in most big cities. In the same period, a third opinion-leading newspaper, *Polityka (Politics)*, a weekly, was issued, which apart from political matters also dealt with economic and social problems, culture and science. It was intended for the educated public and had the freedom to take a relatively broad stance.

In the 1980s, parallel to the official press, underground newspapers were published, representing the outlawed Solidarity movement. They were rather tiny regional newspapers, issued irregularly, and devoted to political matters.

Since the beginning of the 1990s, following the conversion of the economic–political system and the dissolution of the Communist Party, *Trybuna Ludu*, under a new name of *Trybuna (Tribune)*, became just one of many party newspapers, while a new daily representing the new Solidarity political elites, *Gazeta Wyborcza (Election Newspaper)*, with a circulation of 450,000, emerged to soon become an opinion-leader alongside *Rzeczpospolita* (210,000) and *Polityka* (210,000).

The level of newspaper reading in Poland is not particularly high. In the late 1980s and early 1990s, about 90% of adults were daily or occasional newspaper readers. In 1997 the proportion dropped to 72%, but only one in every four read a paper every day, and 15% did so only very occasionally.[4] Most of those who do not read newspapers live in the country, where 40% of adults do not read newspapers at all.[5]

Content analysis

The coverage of biotechnology in the Polish press has been on a smaller scale than in other European countries; the longitudinal study of the Polish media coverage of modern biotechnology presented here accordingly refers to a rather smaller sample of press material. The study is limited to the printed media, because the purely laboratory character of modern biotechnology in Poland and a lack of spectacular achievements abroad until the second half of the 1990s made it insufficiently attractive to make television or radio headlines. Technical difficulties make it practically impossible to demonstrate this claim by analysing the contents of electronic media, but one can reasonably assume that they carried little, if any, material before 1995.

The analysis of the press coverage embraced two groups of newspapers. The first group contained the opinion leaders: *Trybuna Ludu* from 1973 to 1981, *Rzeczpospolita* from 1982 to 1996 and *Polityka* from 1973 to 1996. *Rzeczpospolita* was chosen instead of *Trybuna Ludu* (in the 1980s) and *Gazeta Wyborcza* (in the 1990s), which could also be regarded as opinion-leaders, because *Rzeczpospolita* contained a much better science section than the other papers. Its science section at first appeared only once a week, but in 1996 it became biweekly. So the group consisting of opinion-leaders contained, at any time during the period analysed, one daily and one weekly newspaper. Altogether 254 articles from opinion-leaders were selected and coded.

The other group of newspapers consisted of two national weeklies: *Przegląd Techniczny (Technical Review)* from 1973 to 1982 and *Wprost (Point-Blank)* from 1983 to 1996, in which 135 articles were identified, and a regional daily paper *Głos Wielkopolski (The Great Poland's Voice)*, with 37 articles.

All articles were found by means of a manual search. In the case of the weeklies (*Polityka, Przegląd Techniczny* and *Wprost*), all the issues were taken into consideration. Analyses of the dailies were based on samples of one issue per week: a random sample for *Trybuna Ludu* and *Głos Wielkopolski*, and a purposive sample of those issues with a science section for *Rzeczpospolita*.

The analysis focused on the articles from the opinion-leading press; with the other newspapers serving as complementary material, showing how the general and regional press reflects opinion-leading newspapers. The frequency of articles in the opinion-leading press is shown in Figure 2. The break in the curve indicates a change from random sampling to purposive sampling of the dailies.

The graph justifies a division of the whole period in question into three phases, which are defined by noticeable drops in the curves in 1981 and 1989. The boundaries of these periods, which coincide with the division into decades, correspond with two political breakthroughs in Poland. Differences in frequency of information about biotechnology between these periods are particularly visible in the newspapers, which contain the majority of the articles. The number of articles – i.e. the amount of information on biotechnology – increased clearly only in the last decade. The vast majority of these articles focused mainly on biotechnology: there were three times as many of them (73%) as there were articles about other issues that merely mentioned biotechnology. This ratio was higher in the newspapers (4:1). In the weekly, the number of focused articles exceeded the less-focused articles by only 10%. In fact the proportion of these less-focused articles may be higher – it is sometimes difficult to spot passing references in a manual search.

The press coverage of biotechnology in Poland has some characteristics that extended through the whole period of the study (Table 1). Medicine was the dominant theme (though this receded slightly in the 1980s). Modern biotechnology was presented as uncontroversial, or at least as having more benefits than risks. Most of the benefits mentioned were medical, as were the risks. Modern biotechnology was perceived most often within a frame of progress, and the most frequent actors were scientists. Lastly, the narrative was free of metaphors, and rather matter-of-fact.

Phase 1, 1973–79. The first phase was represented in the opinion-leading press by 39 articles. Although relatively rare, they were often long. Apart from medicine, other important themes were agricultural applications of biotechnology and biotechnological research. Medical benefits were accompanied fairly often by economic and to a lesser degree consumer benefits. Specific to this phase was a comparatively important role for the moral aspects of biotechnology, and its moral risks. Poland and the USSR were locations of biotechnological events.

In the newspapers belonging to the second group, 26 articles were identified: 13 in the weeklies and 13 in the regional daily. The thematic structure and the frequency of benefits and risks in the general weekly were somewhat different from those in the opinion-leaders. Biotechnological research was almost three times as frequent as health and economics. In the regional newspaper (13 articles) the applications of biotechnology in agriculture and animal breeding were the most frequent themes, and the most important benefits were to the economy and the consumer. The reason seems to be the specificity of its readership: *Głos Wielkopolski* is a paper of the most important Polish agricultural region. Risks were hardly ever mentioned.

Phase 2, 1980–89. The total number of articles in the opinion-leaders in this phase was 84. The phase itself had no specific profile. Biotechnological research was the main theme, while the economic theme and the role of economic benefits receded in both newspapers, although less in the daily than in the weekly. The frequency of the medical theme dropped slightly too, but it was still twice as frequent as the economic theme. Industrial actors appeared among the dominating scientific actors. This period in the opinion-leading press was practically an ethical vacuum.

In the other group of papers, the general weekly contained 23 articles. Compared to the previous phase, the proportion of medical and economic themes increased, as did the corresponding types of benefits. Ethical questions, absent in the opinion-leaders, found a place here, both as a theme and as a risk. The 'economic prospects' frame of articles was as frequent as the 'progress' frame.

Only seven articles were found in the regional newspaper. As in the first phase, applications of biotechnology to plants and animals appear as very important themes, second to the medical applications. Two articles mentioned risks.

Phase 3, 1990–96. This phase is of particular importance as it represents new conditions for the development of both biotechnology and biotechnology policy, as well as for the functioning of the press. The most fundamental of these new conditions (the market and political democracy) look set to be permanent, so in characterising this phase new traits can perhaps be seen as indicators of possible new trends in the press coverage.

The material collected in this phase amounted to 131 articles in the opinion-leading press. The role of biotechnology research as a theme had receded considerably. So had economic benefits and the 'economic prospect' frame. Health as a theme of articles and as a beneficiary of biotechnology gained in importance. These were in fact not new trends, but extensions of existing features.

New elements of the coverage in this phase compared to the previous one were a growing number of articles with a biotechnology focus (as a result of a large number of small articles without headlines giving the latest information on new developments in biotechnology); a decrease in the role of consumer benefits, especially in the weekly; and the re-emergence of ethical issues.

In the other group, 99 articles were published in the general weekly. The medical focus and medical benefits were the most common. The most frequently mentioned risks were those for health, but in most cases no risks were mentioned.

In the regional paper the sample consisted of 17 articles. Its previous specific agricultural focus had disappeared, and its thematic profile became similar to that of other newspapers. The only benefits mentioned referred to health, and risk was hardly ever mentioned.

Public perceptions

There has never been a nationwide survey, based on a representative sample, of public perceptions of modern biotechnology in Poland. There are, however, a few surveys dealing with related topics, the results of which could serve as indicators of attitudes towards biotechnology. For example, there have been two surveys based on representative national samples on the subject of organ transplantation, that were carried out in February 1995 and April 1997;[6] and a similar survey concerning *in-vitro* fertilisation that was carried out in April 1995.[7] There have also been surveys commissioned by newspapers about sensational news events such as the cloning of Dolly the sheep. In 1997 the first book on public perceptions of genetic engineering in Poland was published: it presents the results of a survey of secondary school and university students.[8]

A study of Polish youth

In 1993–97 we carried out three surveys concerning public opinion in Poland on biotechnology and genetic engineering. The studies were performed in a big Polish city, so that multifaceted comparisons could be performed. Our theoretical base was a theory of social horizon.[9] According to this theory, the scope of individuals' knowledge is determined by their participation in social groups. The groups specify the perspective on reality, thus providing individual group members with categories of thought that frame both the stereotyping of knowledge and evaluations, and generalisations of opinions. Perception is determined by our place in a social structure.

While looking for a suitable population for our investigations, we assumed that a systematic knowledge of biotechnology would reduce the extent to which knowledge would be framed in stereotypical ways. This systematic knowledge would be found only among either professionals or young people (older generations did not come across biotechnology at school). So we decided to focus on a representative sample of the pupils of final grades at secondary schools and university students in Poznań, i.e. a social category aged 18 to 24 years. For our purposes, a secondary school serves as the last stage of a unified education in a big community. We assumed that because similar social horizons would result in all students from the experience of having been a pupil, any differences would be the result of membership of some other social group.

Our questionnaire was in three parts: a knowledge test; questions concerning attitudes towards selected problems of biotechnology and genetic engineering; and socio-demographic characteristics of the respondent. The knowledge test was prepared by biotechnology experts. We assumed it would indicate the extent to which knowledge about biotechnology becomes stereotyped. The test was constructed in the form of a Guttman's scalogram (an incorrect answer to a given question should remove the possibility of giving correct answers to subsequent questions).

The main part of the questionnaire was based on the division of biotechnological applications into four groups: those involving microorganisms, plants, animals and people. The categories reflect a concept of the human being as the last link in the historical development of species, and are consistent with a current ethical–moral pattern of the environmental order of the world. This pattern is based on an anthropocentrism that is characteristic of our cultural sphere, and according to which the categories are respectively more and more remote from humans with regard to biological similarities: they somehow create a symbolic hierarchy of the world on a scale ranging from higher to lower organisms. We assumed that this scale would be reflected in degrees of acceptance of biotechnological applications.

We also distinguished two spheres of biotechnological activity: scientific research, and the mass application of it in everyday life. Such a distinction was a consequence of the over-representation in our study of children whose parents were very well educated (children in Poland often follow the educational pattern of their parents); so we took into account both the scientific and the utilitarian aspects of biotechnology.

Respondents were able to define their points of view as a full and unconditional acceptance, an acceptance with reservations, or a lack of acceptance. They were also able to take no stand at all.

Perceptions of biotechnology among educated youth

Altogether three studies were carried out, in 1993, 1996 and 1997. The first study looked at a random sample of 400 pupils drawn from a population of fourth graders. This sample was big enough for generalisations with 5% confidence semi-interval at the significance level of 0.05. Guttman's scalogram failed to work in the field, given the state and structure of knowledge of our respondents: answers were not consistent. Taking into account general results of the knowledge test, it did not come out very well. After dividing our respondents into four categories according to the percentage of correct answers, we worked out the following schedule:

Score (%)	Respondents (%)
0–25	17.9
26–50	41.5
51–75	34.9
76 and above	5.7

The inconsistency of the scalogram is due to the differentiation of social horizons. Data collected showed the lack of systematic and reliable biotechnology and genetic engineering courses at secondary school level. They also revealed very poor extra-curricular and extra-school education. Superficial acquisition of knowledge from parents and from other sources of information can explain the haphazard and circumstantial character of respondents' knowledge of the subject. Prime among other information sources used were newspapers (37.5%), television (27%), magazines (18.3%), books (17.1%) and the specialist press (5.6%).

The test results falsified our original hypothesis, according to which the scope of acquired knowledge was supposed to differentiate the attitudes of our respondents towards the problems of applications of biotechnology and genetic engineering. The results also enabled us to determine the scope of the answers based upon group stereotypes whose contribution is inversely proportional to the frequency of correct answers.

Answers to the questions in the main part of our questionnaire, which concerned attitudes to biotechnology, presented two constant tendencies. The first was a decreasing acceptance as respondents came to consider more complex organisms; and the second was a statistically significant greater acceptance of the activities that would further science than of practical uses in everyday life. People who fully and without reservation accept the use of all the biotechnological and genetic engineering techniques with reference to microorganisms, regardless of the field of activity, make up 9.2%, whereas with reference only to human beings they make up only 5%. In the activities promoting the development of science, unreserved acceptance came from 61.2% and 8.9% respectively. For practical purposes in everyday life, the results were 58.4% and 6.9%.

There were similar results in the case of individual usage of biotechnology and genetic engineering. Again there are extreme levels of acceptance concerning microorganisms and people. As far as the manipulation of microorganisms to further human well-being is concerned, an unreserved acceptance was expressed: for scientific experiments by 88.3% of respondents, and for their use in practice by 80.8%; for environmental protection by 88.3% and 78.2%; and for the improvement of food quality by 64.6% and 60.5%. Regarding humans, the highest level of unreserved acceptance was for activities aimed at the elimination of genetic diseases – the percentage here was 61.4% for scientific activities and 49.9% for their practical application on a mass scale. The lowest acceptance was for the creation of GMOs: 8.9% and 7.1%.

So far we have been dealing with declarations of unreserved acceptance. However, there were also respondents who had a favourable approach to biotechnology, yet certain aspects of its use raised concerns for them. Their final position on the problem resulted from a more sedate balance of perceptible characteristics of particular applications with regard to the individual hierarchy of values, usually the ethical–moral ones.

Results show that secondary school children hold polarised attitudes. Differentiation is not explained by the results of knowledge tests, nor by the social characteristics of our respondents (statistically significant correlations did not occur). The opinions presented by the youngsters are simply a reflection of common knowledge and of the knowledge belonging to their social horizon, including its stereotypes and representations (and also the anthropocentric point of view and its consequences). The lack of dependencies explaining the differentiation results both from many demographic and social similarities and distinctions that were common in the sample, and from the inefficiency of the system of popularisation of knowledge. Such a system would be able to stimulate individual interest as well as to form the conditions required for the creation of evaluations, opinions and points of view.

Attitudes to biotechnology among medical students

These results led us into a second study, in 1996, among third-year students from the Academy of Medicine in Poznań. The specificity and advanced level of their of studies ought to have had a favourable influence, limiting the impact of representations and stereotypes on the students' images of biotechnology and resulting in a favourable attitude to it.

The scores in the 'knowledge test' were much higher this time. The structure of the basic sources of information used by the respondents was different. They used books (22.4%), the specialist press (17.1%) and television (12.4%) as sources of information. Of the various institutions and organisations concerned with biotechnology, the highest reliability was achieved by the universities (supported by 69.4% of respondents), whereas ecological organisations were rated relatively low (8.8% of support); the remaining part of support is within the range of an acceptable error margin.

The greater range of knowledge acquired by the medical students, the change of social horizon (which has to do with their studies), the different sources of information, and acknowledged authorities brought about a decrease in the level of acceptance with regard to particular biotechnological practices. More frequently than the pupils, they acknowledged the risks of biotechnology.

Similar results emerged in cases of particular applications of biotechnology and genetic engineering. Again there are the extreme levels of acceptance concerning microorganisms and people. As far as the manipulation of microorganisms to further human well-being is concerned, a full acceptance is declared for scientific experiments by 82.5%, and for their use in practice by 68.6%. For the protection of the natural environment, results were 75.3% and 65.7%; for the improvement of food quality, 63.5% and 57.1%. From among the applications in humans, the highest level of full acceptance was awarded to the elimination of genetic diseases – the percentages here were 47.6% for scientific activities and 44% for their practical applications on a mass scale. The sequence of other spheres were as follows: pre-natal research 21.8% and 17.3%; diagnosis of genetic characteristics 13.5% and 11.3%; and, with the lowest level of acceptance, the creation of mutants 9.4% and 8.9%.

Basic attitudes to these questions are identical among students and pupils. Curiously enough, only the creation of 'new and better people' has a greater acceptance among the medical students.

Recent events, particularly the cloning of Dolly the sheep, precipitated our third study, which again concerned third-year students from the Academy of Medicine. Their educational status is identical to that of the participants in our 1996 studies. Because the data at our disposal have been collected very recently, they have not yet been fully analysed. Still, the information we have already indicates the range of influences the Dolly affair had on attitudes towards biotechnology. Compared to our previous studies, levels of acceptance of particular biological applications are lower than before.

The hierarchy of particular sources of information has changed: there is a return to rapid electronic media, with one in three respondents taking advantage first of all of television; but magazines are also important. Books fall to fourth place. The previous evaluations of reliability remain unchanged. The 'science fiction' character of real events has brought to mind other previously unthinkable aspects of risk.

Some results of related research

Two of the other studies mentioned above concerned *in-vitro* fertilisation (April 1995) and cloning (March 1997).[10] Both were conducted by the Centre for Social Opinion Research (CBOS), and were based on representative national samples. Some 65% of respondents accepted *in-vitro* fertilisation in humans, as opposed to 21% who did not; 41% accepted the cloning of animals and less than 5% the cloning of humans, as opposed to 43% and 63% respectively who did not; 16% were in favour of unlimited genetic experiments, but only 5% thought that no control over them was necessary; 78% were against the experiments and 32% wanted them to be banned. As for the character of the control, 25% of respondents had confidence in committees composed of scientists and moral authorities, 21% believed in legal control and 12% were prepared to leave the control to scientists themselves. The opinions are related to differences in educational level which had been dichotomised into non-educated (elementary education or less) and educated (secondary and higher education). Educated respondents more often opposed the cloning of humans and had confidence in legal and moral authorities to control genetic experiments (61%), and less often demanded that experiments be banned (22%). Among the educated, the attitudes of those with university education towards the cloning of humans are even more negative (85%) than the group average.

The results must be treated with some caution, as there is no information on the strength of dependency in the report. Nevertheless, they are from a representative national sample, and they support the hypothesis of our study that the level of education has an impact on opinions on biotechnology and its applications. Similar conclusions were drawn from Piasecki's study of students' opinions of biotechnology, although he differentiates between respondents on the basis of their knowledge of genetic engineering as measured by a test within the questionnaire, rather than on the basis of educational level. His study confirms the role of social milieu (especially the family) in determining opinions, as well as demographic characteristics: males are more liberal. Another interesting conclusion in this study is the identification of a dissonance between declared outlook on world and attitude towards ethical issues connected with biotechnology. These conclusions do not refer to all Poles, but only to youths, and to better educated youths at that, as the study was based on a sample consisting of secondary-school students and university students (four universities of different profiles).

It seems that potential studies of representative samples of the Polish people are likely to reveal the stereotypes that function in Polish communities – considering that the average level of knowledge of biotechnology among the national sample is much lower than that of the students. These stereotypes however, ought to validate indicator-based conclusions concerning attitudes towards, opinions on and evaluations of biotechnology in different Polish communities. They may even mark out the 'line of resistance' against particular spheres of application.

Commentary

Biotechnology first reached the Polish public via the press. While most of the coverage was in the opinion-leading papers, some of it was to be found in other newspapers. Until the mid 1990s, the press and the scientific community were the only actors on the biotechnology stage: in press reports, scientists were the most common actors. At this time there were more reports of foreign biotechnological events and achievements than of Polish ones (Poland as a location is included in 'other Europe' in Table 1).

Throughout the 25-year period, Polish press coverage has been positive and optimistic, focusing on medical applications of biotechnology and neglecting ethical issues. The growing dominance of the progress frame was accompanied by a clear decline, from the 1970s, of the economic frame, which had almost disappeared by the 1990s.

One of the new phenomena in the press coverage of modern biotechnology after 1989 was the rise of the role of ethical issues in the opinion-leading press. Although noticeable (6%), this emergence of ethical issues does not seem to correspond with the role of the ethical factor in the media in general and in public debates of the 1990s. If moral attitudes may be treated as one of the barriers to public acceptance of the development of biotechnology, then the opinion-leading press has not reflected it adequately.

An important cause of the exclusively positive attitude of the press towards biotechnology until 1980 was the fact that freedom of opinion, especially political opinions and those concerning matters involving public risks, was seriously restricted. Issues such as nuclear energy, the environment, pollution, the pathogenic impact of industry, abortion, and so on were regarded as relatively unproblematic. The press revealed some of their risky or dangerous aspects, but the scope of that sort of information was limited, because the political authorities were afraid of social discontent or panic that they could not control. There were no opportunities for the public to organise themselves into groups to fight particular social risks, or to support specific citizens' initiatives. The public was atomised, and could not form collective opinions. The Church, aware of the authorities' over-sensitive attitude to spontaneous social criticism, kept a neutral position. The situation changed to a certain degree in the 1980s, but organised public activity had an exclusively political character.

In these circumstances, the optimistic image of modern technologies in general and biotechnology in particular could contribute to positive attitudes among the public, especially towards those technologies they did not know from their own experience (such as nuclear energy and *in-vitro* fertilisation).

An optimistic image of biotechnology in the period after 1989 persisted in spite of the changed social and political conditions, the new freedom of opinion, and some new forms of public self-organisation. One of the reasons, it seems, was the fact that the opinion-leading press consisted of fairly liberal newspapers – the ecological press does not play an important role and the conservative press clearly serves a minority. Until 1996 biotechnological issues were, for the broad, even the educated, public, curiosities with little practical importance. The attitudes towards technological innovation coming from the West were co-formed by a wave of pro-Western enthusiasm at a time when Poland aspired to join Western social and political institutions. Only in recent years has a current of criticism towards the West become manifest. One of its sources is the conservative attitude of the Church; another is a unification and strengthening of political parties claiming right-wing ideologies. It is from among these social circles that a morally justified criticism of modern biotechnology may be expected. A critical potential is also to be found in the ecological movement as it becomes stronger.

But their adversaries, the promoters of modern biotechnology, are getting stronger, too. Their base is modern biotechnological industry and international capital, which is investing rapidly in the food and pharmaceutical industries and implementing modern technologies. The investors include the following companies which are quoted on the Warsaw Stock Exchange or are ready for public trading: in the food industry, Heineken Int. Beheer BV (Żywiec), Carlsberg (Okocim), Holsten-Brauerei AG (Brok), PepsiCo (Wedel), Nestlé (Goplana), Eridania Beghin-Say (Zakłady Tłuszczowe Kruszwica), Jerome Foods Inc. (Indykpol), and Conrad Jacobson GmbH (FarmFood); and in the pharmaceutical industry, Glaxo (Polfa Poznań), PPE Fund, and PAE Fund (Polfa Kutno).

Acknowledgement

The authors acknowledge the assistance of the State Committee for Research (KBN, SPUB).

Figure 1. A regulatory profile for Poland

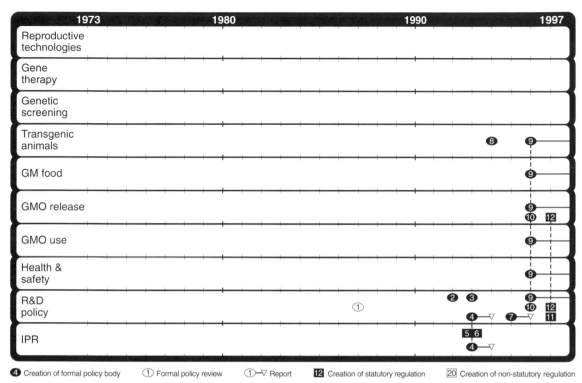

	1973	1980	1990	1997
Reproductive technologies				
Gene therapy				
Genetic screening				
Transgenic animals			⑧	⑨
GM food				⑨
GMO release				⑨ ⑩ 12
GMO use				⑨
Health & safety				⑨
R&D policy		①	② ③	⑨ ⑩ 12 / ④–▽ ⑦–▽ 11
IPR			5 6 / ④–▽	

● Creation of formal policy body   ① Formal policy review   ①–▽ Report   12 Creation of statutory regulation   20 Creation of non-statutory regulation

1 *1987.* Polish Biotechnology Committee at the Polish Academy of Sciences (chairman: Prof. Włodzimierz Ostrowski) was established.
2 *1992.* Subcommittee on Molecular Biology and Biotechnology of the State Committee for Scientific Research was established.
3 *December 1993.* Membership of the Australian Group and formation of Biotechnology Group in Ministry of International Cooperation for Economy.
4 *1993.* Working group established on intellectual property rights, public perception of biosafety and legislation (chairman: Tomasz Twardowski) within the Polish Biotechnology Committee.
5 *September 1993.* Signing of the Budapest Treaty concerning the deposition of microorganisms.
6 *October 1993.* Protection by patents of drugs, chemical compounds, food and food additives techniques of isolation and identification of natural compounds, gene technology.
7 *June 1995.* Unsuccessful initiative of State Committee for Scientific Research to establish an interministerial committee for legal aspects of biotechnology.
8 *1994–95.* First reports of Polish experiments with transgenic plants (wheat, triticale, potato, lupine) and animals (cow, goat): exclusively contained use under strictly restricted regime.
9 *June 1996.* Formation of the experts committee for the elaboration of 'Technical Guidelines' under the auspices of the Ministry of Agriculture; the inter-ministerial

counselling board for GMOs was established (Ministry of Agriculture, Ministry of Health, Ministry of Environment Protection and State Committee for Scientific Research).
10 *July 1996.* Formation of the experts committee, under Ministry of Agriculture supervision, for registration and permits for agricultural use of herbicides and transgenic plants (in cooperation with the above-mentioned committee).
11 *March 1997.* Meeting of Ministries of Environment and Agriculture with several high-level officers of administration and scientific advisors; preliminary acceptance of GMOs released for experiments in Poland under strict supervision of the competent authorities (14 March). Embargo on this information until official notification from the Ministry of Agriculture granted on 18 April 1997.
12 *February 1997.* A new law concerning environmental protection, including several aspects of biodiversity and introduction of GMOs to the environment (Article no. 37a of the proposal), passed through the government in February 1997. The law was passed and signed by the President at the end of 1997. The law concerns environmental protection in general and a separate paragraph is dedicated to biotechnology. It details the requirement for a competent authority to grant GMO release permits. This new body will be jointly coordinated by the Ministry of the Environment and the Ministry of Agriculture.

Figure 2. Intensity of articles on biotechnology in the Polish press

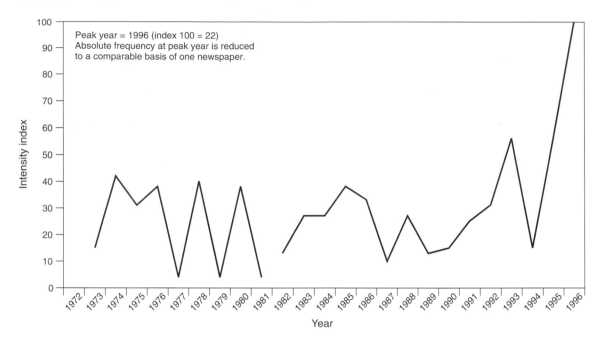

Peak year = 1996 (index 100 = 22)
Absolute frequency at peak year is reduced
to a comparable basis of one newspaper.

Table 1. Polish media profile (for an explanation of terminology please see Appendix 5)

Phase	Freq.[a] (%)	Frame (%)		Theme (%)		Actor (%)		Benefit/risk (%)		Location (%)	
1. 1973–79	15	**Progress**[b]	**56**	**Medical**	**36**	**Scientific**	**77**	**Benefit only**	**49**	*Other Europe*	*36*
		Ethical[c]	15	Agriculture	26	Political	8	Both	41	USA	18
		Economic	13	Basic research	18			Neither	8	USSR	13
		Nature/nurture	8	Ethical	8					UK	10
		Pandora's Box	5	Animal	5						
				Regulations	5						
2. 1980–89	33	**Progress**	**75**	*Basic research*	*39*	**Scientific**	**82**	*Benefit only*	*66*	**USA**	**26**
		Economic	12	Medical	27	*Industry*	*11*	Both	27	Other Europe	21
		Runaway	5	Animal	8			Neither	5	*France*	*8*
				Political	7					UK	7
				Agriculture	6						
3. 1990–96	52	*Progress*	*79*	*Medical*	*51*	*Scientific*	*83*	*Benefit only*	*66*	*USA*	*36*
		Economic	7	Animal	15	Industry	9	Both	28	Other Europe	17
		Ethical	6	Basic research	14			Neither	5	UK	10
				Agriculture	5					France	5
				Identification	5						

a Perentage of corpus in the period; total *n* = 208.
b Bold indicates highest frequency within phase.
c Italics indicates highest frequency within category.

Notes and references

1 Węglewski, O and Fikus, M, *Raport o stanie biotech-nologii – badania i zastosowania* (Warszawa: PAN, 1984).

2 Zabża, A and Ułaszewski, S, 'Raport o stanie polskiej biotechnologii', *Biotechnologia*, 4 (1995), pp13–58.

3 Twardowski, T (ed), *Rozwój biotechnologii. Projekt rozwiązań prawnych dotyczących stosowania genetycznie modyfikowanych organizmów* (Poznań: PAN, 1997).

4 Falkowska, M, 'Czas wolny, orientacje kulturalne Polaków', in Falkowska, M (ed), *O stylach życia Polaków* (Warszawa: CBOS, 1997).

5 Bukraba-Rylska, I, 'Wieś czyli paradoksy transformacji', in Falkowska, M. (1997).

6 CBOS, *Stosunek społeczeństwa do transplantacji narządów* (Warszawa: CBOS, 1995); CBOS, *Postawy wobec przeszczepiania narządów* (Warszawa: CBOS, 1997).

7 CBOS, *Dziecko z probówki* (Warszawa: CBOS, 1995).

8 Piasecki, K, *Inżynieria genetyczna w oczach młodzieży studenckiej* (Warszawa: Instytut Archeologii Uniwersytetu Warszawskiego, 1997).

9 Szczurkiewicz, T, *Studia Socjologiczne* (Warszawa: PWN, 1970).

10 CBOS, *Dziecko z probówki* (Warszawa: CBOS, 1995); MAJ, A J, 'Jak daleko nauka może się posunąć, *Gazeta Wyborcza*, 28 March 1997, p8.

Address for correspondence

Prof. Tomasz Twardowski, Institute of Bioorganic Chemistry, Polish Academy of Science, ul. Noskowskiego 12, 61-704 Poznań, Poland. Tel. +48 61 852 85 03 ext 133 or 134, fax +48 61 852 05 32.
E-mail twardows@ibch.poznan.pl

Sweden

Björn Fjæstad, Susanna Olsson, Anna Olofsson and Marie-Louise von Bergmann-Winberg

The history of industrial development in Sweden

When gene technology was invented in 1972–73, the prospects for brisk scientific and commercial development of this field in Sweden must have seemed rosy. Sweden was at the time one of the richest countries in the world measured by income per capita, and also one of the most industrially developed, having experienced a period of unprecedented economic growth over the past hundred years. The average annual growth rate since 1870 had been about 4%, the highest in the world. A number of Swedish technological inventions during this period gave rise to several large international Swedish-based companies, which had captured substantial global market shares. The mid 1960s were known as 'the record years', because they were characterised by technological and economic optimism and confidence. Later, from the mid 1970s, the same label gradually came to be used ironically.

Among the successful companies were several pharmaceutical firms: Astra, Pharmacia, Leo, Kabi, Hässle, and others. They were all well connected to the academic world, but few of them realised the importance of the progress being made in genetics and molecular biology in the USA and elsewhere. The exception was Kabi, which had business contacts with Genentech by 1977: the two companies signed a contract for the development of bacteria-made human growth hormone as early as 1978. There were, and still are, also some scientifically advanced Swedish companies involved in plant breeding, although not of the same global reach. Agriculture is a very small sector nowadays (1998), employing only about 2% of the workforce.

All through the postwar period, and up to the early 1990s, unemployment was significantly lower in Sweden than in most of Western Europe. There were very few serious labour market disputes, the population was homogeneous, and the people were looked after by the socially engineered high-tax welfare state. The country was militarily neutral but not silent: the political leaders spoke up for the poor and disadvantaged all over the world, and Sweden gave the highest relative development aid in the world – aid that was well meant, but that, it later turned out, sometimes produced no benefits in the countries that received it.

The Social Democratic Party came to power in 1932 and remained in power for the following 44 years. In few other countries has a democracy seen the same party re-elected for so many consecutive years, denoting an unprecedented stability and a popular wish for the status quo. Since 1970, Sweden has had a one-chamber parliament with proportional representation for all parties receiving at least 4% of the vote. For the past several decades, this has resulted in a multitude of parties in Parliament (now seven) and in minority governments in need of support for all government bills from at least one of the opposition parties – a situation characterised by political scientists as not very conducive to the passing of consequent or unpopular policies. A Swedish politician, tongue in cheek, coined the so-called Christmas Eve Theorem, which states: 'When they have a choice between another paid vacation week and higher pensions, Swedes take both.'

However, at the same time, the successful post-war period was slowly coming to an end, and the general picture was imperceptibly changing. The yearly growth rate in the economy moved into a lower gear around 1970, to 2% rather than 4%. By then, the rest of the Western world had overcome the effects of the Second World War, from which Sweden had been spared. The 1968 movement changed the public's perception of industry and politics, most notably during the 1970s. The environmental activists had successfully lobbied against nuclear power and finally made the unwilling political establishment call a national referendum in 1980 – which, by offering three alternatives instead of two, gave an ambiguous result (which the cynics say was planned). It has been echoed many times by socio-political analysts, in incisive terms, that the Swedes, after the still-unsolved murder of their Prime Minister in 1986, realised that they, for better or for worse, live in just another, small, peripheral, insignificant, Western European nation that had, after the 1991 recession, 10% unemployment or more, and a far from perfect harmony between the 'Swedes' and the now as many as 15% of the population defined as 'immigrants'.

However, the large Swedish international industries – Volvo, Saab, Ericsson, ABB, SKF, Electrolux, AGA, Stora, Scania, Skanska, Sandvik, Atlas Copco, Tetra Laval, IKEA, and so on – were never really caught up in these attitudinal and structural transitions. They stayed out of the social debates and of politics (when they occasionally meddled, they tended to lose) and continued to do what they did best: earn money for their shareholders. To this day, there is a disproportionately large number of sizable international industrial corporations based in Sweden, which are well regarded by governments of all colours. The government has done at least part of its share of the work: it has provided a fairly attractive corporate tax system, and has funded basic research generously. State scientific and technologi-cal research councils were set up by the late 1940s and have spent large sums on science ever since. About 3% of Sweden's GNP is allocated to research, which is among the highest shares in the world. Whether it is efficiently spent is another matter. Sweden has very few research institutes of the German type, and the universities are supposed to fill the analogous role. However, most academics do not see it as their goal to further the interests of industry and business, and it is a common truth that this university–company cooperation scheme works superbly only in a few areas, notably medicine/ pharmacology and informatics/electronics.

For many years Sweden stayed outside of the EU. This was not due to the wish of a majority of politicians or of industry, but to voter attitudes. Eventually, in 1995, Sweden joined the EU after a referendum showed an even split of opinion. The debate about the membership has since waned, but there is still a lot of scepticism, mainly about centralised bureaucracy and misdirected subsidies. Opinion polls show that a majority is opposed to the monetary union.

Public policy

The following phases can be distinguished in Swedish policy on biotechnology and genetic engineering.

Phase 1, 1974–79: problem formulation and policy initiation. Awareness at the highest political level in Sweden of the economic potential and possible hazards of gene technology was created in 1974, when the Swedish Cabinet was briefed by a group of distinguished scientists. In 1975, the Royal Swedish Academy of Sciences appointed a working group to study possibilities and risks associated with the recent advances in molecular genetics. Also in 1975, the Natural Science Research Council appointed a committee to review research proposals and to inform the general public about recombinant DNA research.

One or two of the Swedish pharmaceutical companies recognised early on the business opportu-nities of biotechnology, and this period of non-regulation was favourable to the creation of infant enterprises in the field. For various reasons, Swe-den's introductory ventures into industrial applica-tions of gene technology happened to be made by a small state-owned company (KabiGen), and to be financed by a government industrial fund.

In 1978, the party congress of the Centre Party decided to call for a moratorium both on recombinant DNA research in Sweden, and on the construction of laboratories, until the issues had been debated more thoroughly and regulatory legislation put in place. In the same year, the first government public inquiry into gene technology was appointed. The report from this one-person committee, *Recombinant DNA Technology under Control*, was published later the same year.[1] Its main thesis was that it is of vital importance that such research can be performed in Sweden, but that it should be controlled. The report proposed a permanent committee within the Agency for Occupational Safety. In 1980, the Recombinant DNA Advisory Committee was established by Parliament. Genetic engineering experiments were brought under existing legislation for industrial safety and for environmental protection (see Figure 1). At the same time, Parliament asked the Cabinet to appoint another two public government inquiries: one to evaluate the pending legislation after it had been in effect for a period of time, and one to look into ethical and social questions of gene technology.

In 1979 Prime Minister Ola Ullsten (Liberal Party), three weeks before a general election, publicly demanded that industry refrain from recombinant DNA experiments until the legislation had come into force three months later. The reason seems to have been to silence the loud anti-genetic-engineering lobby group within the Centre Party, but this initiative probably did lasting harm to relations with leaders of the biotechnology industry.

The Agency for Technical Development, working independently, ran a framework programme for financing applied genetic engineering from as early as 1979 up until 1987. The Natural Science Research Council soon started to earmark special funds for basic research in gene technology.

In terms of political power, this first phase of policy initiation was dominated by the researchers, with some representation of government agencies; but few parliamentarians were involved, and almost no-one from environmental or animal-rights NGOs. The single-person government investigation was also a notable feature of this phase, mainly because of its policy-initiating and agenda-setting effects, e.g. passing over most animal applications and establish-ing that ethical considerations were not relevant to microorganisms and plants. A dividing line between a very few involved politicians and the vast majority of non-involved politicians indicated a pattern that would hold in the years to come. The legislation that took effect in January 1980 drew up a certain division of power between the government and the loose coalition of academic and industrial scientists which somewhat limited the freedom of the re-searchers.

Phase 2, 1980–86: low political activity. One effect of the regulation was to considerably reduce the intensity of public debate and policy activity. Policy-making in this second phase was characterised by the implementation of the regulation of 1980, which inspired very few parliamentary initiatives. Most notable was an MP's bill to ease the restrictions in the recombinant DNA legislation, so as to avoid the emigration of Swedish biotechnology firms and researchers. Industries and scientists involved in biotechnology backed this initiative. The Recombinant DNA Advisory Committee recommended to the government that only larger genetic engineering plants should need special permission. It was decided accordingly.

It took three years for the Commission on Gene Ethics, which had been requested by Parliament in 1979, to be appointed by the Cabinet. Its report, published in 1984, dealt with animal trials, genetic diagnostics, gene therapy, insurance matters and biological weapons. Applying gene technology in plants and microorganisms was seen as relatively unproblematic from an ethical point of view. This, in turn, meant that environmental organisations were not really involved in the committee work.

Phase 3, 1987–91: less regulation and renewed debate. Another government committee report was published in 1984: *Is There a Need for Control of Recombinant DNA?*[2] The report concluded that the risks of gene technology had been overstated. This led, in 1987, to Parliament abolishing the special legislation on advance examination of recombinant DNA activities. It also made the Recombinant DNA Advisory Committee an agency department in its own right.

After the partial deregulation, the debate was reignited. However, it seems that external events played a more powerful role in this than the relaxation of regulation. For example, the first real public debate in Sweden on the consequences of gene technology for the environment took place during the spring of 1987. Gene technology and animal rights also became an issue, and resulted in 1988 in a supplement to the law on control of animals in domestic production, specifying that transgenic animals were allowed for use in pharmaceutical production, but not in agricultural production. In the pre-election debate in 1988, the Centre Party tried in vain to make gene technology an important issue (which they had successfully done with nuclear power ten years earlier).

In 1988, the Recombinant DNA Advisory Committee wrote to the Cabinet to say that the release of GMOs was not satisfactorily regulated. This initiative, together with MP's bills in Parliament and actions of the Centre Party, led to the appointment in 1990 of the Gene Technology Inquiry.

Besides this, a working committee of the Ministry of Agriculture investigated genetic engineering on plants and animals. Their report, published in 1990, concluded that a ban on genetic research on animals, plants and microorganisms would impair Swedish research capabilities and industrial competitiveness.[3]

At the end of the 1980s, more MP's bills than ever before appeared, most of them – but not all – calling for more comprehensive legislation. Also, legislation prohibiting gene therapy on sex cells was put in place in 1991, although the issue was discussed in a government committee report as early as 1983.[4]

Many MP's bills concerned 'patents on life'. As genetic engineering is a costly undertaking, the question of patenting had to be solved. Newspaper articles about patenting and the ensuing debate made some environmental organisations put gene technology on their agenda for the first time. Some NGO representatives participated actively in the public debate. On the other hand, the two largest political parties, the Social Democrats and the Conservatives, were hardly engaged at all in the regulation of genetic engineering. In the pre-election debate of 1991, the Centre Party again tried to make gene technology an election issue, as did the Christian Democrats and the Environmental Party. Genetic engineering, however, was even less visible than in the election debate three years previously. Policy initiatives increased in number and strength during this phase, coinciding with a vigorous media debate, but new regulation was not introduced.

Phase 4, 1992–96: putting new regulation in place. The results of two major public inquiries were published in 1992 and 1993, which calmed the public debate. *Gene Technology – a Challenge* from the Gene Technology Inquiry covered three areas: the patent question, ethical issues regarding genetic manipulation of animals and plants, and ecological considerations (including release of GMOs).[5] The conclusion was that special legislation on gene technology is not necessary. The Inquiry proposed instead that the various product areas be regulated separately, and that changes in existing laws would suffice.

The other parliamentary investigation committee report was produced in 1993.[6] Contrary to the Gene Technology Inquiry's recommendation, this report called for biotechnological products to be regulated in a dedicated chapter in the existing Environmental Act. Both reports must be seen in the light of the European Economic Area agreement, the result of which was the forthcoming (in 1995) implementation of 1990 EU directives. The Swedish legislative work in this field now started to be more in line with international regulations, after having been quite domestic in character for 15 years.

The main policy event of the fourth phase was

putting the Gene Technology Law in place. It was finalised in 1994 and came into effect on 1 January 1995, since when only minor legislative supplements have been approved so far (1998). In the work on the government bill, more than 50 agencies and organisations were heard, including many NGOs and a wide variety of interest groups.

The final legislation was a compromise in two respects: first, between the four coalition parties in government – Conservatives and Liberals were in favour of biotechnology and industry, and the Centre Party and Christian Democrats were more negative. Second, the new legislation is a special law for gene technology, not a chapter in the Environ-mental Act. The Gene Technology Law thus expresses the conflicting thoughts of the 1992 and 1993 government inquiry reports.

The legislation consists of three pillars regulating genetic engineering in Sweden as far as microorgan-isms, animals and plants are concerned: a framework law on genetically modified organisms; a govern-ment ordinance on genetically modified organisms which distributes among seven different government agencies the responsibility for detailing and supervis-ing the law; and detailed regulations from the seven agencies. A new element in the Swedish law, which is not present in the rest of the EU nor in any EU directive, is that it requires that consideration be paid to ethical aspects, in particular those concerning transgenic animals.

Many proposals for tighter control mechanisms were rejected with reference to present and forth-coming EU directives. The Recombinant DNA Advisory Committee was reconstructed as the Gene Technology Board in 1994. The importing, labelling and marketing of novel foods were new issues for debate and for MP's bills. The suggestion that Sweden should lobby for stricter EU regulation, prohibition and/or compulsory labelling was also rejected by Parliament.

Human gene technology is regulated quite separately, because this field comes under another ministry. There are two relevant pieces of law. One prohibits gene therapy on sex cells and the transfer-ral back into the womb of genetically modified embryos. The other law deals with genetic testing (screening). The administrative authority in both cases is the Agency for Social Affairs and Health, which acts independently of other governmental agencies.

A notable feature during the whole 23-year period is the gap between the experts and laymen and between proponents and opponents. Scientists, industrialists, civil servants, and politicians from the two or three leading parties have been, on the whole, favourable to the new technology. They have had an almost exclusive influence on the naming of members of and experts in government inquiries.

The more negative views of the sceptics – politicians, environmentalists and the general public – have attracted a lot of publicity, but have not been very influential in the legislation. This is noteworthy, since the Swedish type of all-encompassing legisla-tion and the long tradition of government inquiries usually take public opinion into consideration. The discrepancy here is to some extent reflected in the fact that a number of MP's bills calling for more restrictive regulation have continued to be intro-duced in Parliament (without success) even after new legislation was put in place.

In February 1998, the Minister for Education appointed a government commission whose members are parliamentarians from all seven parties. It was charged with analysing the possibilities and risks of modern biotechnology, including scientific developments, ethical considerations, the formation of public perceptions, the need for popular educa-tion, industrial applications, and proposals for changes in regulatory content and structure. The commission has been asked to report to the Cabinet on 1 June 2000.

Media coverage

Sweden has a long tradition of freedom of expres-sion, and the constitutional law of 1766 giving freedom to the press is the oldest of its kind in the world. Newspaper readership is among the highest in the world (together with Norway, Finland and Switzerland). In these four countries, on an average day about 85% of the population aged 9–79 reads a daily newspaper. In Sweden, this high figure is due to a number of factors: the general equality in education; a long tradition of newspaper reading which in turn is connected to the long period of meagre choice among broadcasting media; and the press structure – there is no real distinction between the quality and popular press (except for a couple of afternoon tabloids). The daily press is heavily subsidised by the state. The morning newspapers are delivered to subscribers' homes before 6 o'clock in the morning; and small towns of 20,000 inhabitants or even fewer have their own daily newspaper supplying local news and advertisements, making a total of almost 100 newspapers that are published in Sweden at least four times a week.

Like the other Nordic countries, Sweden is characterised by a very high degree of literacy, high levels of book and newspaper reading, and by a vigorous public service broadcasting system. *Dagens Nyheter* is Sweden's largest morning newspaper and also by far the most influential newspaper in the country.[7] Its political affiliation is 'independent' (formerly 'liberal'), and its audited daily circulation is approximately 400,000. About 95% of the circulation in 1996 was subscriptions with home delivery, and the majority of its readers live in Greater Stockholm. *Dagens Nyheter* has a total daily readership of slightly more than a million people, corresponding to 17% of all adults in Sweden.

The empirical material reported here is derived from a content analysis of stories related to genetic engineering published in *Dagens Nyheter* during the period 1973–96. All relevant stories during the period were examined, and all articles dealing with 'the intervention, handling and/or analysis at the level of the gene(s)' have been used as input in the media analysis. This can be described as a total sample or, rather, as a certain population subset of articles. The total number of articles during the period is 734, and the frequency over time is presented in Figure 2 (in 1973 the number was 0). The method of collecting stories changed from manual to electronic in 1992 (when both methods were used), and this shift is marked by a break in the graph.

The total time period has been divided into four phases based on the frequency and the content of the articles. Each phase represents a rise, a peak and a decline in the number of articles. The four phases have been named after the dominating issues in the debate: safety, ethics, regulation and health; their characteristics are presented in Table 1.

Phase 1, 1974–80: safety. Only a few articles were published in the beginning and middle of the 1970s.

Most of the few articles in Phase 1 were published in 1978 and 1979. The predominant debate concerned proposals to build 'risk laboratories' in Uppsala and in the middle of Stockholm. One consequence of this was that most articles described national issues and events, and another was that almost all articles were published on the debate page. Most of these articles were written by non-journalists, such as scientists and politicians. Notably, all scientists described benefits, and all politicians described risks, although scientists did appear as actors in stories describing risks. Although the 'progress frame' was the most common, many articles described genetic engineering as 'opening Pandora's Box': the consequences of the new technology were truly unknown. This phase was characterised by a distinct debate concerning almost exclusively one issue: the building of risk (or 'safety') laboratories, in which categories of actors took opposite positions: scientists vs politicians, with most of the journalists on the politicians' side.

Phase 2, 1981–85: ethics. This phase had its peak in 1983. The main issue was ethical considerations in connection with human reproduction and individual integrity.

The ethical debate concerned human reproduction and diagnoses on human foetuses, which were coded as medical and ethical themes. Articles published on the debate, editorials, and culture pages often covered ethical and health risks, while articles on the science pages described health and economic benefits. There were differences regarding the kinds of actors that were pictured in different articles. Government actors, politicians, and ethical committees were in focus in articles concerning ethical questions and risks, and scientists were in focus in stories on medicine and basic research. Industry was not mentioned in a single story about ethical considerations, but appeared instead, like the scientists, in stories about medicine and basic research.

Phase 3, 1986–90: regulation. This phase peaked in 1989. Regulatory questions were dominant, and gene technology became a political question.

The regulation phase contained about the same total number of articles as the ethical phase, but had a different structure. There was a large number of articles in one particular year, 1989. During this phase, there were several debate topics, but patenting and regulation were on the agenda throughout the period. Owing to both the scientific developments and the relatively low level of regulation, there was ample room for political actors to acquire a dominant role, alongside the scientists. The latter group was still in focus in articles concerning not only progress and medicine but also ethical questions. Public accountability was the frame of many articles in the sense of Sweden lacking or not having enough regulation in the area. Most articles about gene technology were published on the general news pages.

Phase 4, 1991–96: health. Research concerning health and genetic diagnoses was in focus, framed as progress in the medical field.

This phase included many more articles than the earlier phases, especially in 1995 (the peak year), and the greatest increase occurred in articles concerning benefits. One of the reasons for this was that, during this phase, several applications of gene technology were put into practice, especially in health care (diagnosis, gene therapy) and in criminal investigation (genetic fingerprinting). These innovations were often described in the newspaper as benefits of the new biotechnology. As shown in Table 1, this entire phase was characterised by progress, mainly in health care due to new drugs and diagnostic possibilities. Scientists were the most dominant actors in the stories, and the articles were often published on the general news pages and the science pages. Politicians were also quite frequent actors in articles concerning regulation and genetic screening and identification. Increasingly, the stories covered issues and events outside Sweden, and even though Sweden still was the most common location, the USA was gaining space.

Press coverage of soya and maize

In addition to the longitudinal study, a case study on genetically modified soya and maize was conducted during the end of 1996 and the beginning of 1997. With the assistance of a press-clippings agency, all newspapers and magazines in Sweden were scanned for articles concerning the topic for a period of three months, and a total of around 350 stories were found. Most of the articles covered action taken by the EU and national actors' responses to this. Very few articles reported on soya and maize within the general debate about gene technology, or put forward ethical considerations. The publicity was, instead, clearly focused on health and ecological risks, regulation, product development and consequences for the consumers. This manifests an instrumental view from the mass media, focusing on genetically modified soya and maize as products with technical characteristics, and ignoring moral and ethical aspects. One obvious explanation is that most of the articles were written by news agencies and published without much rewriting or comment.

Some general patterns occur during the 24 years of study. There were many articles covering medical and pharmaceutical research and production. Scientists were often actors in these stories, and the frame was 'progress'. These articles were published during the whole period of study.

Turning to changes during the period, we find that in the beginning there was a lot of debate about risks and ethical considerations published on debate and editorial pages, while later on, benefits and medical findings became more dominant. The articles increasingly were published on the general news pages (Table 2). There was also an internationalisation of the debate over time.

It is notable that relatively few articles cover risks and hazards: the reason is probably the dominance of a theme that is almost always framed as progressive and/or positive: medical research. Although medical research was a frequent topic and Sweden has a quite large biomedical industry, the industry has been a silent actor since the beginning of the 1980s. When the industry has been in focus, it has been on the initiative of journalists – the industry itself has not tried to take part in the public debate, and has in fact actively stayed out of it.

Another interesting result is that the Swedish mass media have not focused much on food and crop plants over the years. During the regulation phase (1986–90), this issue was put on the agenda, but the question that was raised concerned the patenting of life, and not whether or not it was risky to eat genetically modified food. The soya and maize case indicated a new phase in the debate, where novel foods and risks associated with them were put on the agenda. Interestingly, this still did not evoke a lot of ethical questions, but instead the media described the foods as products that were potentially hazardous to consumer health. Dolly, the cloned sheep, did catch the media's attention during a few weeks in the spring of 1997, as she did in the rest of Europe and the Western world, mainly as a precursor to the possibility of cloning humans. But she disappeared from the news as suddenly as she had arrived. So, during most of 1997 and thereafter, it was still questions about genetically modified food that characterised the rather calm media debate after this study was ended.

Public perceptions

Over the years, there have been several surveys in Sweden of various aspects of public perceptions of modern biotechnology. The first survey with a few questions dealing with gene technology was carried out in 1978, and after that there are examples from 1989, 1990, 1991, 1992, 1994, and 1995; and finally there is the Eurobarometer survey of 1996.

Surveys of public perceptions 1977–96

In 1977–78, a lot of media attention was given to a group of scientists at Uppsala University who wished to build a so-called risk laboratory. In a general survey about attitudes to technology, Fjæstad found that 58% of the Swedish public reported that they had heard about the planned risk laboratory.[8] In a follow-up question, as many as 45% of them wanted the research to be prohibited by law, 37% thought it should be permitted and 18% did not know. Women were more sympathetic to a research ban, as were younger people.

The Recombinant DNA Advisory Committee conducted three surveys in 1989, 1990 and 1992 (these have not been published). In 1989, 81% of the public said they had heard about gene technology and in 1990 this had risen to 84%. Some 58% were worried about genetic engineering in 1989, and 36% in 1990. Slightly more men had heard about gene technology than women, but women were more worried than men.

In 1989, people were most worried about 'tampering with nature' (89%), but they also feared 'new unknown diseases' (63%) and 'new unknown organisms released into nature' (61%). The pattern was the same in 1990 but the percentages were slightly smaller. In 1992, the respondents were asked an open-ended question about the associations they made with gene technology. In response, 28% answered in the category 'genetic manipulation of humans', 17% 'tampering with nature', 14% 'research' and 9% 'genetic manipulation of animals'. As many as 25% did not know or did not answer the question.

A private research institute, FSI, included gene technology in its 1991, 1994 and 1995 omnibus surveys about issues that might be troubling people. As many as 30% in 1991, 26% in 1994 and 33% in 1995 perceived genetic engineering as one of the major threats to the world. Women were consistently much more negative than men.

Survey results: Eurobarometer 46.1, 1996

To monitor the public perception of genetic engineering in Sweden, data from the 1996 Euro-barometer on biotechnology were used. The Swedish survey consisted of 1008 interviews, representing the population over the age of 15, and was conducted in November 1996. General questions about biotechnology and questions about specific applications were asked. There were also questions about attitudes, knowledge, trust and regulation, as well as an open-ended question asking what comes to respondents' minds when they hear the expression 'modern biotechnology including genetic engineering'.

In reply to this open-ended question, as many as 12% of Swedish respondents mentioned the Belgian Blue cattle or synonyms like 'monster bull', or 'deformed beef cattle'. This was, by far, the single most mentioned association with modern biotech-nology, and was consistent with earlier focus-group interviews.[9] The respondents unanimously associ-ated genetic engineering in a negative way with Belgian Blue. Some stated explicitly that gene technology had gone way too far in creating such monsters. This is a provocative result, because this specific breed of cattle is not a product of genetic engineering or other forms of modern biotechnol-ogy, but a decades-old spontaneous mutation produced within a programme of traditional breeding methods. The Belgian Blue had recently been introduced in Sweden; and it had attracted a lot of attention, in the press and on television, that had focused on the seemingly abnormal physical characteristics of the cattle: the giant body has double muscles in certain places and weak legs, and the cows have problems calving.

Other common associations were food and animals in general, usually followed by a rejection of modern biotechnology. Answers expressing a favourable opinion on modern biotechnology including genetic engineering almost exclusively referred to medical applications and research. As much as 14% of the public did not associate anything with modern biotechnology, or answered 'don't know'.

We also constructed four clusters of images from the open-ended question. None of these clusters of images was more frequent than the Belgian Blue (12%): they were 'tampering with nature' (11%),

'progress' (9%), 'human monsters' (7%) and 'eugenics' (4%). It seems that a significant propor-tion of the Swedish public perceives modern biotechnology as something unnatural and unwanted that can create different kinds of monsters.

Before the open-ended association prompt, a general question was asked about the way science and technology change lives. Six different areas of science and technology were listed: solar energy, computers and information technology, space exploration, telecommunications, new materials or substances, and biotechnology/genetic engineering (split ballot). For each of the areas, the respondents were asked if they thought it would improve our way of life in the next 20 years, if it would have no effect, or if it would make things worse.

Optimism about genetic engineering is clearly lower than for all the other technologies offered. This list is headed by telecommunications, where 84% think it will improve their way of life, followed by 80% for computers and IT, 77% for solar energy, 72% for new materials or substances, 57% for biotechnology, 51% for space exploration, and 42% for genetic engineering. There is a gender difference in all areas, women being less optimistic than men. Correspond-ingly, the pessimist group is by far the largest for biotechnology and genetic engineering: 18%.

Attitudes to applications

Turning to six different applications of gene technology – genetic screening for diseases, produc-tion of pharmaceuticals, creation of special breeds of research animals, xenotransplants, food production, and crop plants with certain characteristics – the picture is more ambivalent. Respondents were asked whether they thought the different applications of biotechnology were useful, risky or morally accept-able, and whether they should be encouraged or not. In Figure 3, mean scores are given on a five-point scale, where 0 is the neutral point. The Swedish public is, on average, positive towards medical applications and negative towards animal and food applications. The Swedes are more negative to animal and food applications than Europeans in general: Sweden is one of the most negative coun-tries in Europe, together with Austria, Norway, Denmark, Switzerland and Germany.

How, then, can the public be segmented regarding attitudes to different applications of biotechnology? In multiple regression analysis, the following variables contributed significantly to the rather weak regressions (multiple R varies between 0.13 and 0.25): gender (the strongest predictor for all six applications), knowledge (all applications), interest and education (for crop plants and food produc-tion), and religiosity and age (for the two medical applications). Across all applications, men were more willing to encourage further development of gene technology than women. They also saw the applications as more useful and more morally acceptable, but there was no gender difference on perceived risk.

Previous research shows that people in general do not know very much about the details of modern biotechnology. The Swedish public knows slightly more than the citizens of most EU countries: a mean of 5.4 on a nine-point scale makes Sweden one of the top three nations in the EU. People with higher scores on the knowledge scale tended to be more encouraging of the six different applications. Older people were more positive about genetic testing and drug production by GMOs than were younger people, while religious people were less positive than others about these applications. The picture looks a little different for crop plants and food production: besides gender and knowledge, interest (if the respondent recalled hearing about and/or had discussed biotechnology) and education play a role – people with higher education were less negative, as were those who have not participated in the debate.

A question of special interest is the degree to which the respondents' propensity to encourage the different applications is explained by their views on usefulness, risk and moral acceptability of the very same applications. The results of such an analysis show the same pattern for all applications: moral acceptability is the strongest predictor (average β = 0.52), followed by usefulness (average β = 0.35). Risk, which has been a strong factor in the public debate, has, quite surprisingly, a very low and, for some applications, a non-significant predictive value (average β = 0.06). This is not a unique finding for Sweden; the pattern actually looks the same for the whole of the EU. In a factor analysis of the four response variables (usefulness, risk, moral acceptability and encouragement) across all applications, the risk items form a separate factor of their own and do not behave in the same manner as the other items, which form factors according to application. This indicates that the risk dimension is quite independent of the other three response variables.

Regulation

Who, then, should regulate gene technology? The respondents were offered seven organisations or types of organisation and asked who they thought best placed to undertake this function. As many as 34% said that international organisations, such as the UN or the WHO, should regulate modern biotechnology. Self-regulation by scientific organisations was also rated highly, 32%. Remarkably fewer, 8%, chose government agencies in Sweden, 6% the national Parliament, and as few as 2% the EU. This can be

interpreted in two ways: people do not trust national and EU political institutions; and/or they see gene technology as having global implications.

The survey also addressed some public concerns. A large majority, 82%, of the Swedish public thought that genetically modified food should be labelled as such, 49% that current regulations are insufficient to protect people from the risks of biotechnology, and 69% that there should be public consultation about new developments in biotechnology.

People were also asked which of seven organisations or groups they trust to tell the truth about two areas of modern biotechnology. For genetically modified food crops, people reported most trust in environmental organisations (29%), followed by farmers' organisations (23%), consumer organisations (12%) and universities (12%), whereas for xenotransplants people chose the medical profession (37%), followed by animal welfare organisations (19%) and universities (12%).

Cluster analysis of the group of Swedes expressing pessimism about modern biotechnology and genetic engineering (18% of the sample) revealed two distinct groups: 'blue' scepticism and 'green' critique.[10] The blue scepticism is formed by a traditionalist, conservative opposition based on moral values, while the green critique is a modern opposition based on uncertainty and risk. The blue group comprises 9.2% of the sample and the green group 8.6%. In the blue group, the gender distribution is about even (54% male and 46% female), but in the green group 76% of the members are women. The blue group consists mainly of middle-aged and older people (mean age 55 years) with low levels of education, living in small towns or in rural areas, somewhat materialistic, having low knowledge of biotechnology and not perceiving the risks as high. The green group, on the other hand, is made up of younger people (mean age 36 years) with high education, living in larger cities, voting left of centre, holding post-materialistic values, having relatively high knowledge of biotechnology and perceiving the risks as high.

To summarise, the Swedish public is relatively negative, and has been consistently so over the years, towards modern biotechnology including genetic engineering, especially regarding animals and food production. This resistance is mainly based on moral/ethical values and not on assessment of physical risks. Women are generally more negative than men towards modern biotechnology. The Swedish public knows relatively more about genetics and biotechnology compared to the rest of Europe. The Swedes do not trust their politicians to give correct information, but instead turn to environmental organisations; and they prefer the regulation to be on an international level or handled by the scientists themselves.

Commentary

A number of observations can be made from the Swedish data. From the viewpoint of the proponents of gene technology, the Swedish debate started on a bad note. From 1977 to 1979, a few very active politicians and a number of freelance social commentators wrote a number of newspaper debate articles that were very negative about recombinant DNA research, carrying headlines such as 'God Knows what Monsters You Have in Your Test Tubes'. There are probably a several different reasons behind this start to the debate. Nuclear power had been successfully introduced into political discourse, and it was generally the same forces behind the anti-recombinant DNA research initiatives. These politicians and critics had probably been alerted by the 1974–75 research moratorium that led to the Asilomar conference. Another important factor was two early proposals to build contained recombinant DNA laboratories in Uppsala and Stockholm.

When legislation took effect on 1 January 1980, debate waned. The number of stories in *Dagens Nyheter* dropped by half between 1979 and 1981. This pattern was, with a slight variation, repeated 15 years later when new legislation was put in place, and the number of articles fell by about 40% from one year to the next. (The same was true of nuclear energy: after the referendum in 1980, and after having been the top political controversy for a number of years, the issue suddenly died for more than 15 years as a major subject for publicity, debate or policy initiatives.) The lesson to be learnt here for proponents of new technologies is probably that the very existence of public regulation calms debate. The content of the regulation seems, ironically, to be of less importance.

The role of the media

All in all, the positive stories in *Dagens Nyheter* outnumbered the negative ones. Still, we have interviewed a number of the individual actors involved in research or policy-making, and the general impression seems to be the opposite, irrespective of whether the interviewee is generally for or against genetic engineering. The reason for this is possibly that the stories were unevenly divided between categories. The positive stories consisted of mostly factual and non-emotional reports on scientific and/ or economic progress – stories without pleas in any direction. These articles, often without pictures, were placed on the business or the general news pages. They were largely written by wire agencies or in-house news journalists. The negative and critical stories, on the other hand, were more emotional, they were placed on the op-ed pages, and they were to a large extent written by external persons, quite a few of whom were well-known politicians, debaters and columnists. A special case was the series of large, critical articles in the spring of 1989 on biotechnology, written by in-house journalist Thomas Michélsen. The articles were mostly quite factual in their assertions, but the total picture emerging was seen as clearly negative toward both science and industry. The series attracted a lot of attention. The scientific community was quite upset, but industry was not visibly moved. These articles made some environmental NGOs put gene technology on their agendas for the first time.

The attempt in 1988 by the Centre Party to put biotechnology on to the general election agenda failed. Perhaps it was just one environmental issue too many – the 1988 election came to be known as the 'environmental election': coinciding with the election campaign (in August and September) there was a strong media focus on a sudden explosion of algae in the coastal seas, caused by leakage of fertiliser from agriculture, that was believed to be the cause of a simultaneous dying-off of seals around the Swedish coasts (later it was shown that the deaths were caused by a virus). This chain of events helped the Environmental Party into the Parliament for the first time. Perhaps gene technology was too difficult to communicate. At the time there was no such thing as 'gene food', just mainly medical applications.

The intersection of policy and media phases

The four discernible phases of the media study are almost the same as the phases in the policy study:

Media study	Policy study
1974–80: safety phase	1974–79: problem formulation and policy initiation
1981–85: ethical phase	1980–86: low political activity
1986–90: regulation phase	1987–91: less regulation and renewed debate
1991–96: health phase	1992–96: putting new regulation in place

This may not seem a coincidence, but the division into phases was nevertheless based on different perspectives relevant to each study: the amount of publicity and its main contents in the media study, and forum and momentum of regulation in the policy study. It is possible, though, to hypothesise about causal connections in both directions. For example, initial media attention on safety (for workers and for citizens living near the risk laboratories) may have alerted politicians to the need for regulation. Later, when the safety problem was taken care of with the legislation of 1979–80, the media were free to move on to ethical issues. And with less media attention on physical and biological risks in the beginning of the 1980s, there may have been fertile ground for the deregulation of the mid 1980s. From 1987, with less regulation, a new media debate about the need for tighter regulation started which may have influenced the new round of political interest in regulation that eventually led to the Gene Technology Law.

Moral acceptability

In the 1996 Eurobarometer, Sweden stands out as one of the most sceptical countries of the EU. As in all the other countries in the survey, one of the major findings was that the encouragement of gene technology applications does not co-vary with the perceived risk; and varies only about half as strongly with utility as with moral acceptance of these applications. From the open question, from our focus groups, and from other current research in Sweden, we conclude that the main constituents of 'moral acceptability' (which was not defined in the Eurobarometer questionnaire) can be described as follows in the Swedish case:

1 Many Swedes think genetic engineering is an inappropriate tinkering with life, and, more specifically, with the meaningful order and ecology of nature where every species has its place and its purpose and where natural boundaries should not be transgressed by unnatural means.

2 Gene technology is seen by many Swedes as having existential, far-reaching, unknown, and not-easily-observed consequences – not very different from radioactivity. The fact that smoking or travelling by car kills many more people is not important since these are perceived to be risks that can be estimated and weighed. The importing of genetically modified soya or maize is described both in the media, by the opponents, and by large food retail chains in terms that are reminiscent of drug smuggling: the food products are referred to as 'leaking uncontrollably over the borders'.

3 The argument that gene technology would help the Third World to a second green revolution that could counteract hunger appears not be thought valid by a majority of Swedes, and is, thus, not a factor that improves general attitudes toward genetic engineering.

4 Genetic engineering is perceived by many people as not doing very much good for the average citizen (except for the medical applications), but is instead regarded as a top-down technology. Who cares if the producers of corn or cotton or ham can make the same products a little more cheaply? The advantages for the consumers have not been very clearly stated. For them, gene technology appears to be just

another questionable production method designed to help boost commercial interests and big business – not very different in the minds of consumers from spraying poisonous insecticides and herbicides on food plants, feeding livestock with naturally dead animal corpses, or injecting cattle with growth hormone or penicillin that may leave residues in the meat.

Interestingly, the most common answer to the open question 'what comes to mind when the term biotechnology including gene technology is mentioned?' is Belgian Blue, a 'monster' cattle breed. Not a product of gene technology but of traditional breeding methods, it looks unnatural, and it is infamous in Sweden for hardly being able to carry its own weight, and for needing caesarian births. Belgian Blue is seen as an EU project (since it was allowed into Sweden under EU rules). As mistrust against central authority, including the EU, seems more widespread than ever in Sweden, this does not help the acceptance of genetic engineering. The very fact that gene technology is construed by many as an EU project (e.g. because of the EU's approval of the modified soya and maize), may result in opposition.

There are no obvious signs that genetic engineering will be accepted to any great extent in the near future. On the contrary, genetically modified food came into focus only at the end of 1996, and the publicity during 1997 in the popular media, rather more extensively about genetically modified food than about Dolly the sheep, has hardly been supportive. The scientific community and its various branches (academies and others) now have seen the need to address the public, and there have been series of lectures in many cities and other initiatives. However, in spite of the main results of the Eurobarometer survey being published in Sweden in August 1997, emphasising the fairly weak connection between knowledge and attitudes and the strong linkage between morality and attitudes to applications, these initiatives have not aimed at discussing bioethics, but at teaching people about genetic engineering, assuming that this will make the public take on 'sounder' attitudes. The retail food business has not been supportive of science and industry, but has sided instead with the opponents in an attempt to not alienate the consumers. For this, the retail chains have lately been accused of not daring to say what they believe; it is thus taken for granted that they, deep down, do realise that genetically modified food is no more harmful than other food – a position publicly taken by the Government Food Agency. Since medical applications are accepted, another lesson to learn for the proponents of genetically modified food is probably to emphasise the continuity from traditional plant and animal breeding, and to stress consumer rather than producer utility.

Acknowledgement

The Swedish national research was funded by: the Freja Foundation, the Mid Sweden University Research Council, the Erinaceidæ Foundation, the Magn Bergvall Foundation and the Swedish Natural Science Research Council.

Figure 1. A regulatory profile for Sweden

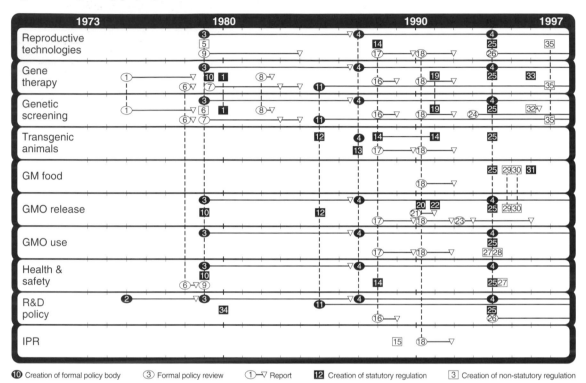

1 *1975.* The Swedish Academy of Sciences appointed a working group for possibilities and risks associated with the recent advances in molecular genetics on 3 March (1980, legislation on genetic technology).

2 *1975.* A committee on matters relating to recombinant DNA research created at the Natural Research Council and recognised by the Medical Science Research Council.

3 *1979.* According to a government ordnance 1980, the establishment of the Swedish Recombinant DNA Advisory Committee was agreed in parliament (13 December).

4 *1987.* Government decision (parliamentary procedure) converted Swedish Recombinant DNA Advisory Committee to a government agency. On 1 July 1994 the agency was re-established as the Gene Technology Board.

5 *1979.* Prime Minister Ola Ullsten of the Liberal Party publicly demanded that industry refrain from recombinant DNA experiments pending forthcoming legalisation.

6 *1978.* The first government public investigation into genetic technology appointed (16 February). Report DsU 1978:11 published: 'Recombinant DNA Technology under Control'.

7 *1979.* Parliamentary commission on genetic ethics appointed. First report: DsS 1983:12 'The Application of Genetic Technology on Humans'. Second (main) report: SOU 1984:88 'Genetic Integrity'.

8 *1982.* Symposium on genetic ethics organised by the Swedish Recombinant DNA Advisory Committee and the Agency for Technology Development (STU). Report published by STU: 'Ethics and Gene Technology'.

9 *1979.* Government investigation committee for the evaluation of genetic engineering legalisation appointed in 1980. Report published in 1984: SOU 1984:5 'Is recombinant DNA technology necessary?'.

10 *1979.* Government ordnance (20 December). Parliamentary report (rskr 1979-80: 107): **1** change in instruction for the agency for industrial and occupational safety (SFS 1970:1172); **2** Change in instruction for the Swedish Recombinant DNA Advisory Committee (SFS 1979:1173); **3** preliminary

examination according to the law for work environment for Recombinant DNA questions (SFS 1979:1173); **4** change in Environment Protection statement (SFS 1979:1175), changed to environment protection ordnance 1981:574, changed 1981:1074.

11 *1985.* Foundation of the State Medical–Ethical Council in 1985.

12 *1985.* Law 1985:342 on control of animals in domestic production. Transgenic animals are not considered as food production, but are approved for pharmaceuticals.

13 *1987.* Law on hunting 1987:259 (§ 36) + regulation 1987:905 (deliberate release of non-naturalised animals).

14 *1988.* Law 1988:534 on animal protection includes a passage on gene technology which states that the government may regulate animal gene technology; §19, referring to commercial production and general research. New law on ensuing exceptions to §19 (1991:404) granted to Agricultural Agency.

15 *1989.* Patents and Register Agency: no patents on living animals (with the exception of patents on microorganisms). There is also a difference between discovery and invention (cf. EU 1990 'biotechnical packages' and secret inventions (cf. also Swedish law 1990:409).

16 *1988.* Investigative committee on research ethical control is appointed, the report of which is published 1989 (SOU 1989:74 'Research–Ethical Examination').

17 *1988.* Working group set up under the Ministry of Agriculture, which in 1990 (DsU 1990:9) publishes its report 'Gene Technology: Plants and Animals', dealing with risks and ethical questions.

18 *1990.* A government investigation on biotechnological patents, ecology and ethics is appointed(March 29). Gene Technology Committee publishes report SOU 1992:82 'Gene Technology – a Challenge'. No suggestion on patents however.

19 *1991.* Law 1991:114 on the use of special gene technology at general health checkups, and Law 1991:115 on action on human fertilised eggs for research and therapy purposes.

20 *1990.* Law on plant protection, giving the Agricultural Agency the right to regulate the setting out of transgenic plants in greenhouses or outdoors.

21 *1990–91.* Chemicals Inspectorate policy review on biological insecticides.

22 *1991.* Law 1991:639 on on preliminary inspection of biological insecticides. The law is to be administered by the Chemicals Inspectorate.

23 *1992.* Parliamentary investigation committee established. Report: SOU 1993:27 'Environmental Acts'. The main report followed in 1996: SOU 1996:103 'Sustainable Development'.

24 *1993.* An investigative committee (chair: Stig Larsson) is set up to handle the question of access to genetic information.

25 *1994.* Framework legislation: **1** Law SFS 1994:900 on genetically modified organisms, **2** government ordnance on GMOs, SFS:1994:901 (change to SFS 1994:1515), **3** government ordnance on instruction for the gene technology board: SFS 1994:902 (change to SFS 1995:365), **4** government ordnance on change of the government ordnance on secrecy, SFS 1994:1030.

26 *1994.* The Office of the Auditors of the Parliament appointed to investigate and evaluate biotechnical legislation and its effects on science and industry in 1994.

27 *1994.* Authority regulation for various agencies (non-statutory regulations): contained use of GMOs (AFS 1994:46). The Fisheries Agency (FISF 1995:10) established regulations on genetically modified marine organisms.

28 *1994.* The Chemicals Inspectorate (KIFS 1994:11) established regulations on GMOs.

29 *1995.* The Agricultural Agency established regulations on GMOs (SJVFS 1995:33), and regulations on genetically modified plants (SJVFS 1995:45).

30 *1995.* The Swedish National Food Administration (SLV FS 1995:3). Regulations and general advice on permission for deliberate release of such food that contains or consists of genetically modified organisms.

31 *1996.* Law on labelling of genetically modified food and law on labelling of country of origin on genetically modified food.

32 *1996.* Report on genetic testing (DS 1996:13) and 'Medical Testing in Working Life' (SOU 1996:63).

33 *1996.* Government ordnance (1996:608) with instruction for the State Working Committee for the Evaluation of Medical Methods.

34 *1980.* Requirement for preliminary scrutiny of working methods for new adaptations of Recombinant DNA, concerning research or industrial application of new techniques. Permission granted by the Agency for Industrial and Occupational Safety.

35 *1997.* Proposed bill on genetic integrity (1997:60).

Figure 2. Intensity of articles on gene technology in the Swedish press

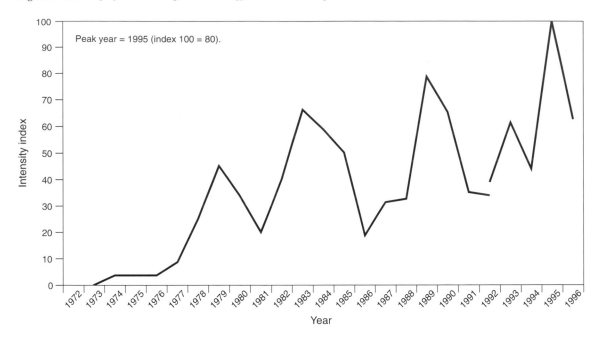

Figure 3. Attitudes to applications of biotechnology in Sweden, 1996

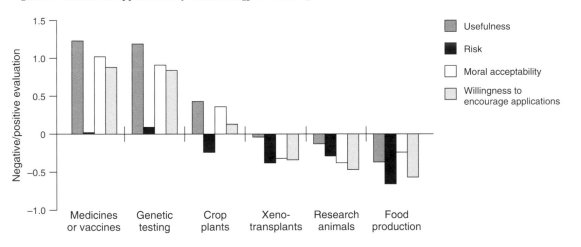

Table 1. Swedish media profile (for an explanation of terminology please see Appendix 5).

Phase	Freq.[a] (%)	Frame (%)		Theme (%)		Actor (%)		Benefit/risk (%)		Location (%)	
1. 1974–80	13	**Progress**[b]	**46**	*Safety & risk*	*31*	Scientific	41	*Both*	*34*	Sweden	83
		Pandora's Box[c]	19	Basic research	30	Political	20	*Neither*	*30*	USA	9
		Accountability	18			Industry	13	Benefit only	18		
		Economic	12			*Media/public*	*10*	Risk only	17		
2. 1981–85	26	**Progress**	**44**	**Medical**	**27**	Scientific	43	**Benefit only**	**34**	Sweden	74
		Ethical	36	Basic research	19	*Industry*	*17*	Neither	23	USA	15
		Economic	13	Ethics	18	Political	15	Both	22		
						Ethics	10	*Risk*	*21*		
3. 1986–90	24	**Progress**	**43**	**Medical**	**30**	Scientific	47	**Benefit only**	**32**	Sweden	63
		Accountability	26	Regulation	14	*Political*	*23*	Both	29	USA	18
		Ethics	16	Ethics	12	Industry	15	Neither	22		
								Risk	16		
4. 1991–96	37	*Progress*	*58*	*Medical*	*44*	*Scientific*	*54*	*Benefit only*	*46*	Sweden	53
		Accountabilty	14	Regulation	12	Political	18	Neither	25	*USA*	*21*
		Ethics	12					Both	18		
		Pandora's box	10					Risk	12		

a Percentage of corpus in period; *n* =734.
b Bold indicates highest frequency within phase.
c Italics indicates highest frequency within category.

Table 2. Type of article in the Swedish opinion-leading press

	1974–80 (%)		1981–85 (%)		1986–90 (%)		1991–96 (%)	
Newspaper sections	*Debate/editorial*	*45*	Debate/editorial	35	General news	34	*General news*	*54*
	Science	10	*Science*	*26*	Debate/editorial	30	Science	24
	General news	10	*Culture*	*19*	Science	13		

Notes and references

1 Ds U 1978:11, Hybrid-DNA tekniken under kontroll.
2 Ds A 1984:5, Behövs Hybrid-DNA kontrollen?
3 Ds Jo 1990:9, Genteknik, växter och djur.
4 Ds S 1983:12, Genteknikens tillämpning på människa.
5 SOU 1992:82, Genteknik - en utmaning.
6 SOU 1993:103 Miljöbalken. Huvudbetänkande av miljöbalksutredningen.
7 Petersson, O and Carlberg, I, *Makten över tanken* (Stockholm: Carlssons, 1990).
8 Fjæstad, B, 'Människor och den tekniska utvecklingen: allmänhetens attityder', in Sörbom, P (ed),

Attityder till tekniken (Stockholm: Bank of Sweden Tercentenary Foundation and The Royal Swedish Academy of Engineering Sciences, 1978), pp31–45.
9 Olofsson, A and Olsson, S, 'The new biotechnology: media coverage and public opinion', in Fjæstad, B (ed), *Public Perceptions of Science, Biotechnology, and a New University* (Östersund: Mid Sweden University, 1996), pp48–87.
10 Hviid Nielsen, T, 'Behind the color code of "no" ', *Nature Biotechnology*, 12 (1997), pp1320–1.

Address for correspondence

Prof. Björn Fjæstad, Department of Human Resources, Management and Environment, Mid Sweden University, SE-831 25 Östersund, Sweden. E-mail Bjorn.Fjaestad@mam.mh.se

Switzerland

Heinz Bonfadelli, Petra Hieber, Martina Leonarz, Werner A Meier, Michael Schanne and Hans-Peter Wessels

The economy

Switzerland is among the most highly industrialised countries. In 1995, 4.2% of the population of around 7 million people were employed in the primary sector (agriculture and forestry), 29% in industry and 66.8% in the service sector. Unemployment rose from under 1% in 1991 to 4.7% in 1996. Western Switzerland and the canton of Tessin were especially affected with unemployment at 8–9%.

The branched structure of the Swiss economy has changed fundamentally since the late 1980s. Traditionally economically strong branches of industry, such as mechanical engineering, have suffered considerable job losses, while branches of the service sector, such as health, leisure and teaching, have proliferated. However, employment in other service branches, such as retail and banking, has also diminished since 1991.

A large proportion of Switzerland's GNP comes from the export trade, mostly with the OECD countries. In 1996, they accounted for 78% of exports and 90% of imports. The EU is particularly important in this respect (61% of exports and 79% of imports). The mechanical engineering and electronics industries as well as the chemical–pharmaceutical industry produce the majority of exports.

The largest industrial enterprises in Switzerland include Nestlé and Kraft Jacobs Suchard (food); Asea-Brown-Boveri (machinery); and Novartis (established in 1996 following the merger of Ciba and Sandoz) and the Roche group (pharmaceuticals/chemicals).

Research and development

The Swiss federal state finances both basic research and applied and application-oriented research projects. However, the main emphasis is on basic research, while applied research is generally left to the private sector – the state limits itself to providing a suitable economical–political framework for research and industry.[1] Federal R&D costs remained stable between 1989 and 1993. Between 1994 and 1996 they slightly decreased for the first time. Universities spent the most on the medical, pharmaceutical and natural sciences.

The private sector's spending on R&D activities was concentrated in the chemical–pharmaceutical (38%), machinery (19%), and electronics industries (17%) in 1996.[2] Between 1975 and 1989 this expenditure had been rising, but it fell by 10% between 1989 and 1992 and rose again by 6% between 1993 and 1996. Development expenditure by Swiss subsidiary companies abroad exceeded expenditure in Switzerland for the first time in 1992 (by about 7.1 million Swiss francs). Between 1993 and 1996 this expenditure rose again by 14%. The companies involved had previously been very active in Switzerland, and were now cultivating new R&D activities in other countries.

The Swiss Priority Programme for Biotechnology (SPP BioTech) has existed since 1992. This programme for application-oriented research was created by the Swiss National Science Foundation[3] to advance biotechnological research and to secure Switzerland's position in the field. The focus is on improving collaborative efforts between the universities and industry. The first phase (1992–95) was dedicated to research and the second phase (1996–99) emphasises the realisation of research findings.

The biotechnology industry

An estimated 200 companies and research institutes are currently active in biotechnology in Switzerland.[4] In 1994, around 6300 people were employed in the area, most of them by the corporate giants of the pharmaceutical, chemical and food industries.[5]

In the 1970s, universities were the first to apply biotechnology (to molecular biology, microbiology, biochemistry and botany). In 1997, Switzerland was ranked first in the fields of immunology, molecular biology, pharmacology and physics.[6]

Industry began applying biotechnology in the early 1980s. However, the most important actors in biotechnology today are Novartis, Roche and Nestlé. Novartis uses biotechnology primarily in therapeutics, diagnostics and agribusiness – in R&D as well as in production. Roche emphasises therapeutics, vitamins and diagnostics; they have been manufacturing products (such as Interferon) with the help of gene technology for many years. Besides in-house research, these companies also support private research institutions.

Since the late 1980s, Swiss pharmaceutical companies have increased their research collaboration with small, foreign biotechnology companies (especially in the USA). Some of these collaborations have led to the acquisition of foreign firms.[7]

Concurrently, the growth of biotechnology in pharmaceutical–chemical companies in Switzerland has halted.[8] Nestlé had broadly ceased its development of biotechnology by the late 1980s. A real shift of biotechnological activities to foreign countries began. However, this did not mean a reduction in biotechnological activity in Switzerland.[8]

Medium-sized and large companies became active in biotechnology at the same time as the international companies.[9] However, medium-sized companies in the food industry were cautious about applying biotechnology. Small and start-up biotechnology companies, on the other hand, first formed in late 1980s (the number of new companies has remained stable in recent years). These small firms are especially active in planning and construction (80–95% exports) and as suppliers of component products for the pharmaceutical and food industries.

A number of foreign firms are also engaged in biotechnology in Switzerland. These include subsidiary companies of European or US corporate giants such as Novo Nordisk and Schering-Plough.

Political system and culture

The regulatory situation in biotechnology in Switzerland is largely influenced by Switzerland's specific and rather unique political system and culture.[10] Switzerland has a non-parliamentary system of government which is characterised as *kollegial* (collegial). This means that the government (Federal Council) is not based on a majority in Parliament but is elected for four years at the beginning of each legislative period. It is a collective of seven Federal Councillors with equal rights; there is no head of state such as a chancellor, prime minister or president.

The Federal Councillors are elected by Parliament (the Federal Assembly). Each of the seven Federal Councillors is head of a department (ministry) and recruits civil servants. The Federal Council also directs and controls administration. The collective of the seven Federal Councillors is composed of members of all four major political parties: two members of the Free Democrats (liberals, traditionally strongly connected with industrial and trade interest groups); two members of the Christian Democratic People's Party (roots in the Catholic states and connected with industrial, trade and employers' associations); two members of the Social Democratic Party; and one member of the Swiss People's Union (advocates for Swiss farmers and small commercial enterprises and businesses).

The Swiss Parliament (Federal Assembly) consists of two houses: the National Council in which the people are represented (analogous to the US House of Representatives); and Council of States in which the cantons are represented (analogous to the US Senate).

Switzerland is a direct democracy characterised by a high degree of federalism: it consists of 26 states (cantons); and there are four official languages (German, spoken by 63.7% of the population; French, 19.2%; Italian, 7.6% and Romansh, 0.6%), which more or less define four different mentalities.

The system of direct democracy affects all levels of political opinion formation and all the actors involved. The referendum and the constitutional (popular) initiative have a particular impact on decision-making. The mandatory referendum deals with amendments to the constitution and with urgent federal decisions. The optional referendum can be used by 50,000 eligible voters to vote against federal laws, generally binding federal decisions, and international treaties which are passed by the councils. All referendums represent sovereign and binding decisions, in that they cannot be overruled except by another referendum. With all acts of legislation except those designated as matters of urgency, direct democracy has a delaying effect. That is, the law in question is not effective until the referendum has been carried out.

The popular initiative, however, has quite different implications. The right of 100,000 eligible voters to demand, with their signatures, a partial or total revision of the constitution is primarily a route for asking new political questions and for exerting pressure on established forces. Any seven Swiss voters can begin the process by submitting a request for an initiative and a description of the desired change in the constitution. They then have 18 months to collect 100,000 signatures in support of their petition. They can either suggest the general terms of the change, or they can present the exact text of the proposed amendment. Most initiatives are specifically worded. Before the proposal is placed on the ballot the government can either endorse it, recommend rejection, or recommend rejection and submit a counterproposal. Typically, such counterproposals accept some of the petitioners' demands while omitting others. It is not uncommon for the entire initiative process to take seven years or more.

For sponsors of many Swiss initiatives today, success at the polls is not necessary. Rather, the initiative is submitted in order that its withdrawal can be offered when bargaining for political change. The initiative has therefore become an instrument of power. Often, initiatives aim to generate political pressure in an area by demanding policy changes beyond those that the sponsors would be happy to accept.

Launching initiatives can also be a highly effective propaganda exercise for minor political parties. It establishes their agenda in the electorate's mind. The direct success of initiatives, modest as it may be,[11] is accompanied by a considerable indirect effect: often they provoke political discussion and thus work as a dynamic element within the political system.

Public policy

The legislative process in Switzerland takes place on two levels: the constitutional and the legal. The constitution thus forms the framework for laws. It formulates objectives which are explicated on the legal level. Because two constitutional initiatives had and have a decisive influence on the political process associated with the biotechnology debate, the following discussion focuses on the regulatory debate on the constitutional level. The legislative process is detailed in Figure 1.

National policy on biotechnology

The federal policy on biotechnology is based on article 24novies of the constitution, which on one hand finely regulates the human domain, and on the other hand sets general goals in the extra-human domain. The objectives stated in the constitution shall, as determined by the Federal Council, be fulfilled through the adaptation and extension of existing legislation. The Federal Council has no intention of creating a specific, superposed law on gene technology.

The legislative process

Figure 1 shows that legal regulation in biotechnology started rather late in Switzerland. From 1975 until 1989 there was no specific federal regulation in biotechnology at all. Nevertheless, there was some regulation, mainly through guidelines initially issued by the Swiss Academy for the Medical Sciences (SAMW) and, after 1986, by the new interdisciplinary Swiss Committee for Biological Safety (SKBS). The first guidelines issued were the NIH guidelines for the use of GMOs, adapted for Switzerland in 1975 by the Arber Committee (established by the SAMW), chaired by Professor Werner Arber, a Nobel-prize-winning microbiologist.

The Committee issued recommendations for 'the use of *in-vitro* fertilisation and the transfer of embryos' in 1985, and medical–ethical guidelines for medically assisted reproduction in 1990. The SKBS succeeded the Arber Committee and dealt with the use of GMOs. As a body for the self-regulation of science and industry, it cooperated closely with the Federal Office for Environment, Forestry and Landscape. Thus, although it was a non-statutory regulatory body, its guidelines were nevertheless decisive for any use of GMOs (controlling almost all federal funding). Only in 1997 was the SKBS replaced by a statutory body, the Federal Expert Committee for Biological Safety (EFBS). This meant that government took a passive role in the

regulation of biotechnology for a long time. This period of little or no federal regulation, and corporate or self-regulation, of science and industry ended in 1990, when a transitional phase began that was characterised by an increased awareness of legal uncertainty and the start of the legislative process in biotechnology.

The 1991 Federal Ordinance on Incidents in Industrial Plants and Installations (SfFV) was the first statutory federal regulation in biotechnology. With the referendum on EC membership coming up in December 1992, Switzerland wanted to have a regulation ready that was compatible with EU guidelines 2020 (1990).

An actual process in the federal regulation of biotechnology was set in motion by the constitutional amendment article 24novies 1–3 in 1992, regulating reproductive medicine and gene technology in humans as well as entitling federal government to regulate the extra-human domain on the constitutional level.[12] It set the frame for all laws and ordinances issued since. This constitutional article was the result of a debate that started in the mid 1980s. Since then the implementation of the amendment on the level of laws and the enactment of federal decrees has dominated the regulatory process in biotechnology. The most important characteristics of this process are as follows.

1 The Federal Council has no intention of creating a specific, superposed law on gene technology, but wants to adapt and extend existing legislation (see Figure 1 for the IDAGEN and Schweizer reports). For example, it revised the law on food (*Neues Lebensmittelgesetz*) in 1992 and enacted the executive ordinance on food in 1995, regulating approvals and declaration of genetically modified food; and it revised the law for environmental protection (*Umweltschutzgesetz*) in 1995, regulating the contained and deliberate release of GMOs.

2 The popular initiative 'for the protection of living beings and the environment from genetic manipulation', which became known as the Gen-Schutz Initiative, was submitted in 1992. It demanded more detailed and restrictive regulation of the extra-human domain. This started a debate on regulation in that area, and set the agenda for recent policy-making in biotechnology (the referendum on the initiative was held on 7 June 1998), focusing the debate on the regulation policy for the extra-human domain. This regulatory process (which is shown in more detail in Figure 1) is part of, and the result of, a political debate.

The political debate

There have been broadly two phases of debate about biotechnology in Switzerland. The first phase started in 1984 and ended in 1987 with the popular initiative of the consumer magazine *Der Schweizerische Beobachter*, 'against the abuse of reproductive and gene technologies in humans', which became known as the Beobachter Initiative.[13] This phase was characterised by the growing awareness of a lack of regulation in biotechnology, and by the focus on preventing abuses, particularly in humans. The previous decade saw no political or public debate whatsoever, although there had been some awareness among experts and scientists at an early stage.

The second phase started when the Beobachter Initiative was submitted in 1987, and has continued ever since. This phase is characterised by a growing polarisation of the debate, and by the focus on regulatory policy-making.

Phase 1, 1984–87: beginnings of the debate. There was no political or public debate on biotechnology in Switzerland for the first decade after 1972. Gene technology was mostly confined to university and research institute laboratories, and there were no publicly known applications, so it was not a political issue. The political debate in Switzerland began in the mid 1980s. It originated from discussions on the new reproductive technologies (*in-vitro* fertilisation, embryo transfer) and their associations with gene technology (such as germline therapy).

Church and conservative groups, but also leftist feminist groups (with altogether different motives), were disturbed by media reports on surrogate mothers and the use of 'a plethora of embryos', as well as by the birth of the first Swiss test-tube baby in 1985. They demanded an end to physicians' and scientists' self-regulation, which they wanted replaced by legal regulation at the federal level. While the cantons had already begun to translate these demands for societal control of the new technologies into laws, the Federal Council still saw no need for action.

In response to the Federal Council's passive stance towards reproductive and gene technologies, parliamentary proclamations calling for federal regulation began to proliferate in 1984. The criticisms emanated from essentially two groups: one consisting of the Church and conservative groups, represented in Parliament by the Christian Democratic People's Party and the Swiss People's Party; and the other of left-wing political parties such as the Social Democratic Party and the Progressive Organisation of Switzerland. The women in these parties were particularly active on the issue. Both groups criticised, though for different reasons, the lack of any state and societal control. A general legal regulation of gene technology had by then already been alluded to, particularly by the left-wing groups. In the centre of the

debate stood abuse prevention.

The debate may have also been triggered by the then new interest of the large chemical and pharmaceutical companies in biotechnology.[14] The media increasingly heralded biotechnology as momentous, and discussed the prospects for applications in various fields.

At the same time, an opposition movement started to form. Not surprisingly, it was set into motion in Basle, in the northern part of Switzerland, with its unique concentration of pharmaceutical and chemical companies (then three multinationals: Ciba, Sandoz and Roche; and in 1988 Novartis, Ciba Specialty Chemicals, Clariant and Roche). The chemical industry was playing a very important economic role in the region.

In 1986, relations between the local people and the chemical industry experienced a severe blow: a fire in a Sandoz chemical plant on the outskirts of Basle created a new climate of opposition and a new awareness of risk. As a direct consequence of this incident, civil movements formed demanding better protection for the population. They profited from another movement that had just been successful: the anti-nuclear movement, which had opposed the building of a nuclear power plant in the very same area. Having fought for ten years, these people knew how to mobilise resistance. Out of this general opposition to the chemical industry, nourished by the anti-nuclear movement, evolved the more radical and, at first, rather regional opposition to biotechnology. The organisation that represented this movement was the Basle Appeal against Gene Technology, which was founded in 1988. It is an important member of the NGO the Swiss Action Group on Gene Technology (SAG).

The Beobachter Initiative acted as another catalyst for the political process: it eventually set in motion a broad public debate on reproductive and also gene technology. The Initiative Committee had already begun to formulate a proposal in 1984, the collection of signatures began in 1985 and, finally, in 1987, the Beobachter Initiative was submitted and deemed valid. The Federal Council now had to act: under the pressure of events, the Council had already installed an expert commission, the Amstad Commission, in 1986 (see Figure 1).

Phase 2, 1988–98: focus on federal legal regulation policy. In this phase the legal regulation of biotechnology stood in the centre of the debate. The report from the expert Amstad Commission precipitated the Federal Council's decision in 1989 to formulate a counterproposal to oppose the Beobachter Initiative. This counterproposal distinguished itself from the text of the initiative in two points: it no longer included any prohibitions, but did incorporate an article on the extra-human domain.

The Federal Council thus essentially continued to pursue its previous policy of little or no regulation. Concurrently, the regulation debate had expanded to include the extra-human domain. The Federal Council's counterproposal was unacceptable to both Parliament and the Initiative Committee. Moreover, the passive role of the Federal Council and administration was not essentially changed by the counterproposal; they played only a marginal role in the debate about the details of the regulation.

The real agents in the debate were the state councillors and the representatives of the Initiative Committee. At the same time, political actors outside Parliament gained influence: environmental and animal protection organisations, organisations aiding developing nations, as well as technology critics (preceded by the anti-nuclear movement) consolidated at that time to form a distinctive opposition to gene technology.

In 1987 the most important actors in this opposition formed the Swiss Action Group on Gene Technology (SAG), which was established as an association in 1990.[15] During the course of the debate, these extra-parliamentary actors gained influence and introduced new topics: the release of genetically modified products, transgenic animals, hazards to the environment, and the patenting of life forms.

The parliamentary debate on the Beobachter Initiative ended with Parliament's counterproposal, which was finally put to a vote on 17 May 1992. The counterproposal contained a constitutional article which endeavoured to regulate reproductive and gene technologies in the human domain with guidelines and prohibitions, and in the extra-human domain with competence standards. The public debate on gene technology intensified in parallel with the parliamentary debate. Since 1990, the first acceptance problems had manifested themselves in public: there had been resistance to the construction of Ciba-Geigy's biotechnological production plant in Basle, and to the release of genetically modified potatoes in 1991.

With the vote on 17 May 1992, the Parliament's counterproposal, in the form of the constitutional article 24novies, was accepted by 73.8% of Swiss voters, with the usual low turn-out of 38.6%. There were no notable differences in voting results among the three main language and culture regions.[16] Advocates as well as opponents of the counterproposal were primarily motivated by a distrust of the new technologies and a desire for societal control. The advocates were also motivated by an interest in regulating biotechnology at the constitutional level, although the legal context should not hinder research in this new scientific discipline. The opponents, with their rejection of the counterpro-

posal, expressed a general distrust of the new technology, associated with doubts about the authorities' determination and ability to legally regulate biotechnology.[17] The vote illustrated a broad consensus among the public on the necessity of legal regulation of biotechnology in Switzerland. But the increasingly prominent polarity of the political debate was not yet evident in the voting results.

This situation changed with the initiative for the protection of living beings and the environment from genetic manipulation – the Gen-Schutz Initiative. Out of a dissatisfaction with the regulation of the extra-human domain in article 24novies, the action group SAG had already launched the Initiative in April of 1992, shortly before the vote. The Initiative demanded prohibitions in the extra-human domain; consequently, the polarisation of the debate began. On one side stood the supporters of the Initiative, primarily the NGOs, and, on the parliamentary level, especially the Green Party and many Social Democrats. On the other side stood primarily the actors: academics and industry, who viewed the acceptance of the Initiative as a threat to research and to the economic standing of Switzerland; they were represented in Parliament above all by the Free Democratic (liberals). Apart from the patenting of life forms, biotechnology and food were increasingly among the most controversial issues. As a result, the food industry and retail trade as well as consumer and agricultural organisations also joined the debate.

The growing polarity also manifested itself in the parliamentary debate: Parliament was not able to formulate a counterproposal, as it had done in 1991 with the Beobachter Initiative. Not until 1996 did Parliament demand accelerated legislation in the extra-human domain, with the Gen-Lex Motion. This motion called for the largest possible segment of the constitutional mandate on the extra-human domain to be regulated before the vote on the Gen-Schutz Initiative on 7 June 1998. In December 1997 the government unveiled to the public its draft proposal for the Gen-Lex package. It then was put into the formal consultation process which concluded on 31 March 1998.[18]

The year 1997 again brought an intensification of the debate: in Parliament, the number of proclamations on biotechnology reached a climax, with genetically modified soya as one of the main points of discussion. Biotechnology had also achieved high visibility in the media; public presentations such as lecture series and podium discussions also took place, and were already a part of the voting campaign. In the winter of 1997/98 the government's proposition and the reaction of the political parties as well as of the SAG were broadly discussed in the media.

Relationship with EU policy-making

Switzerland's relationship with the EU influenced the biotechnology debate particularly at the legal level: the revision of the patent legislation, demanded by a parliamentary motion (the Motion Auer, see Figure 1) in 1986, was delayed because Switzerland was awaiting the revision of EU environmental protection law. Compatibility with EU directives 90/220 and 90/219 became a major issue in the debate. This was also related to the vote on EC membership (which the Swiss rejected at the end of 1992). It is difficult to discern the EU's influence at the constitutional level: considerations about compatibility with EU ordinances and international developments may have played a role especially for the political actors (in particular the administration).

Media coverage

Switzerland is a country rich in newspapers. Despite the pronounced trend towards newspaper and circulation consolidation in the last few years, approximately 100 newspapers still appear daily. Yet only ten daily newspapers sell 100,000 copies. These ten newspapers account for slightly more than one-third of the total daily circulation. A typical Swiss daily newspaper has a circulation of 15,000. In the vast majority of the newspapers, only the local sections are independently written; all other sections – home news, foreign affairs, editorials, special interests – are provided by the national news agency.

For this reason the reporting of the national news agency on gene technology was regarded as representative, and was one focus of investigation.[19] This agency provides nearly all Swiss editorial staff (98%) with current news. Because the agency produces news separately in German, French and Italian for the media in the three language regions, cultural differences in the reporting of gene technology can also be directly investigated. The results of this study are shown in Tables 1 and 2.

Distribution of articles

From 1983 to 1996, 943 articles on gene technology were distributed among the three languages of German, French and Italian. A total of 606 representative articles were selected from this statistical population and included in a long-term analysis. Few articles on gene technology appeared up until 1990. Reporting intensified after 1990, and peaked in 1991. Subsequently, this rate of reporting persisted, although fewer articles on this topic were distributed. Reporting on gene technology since 1995 – especially on the Gen-Schutz Initiative – has flourished, increasing again the number of articles produced (see Figure 2).

In the German news service, months in which a greater than average number of articles were produced were ascribed to political regulatory discussions associated with the Beobachter Initiative. However, in the French and Italian services, corporate takeovers and investments in the Swiss chemical and pharmaceutical companies led to the high frequency months.

Differences between the language regions

Reporting on gene technology has different emphases in the three main languages. National data unsatisfactorily reflected these differences in the culturally differentiated reporting: there is no 'national' reporting on this topic.

Reporting by the German service was largely political. The Beobachter and Gen-Schutz Initiatives were extensively covered. The French service frequently reported on the chemical industry in the German-speaking part of Switzerland. The economic context was strongly emphasised. However, a differential examination and fundamental analysis of the cases revealed that – beneath the 'economic surface' – the political pattern constituted a significant portion of the reporting. The Beobachter Initiative was, however, rarely discussed. The Italian service paid the most attention to scientific issues. The portion dedicated to foreign events accounted for more than half of the reporting, much more than in the German or French services.

Differences among the three language regions were also discernible in the costs and risks cited in the articles. The German service described twice as many costs and risks as the French service. The benefits and uses mentioned were related to the economy and health. Their description in all three services were qualitatively and quantitatively commensurate.

Reporting on gene technology in all three services was primarily focused on Switzerland. However, the orientation towards the EU in both Latin services, but particularly in the French service, was significantly more elaborate than in the German service.

Phases of reporting

The frequency of reporting was the *a priori* determining criterion for the formation of phases. It was first established that all below-average annual values fell between 1983 and 1989, and all above-average annual values fell between 1990 and 1996. Maximum values were recorded in 1991 and 1992 due to reports on the Beobachter Initiative, and in 1995 and 1996 due to the Gen-Schutz Initiative. Minimum values were noted in the years 1983, 1985 and 1986. In these circumstances, phases could be determined. They were differentiated by the absolute changes in the number of articles from year to year, which were especially striking. However, it should be noted that a phase cannot be understood as a concluded unit. Rather, a phase is stamped with a dominant pattern of reporting; beside this, other patterns of reporting are also cultivated. The demarcation of phases is thus somewhat blurred.

Four phases were drawn from a frequency analysis and frequency changes from year to year, and from a corresponding analysis of the patterns of reporting (correlation of topics, actors and arguments) (see Table 1).

Phase 1, 1983–86: industrial, economic and political reporting. Economic issues dominated this phase. Economic arguments were used in more than half of the articles, and the actors came from industry. Medical topics were mentioned in 25% of cases. A large part of the reporting concentrated on describing pharmaceutical products and vaccines.[20] The most frequently noted benefits and uses of gene technology were economic. This underscores the dominance of economic reporting in this phase.

Phase 2, 1987–92: political reporting. The issues reported changed with the launching of the Beobachter Initiative. Debates on regulation and political discussions now took centre stage. The large proportion of medical issues was due to the Initiative, which dealt with reproductive technologies.[21]

From 1987, NGOs first became important actors.[22] Environmental organisations were first mentioned then. Reporting in this 'political' phase concentrated even more on Switzerland. The costs and risks of gene technology were mentioned more often, as were its moral and ethical implications. Risks were highlighted in direct quotes from politicians, and were seldom at issue in texts written by journalists.

Phase 3, 1992–93: political reporting. The third phase is difficult to distinguish from the second. After the vote on the Beobachter Initiative on 17 May 1992, the journalistic interest in this topic abruptly faded – the wind went out of the topic. The pattern of reporting changed as environmental organisations lost their relative importance as actors to the scientists, political parties and the government.

Phase 4, 1994–96: industrial, economic and scientific reporting. Economic topics, actors and arguments dominated this phase of reporting. Moreover, scientific arguments were now emphasised. The political discussion, however, had not faded. As well as economic advantages, the benefits and usefulness to the public's health were particularly emphasised in this phase.

The increased reporting on the Gen-Schutz Initiative may possibly be connected with a greater stress again being placed on a political pattern of reporting. This theory must await confirmation.

Neue Zürcher Zeitung: *phases of the debate*

The *Neue Zürcher Zeitung* is a high-quality and elite newspaper with a daily circulation of over 160,000. This circulation ranks it third in Switzerland; its readers are mainly urban, well educated and, above all, earning high incomes. The *Neue Zürcher Zeitung* is one of the few papers with an international and intercultural reputation. We studied its coverage by constructing artificial weeks between 1972 and 1996 with 1026 sampling points. A total of 211 articles were selected and included in the sample. The results are shown in Tables 3 and 4. Reporting in the *Neue Zürcher Zeitung* can easily be divided into a phase before 1987 and a phase after 1987.

Phase 1, from 1972 until and including 1987, was dominated by scientific and research themes. Precise and detailed articles, mainly published in the section 'Science and Technology', covered themes like basic research and medical applications. The frame of many of these articles was that modern biological research is an important step towards progress. If advantages and benefits of this new technology were mentioned, they referred mainly to health issues. Towards the end of Phase 1, political and industrial actors gained influence. The topic became increasingly a political issue at a national level.

Phase 2, from 1988 until and including 1996, was a phase of political news coverage. With the submission of the Beobachter Initiative in 1987, the focus of reporting shifted from science to politics. Regulatory themes gained priority. Medical themes mostly dealt with the issues of the Initiative, such as reproduction and childbearing. These articles dealt with debates taking place in Switzerland. Political actors appeared on the agenda as well as NGOs, which now played an important role. One quarter of all the articles in Phase 2 can be put into the frame of 'public accountability'.

It is significant that, in the mean time, scientific reporting continued. 'Progress' served roughly as a counterframe to 'public accountability'. Both the Beobachter Initiative and the forthcoming Gen-Schutz Initiative intensified the overall reporting on biotechnology. Risks were increasingly mentioned in Phase 2, although indirectly; they were primarily moral and ethical risks.

We also analysed articles in 1997 that were not included in the sample for the profile. They emphasised the importance of political events. Due to the forthcoming vote on the Gen-Schutz Initiative, reporting increased in frequency as well as in variety. Dolly, the cloned sheep, for many the event of the century, was reported, but she never found her way on to the front page of the *Neue Zürcher Zeitung*. More space was given to the issue of genetically manipulated soya – a shipload of which was stopped by Greenpeace in Basle's Rhine harbour, and the withdrawal from grocers' shelves of 'genetically manipulated' Toblerone chocolate (which is something of a national icon). Both stories were discussed in depth in the *Neue Zürcher Zeitung*, under headlines such as 'Greenpeace action against the import of genetically manipulated soya: cargo steamer in Basle seized', 'Toblerone of all things', and 'Special case chocolate'. Risks for the consumer were depicted. In the run-up to the vote on the Gen-Schutz Initiative on 7 June 1998, press coverage of biotechnology intensified.

Two issue–actor–context patterns dominated all phases: one political and the other scientific. The political pattern was associated with the two popular initiatives on the political and legal regulation of gene technology. The scientific pattern, on the other hand, was linked to basic research, medical applications, etc. The actors were mainly scientists and universities. An economic pattern, however, played only a minor role in the *Neue Zürcher Zeitung*.

Public perceptions

Methodological notes

The Swiss survey was carried out between 20 May and 10 June 1997, six months after the Eurobarometer survey. In contrast to the EU results, repercussions from the media's reporting on Dolly the sheep were felt. Because of the three language regions in Switzerland, three versions of the questionnaire had to be developed. These were kept largely identical to the Eurobarometer questionnaires from Germany, France and Italy. However, several new questions were included in the Swiss questionnaires.

The statistical population was composed of resident Swiss citizens aged 15–84 years. A sample of 1033 interviewees (605 German-speaking, 220 French-speaking and 208 Italian-speaking) was formed via a combined random-quota sampling procedure, and each was personally interviewed. In the first step, 100 communities weighted for size were randomly selected, although language region and urbanity/rurality were also controlled for. In a second step, target respondents in each community were selected based on gender, age and occupation quotas. For the analysis of the entire Swiss sample, the collected interviews were re-weighted in terms of the actual proportions of the German-, French- and Italian-speaking populations.

The interpretation of the findings both looked at the position of Switzerland compared to the EU, and investigated the hypothesis that the three language groups in Switzerland orient themselves in cultural respects towards their same-language 'next-door giant neighbours': the German speakers towards Germany, the French speakers towards France, and the Tessin residents towards Italy.

Survey results: Switzerland compared to the EU

The Dolly incident left evident marks on the Swiss citizens' agenda: while in the winter of 1996 only half of EU respondents reported having heard about gene technology in the media within the previous three months, three-quarters of Swiss respondents reported having heard something about gene technology in the media over the summer of 1997 (see Table 5). Comparatively many fewer (27%) maintained that they often or even always kept apace with biotechnology and gene technology, and had followed reports in the media. The influence of the media was pursued in an additional question: 37% felt that recent reporting had influenced their attitude towards biotechnology and gene technology; of these, 12% perceived that the media reports had influenced them in favour of and 26% against gene technology.

Biotechnology and gene technology were significantly meaningful to approximately half of all Swiss citizens; one-quarter found them of intermediate and one-fifth of rather small relevance. These values correspond exactly to the weighted averages in the EU. However, the relevance of gene technology was not homogeneous: the French- and Italian-speaking respondents allotted it a greater significance than the German-speaking respondents. It is notable that these differences were also found in the EU countries that surround Switzerland: gene technology was likewise attributed a greater relevance in Italy and France than in Germany. Apparently, the short but intensive media coverage of Dolly the sheep had no significant lasting influence on the personal relevance of gene technology.

The Swiss public were ambivalent towards gene technology: one-third felt that it would improve life and one-third that it would make life worse in the next 20 years. Only a minority (11%) perceived no consequences. The question 'What is your personal attitude towards modern biotechnology and gene technology?' (on a scale of 1, 'strongly opposed to' to 10, 'strongly in favour of'), which was only asked in Switzerland, showed an even stronger attitude of rejection: 50% were opposed to biotechnology and gene technology (1–4 points), 30% were undecided (5–6 points) and only 20% were advocates (7–10 points). The positive evaluation of gene technology by the Swiss thus fell approximately 10 percentage points below the rest of Europe. This rejection was especially strong in German-speaking Switzerland – their evaluation was practically identical to that in Germany. In parallel, the French-speaking Swiss judged gene technology significantly more positively, like the public in France. However, ratings from the canton of Tessin corresponded not to the positive ratings in Italy but to those in Germany.

When interpreting these findings, the fact that a referendum on gene technology would be taking place in Switzerland in 1998 must be taken into account. Already in 1997, media reports on issues in biotechnology and gene technology had achieved a high visibility – one with a rather negative quality following the controversies over Dolly and genetically modified soya.

The acceptance of biotechnology and gene technology was not only dependent on language region, but varied considerably among different socio-demographic segments of the population: the relationship was stronger not only in men and young people compared to women and older people, but also in respondents with higher education. People with a positive basic attitude towards technology were also more inclined to accept biotechnology and gene technology, whereas those with ecological concerns were more inclined to reject them.

The role of religiosity was unclear: whereas in the western part of Switzerland it tended to be negatively correlated with the acceptance of biotechnology and gene technology, negative correlations were found in the German-speaking part of Switzerland and in the canton of Tessin. Furthermore, a significant relationship was found between acceptance and knowledge; however, as in the EU, opponents of gene technology were found among the informed as well as among the uninformed.

Survey results: attitudes

The use of gene technology to produce medicines and vaccines, as well as genetic testing, was judged as useful and morally acceptable by a good three-quarters of the respondents. Although a certain risk was perceived, approximately two-thirds supported the application of gene technology in these areas, because of its advantages. In comparison, the usefulness of gene technology in increasing the resistance of cultivated plants or in breeding animals for laboratory experiments was claimed to be beneficial by only 50–60% of the respondents. Since approximately the same proportion judged these applications to be relatively risky and were ambivalent with respect to their moral acceptability, it is not surprising that they were also not in favour of them. Finally, a majority was clearly opposed to xeno-transplantation and genetically modified foods: these were considered too risky and morally unacceptable, and claims of their usefulness were also not found convincing; 60–80% therefore did not support these applications (see Table 6 and Figure 3).

In contrast to the Eurobarometer survey, the case of Dolly the sheep in March 1997 led to the expression of public opinion on the 'cloning of life forms, e.g. for the production of human tissues'. Opinion on this question did not fare as well as the six areas of application mentioned above: 72% of those interviewed saw no social use for cloning, 73% perceived a risk to society, and 76% deemed cloning morally unacceptable. It is thus not surprising that almost 80% of respondents believed that the cloning of life forms should preferably not or definitely not be supported.

Comparisons with the average ratings from EU countries revealed that Europeans judged the usefulness, moral acceptability and worthiness of support of biotechnological and gene technological applications more favourably than did the Swiss, whereas the risk to society was judged the same by Switzerland and the rest of Europe. A comparison revealed that judgements on the applications of biotechnology and gene technology in German-speaking Switzerland approximated those of the Germans (see Table 6). In comparison, the judgements of the French-speaking Swiss in practically all areas of application were more positive, and comparable to the of European average.

Survey results: regulation and control

The majority of the Swiss public interviewed did not trust industry, and believed that the current regulation of biotechnology and gene technology was unsatisfactory. They demanded the labelling of genetically modified foods, and a public forum. They rejected having to tolerate the risks merely to remain competitive in the marketplace. The particularly negative attitude of the German-speaking Swiss towards gene technology was also conspicuous in the area of regulation. The Swiss therefore maintained a more sceptical attitude compared to the European average: the call for control was more pronounced (see Table 7).

The regulation of biotechnology and gene technology was most readily entrusted to international organisations such as the United Nations and the World Health Organization, although Switzerland is not a member of the UN. Less confidence was placed in governmental authorities, the Federal Assembly and the EU. Consumer and environmental agencies, and also universities, were the most trusted sources of information. Information from political parties and trade unions, and also from industry and governmental authorities, met with scepticism. A comparison of the evaluations of regulation agencies and information sources revealed comparatively small differences between Switzerland and the rest of Europe.

Commentary

The course of media reporting and political discussion in Switzerland clearly indicates that the

Beobachter Initiative and the Gen-Schutz Initiative exerted a decisive influence on public discussions on biotechnology. Both initiatives originated outside of institutionalised politics and were, at least at the outset, supported by comparatively few influential organisations and individuals.

The first articles on the topic of genetic engineering appeared sporadically in the Swiss media in the 1970s and early 1980s. Most articles focused on the economic aspects and actors. Scientific topics included pharmaceuticals and vaccines. The political debate consisted of a handful of sporadic proclamations. Scientists and industry developed their own standards in this phase of self-regulation, virtually uninfluenced by politics, the public or the media.

The Beobachter Initiative

Developments in reproductive medicine occasioned the Beobachter Initiative in the mid-1980s. Initially, the Initiative was chiefly supported by Catholic and rather conservative circles. However, because a ban on germline manipulation in humans was also called for, the initiators prompted the first broad discussion of genetic engineering, which echoed through the Swiss public. As well as Catholic–conservative organisations, feminist organisations and arguments marked this phase of the debate.

Since the submission of the Initiative in 1987, reporting on biotechnology broadened both quantitatively and in terms of the variety of topics covered. From 1987, NGOs became increasingly important. Environmental organisations emerged, and the debate shifted to a political level. Pharmaceutical corporations became more deeply involved with biotechnology, and heralded it as a key technology.

In 1989, the government published a counterproposal to the Beobachter Initiative, which included a new article on the extra-human application of biotechnology. However, left-wing political parties and ecological groups regarded the counterproposal as unsatisfactory. They were also unhappy with Parliament's revision of the government's counterproposal. The debate increasingly concentrated on the release of genetically modified products, environmental dangers, the north–south problem and the patenting of life forms and transgenic animals. A distinct anti-genetic-engineering movement formed. Growing problems of public acceptance manifested themselves in this phase. In the centre of the vehement discussions in 1990–91 stood Ciba-Geigy's planned biotechnological production plant in Basle (which was eventually built just over the border in France) and, in 1991, the first experimental release of genetically modified plants in Switzerland. Since this first release, no proposals have been submitted for further field experiments.

Swiss companies conduct all such experiments in foreign countries, in some cases only a few kilometres beyond the Swiss border.

The Gen-Schutz Initiative

Following the overwhelming approval in 1992 of the counterproposal to the Beobachter Initiative, public debate noticeably abated. The anti-genetic-engineering movement had launched a new, restrictive initiative a month before the vote, which demanded the prohibition of transgenic animals, biotechnological patents and the release of genetically modified organisms. It proved difficult to collect the required 100,000 signatures for this Gen-Schutz Initiative within eighteen months. In this phase the debate visibly polarised.

The submission of the Gen-Schutz Initiative in 1993 revived the discussion. Industry and research now appeared as more potent actors (industry had already been an economic and a scientific factor in earlier reporting). Economic and scientific arguments were stressed in the media. The polarisation increased; distinctive discussion fronts formed and increasingly migrated towards a classic left–right dichotomy. Industrial associations and civil parties stood on one side of the debate, and opposite them stood the NGOs and, after the Social Democratic Party's proposals for reconciliation had been frustrated, left-wing political parties. The positions on the Gen-Schutz Initiative within the trade unions and agricultural organisations had dispersed.

Since 1994, food produced using recombinant DNA technology has increasingly become an important topic. The retail trade, consumer protection organisations, agricultural organisations and then the food industry appeared as additional actors. Since the end of 1996, when the importing of genetically modified soya beans raised tempers, questions about food have become a priority in the debate. For the first time, a foreign company entered the debate in Switzerland (the US agribusiness corporation Monsanto). While the debate up to this point had been concerned primarily with the activities of Swiss universities and companies, the debate was now international.

Because biotechnology advocates and critics were irreconcilable, Parliament's attempt in 1996 to find an acceptable compromise which could serve as a counterproposal to the Gen-Schutz Initiative was thwarted. The intensity of public debates continued to increase up to the vote on the Initiative, which took place in June 1998. Because the biotechnology industry and researchers in Switzerland believed their existence to be threatened, an unusually intense voting campaign was carried out. The resources pumped into the campaign exceeded normal proportions. The anti-gene-technology movement also had considerable resources at its disposal, and carried out a very professional campaign.

To the surprise of most political observers, the Initiative was rejected by a large majority of Swiss voters (66.6%), with a usual low voter turnout of 40.6%. Independent of the voting result, the relentless discussions, with hardened positions on either side, will presumably continue to resonate in the public's memories for a some time to come.

Cultural differences

It is not a coincidence that both the Beobachter and Gen-Schutz Initiatives were launched in the German-speaking part of Switzerland. The questionnaire on media reporting and its statistical analysis revealed significant cultural differences in the individual and societal perception and judgement of biotechnology. The French-speaking population was significantly less critical of biotechnology than the German- and Italian-speaking populations. Differences between French and German reporting were most conspicuous with respect to the Beobachter Initiative, which was barely reported in the French-speaking part of Switzerland.

The results of the vote on the counter-proposal to the Beobachter Initiative, which were more or less comparable in the language and cultural regions, seem to contradict this evidence. The opposition was heterogeneously composed of Catholic–conservative and feminist groups, who attacked the proposal as too liberal. However, liberal voices were heard, which declared the constitutional article to be too restrictive. Presumably, the opposing actors, interests and motivations in the different language regions neutralised each other and led to the similar voting results.

The political dimension

The fact that the debate concentrates on the application of biotechnology in the extra-human domain is a result of Switzerland's early and, by international standards, rather restrictive regulation of the human domain on the constitutional level, as exemplified by the prohibition of human cloning. However, the debate on applications in the human domain could experience a renaissance in 1998 when the draft of the execution provisions will be introduced on the legislative level (the Reproductive Medicine Law) and discussed in Parliament.

The Swiss public react sensibly to the application of genetic engineering in plants and particularly in animals, as in other countries. The question of how to interpret the norm, firmly established in the constitution, that the dignity of all living creatures should be preserved, is currently the focus of intense discussions, especially with respect to transgenic animals in research and in agriculture. If the Gen-Schutz Initiative is rejected, this controversy may become more heated. The discussion of recombinant DNA technology in food production has also received considerable attention, which has led to a burgeoning of this topic in political and journalistic discourse.

The debates on regulation in Switzerland originated outside institutionalised politics through public referendums such as the Beobachter and Gen-Schutz Initiatives, which at first were supported by comparatively few influential organisations and individuals. The intensity of the debate on extra-human applications can certainly be attributed in part to the fact that industry, science and the authorities held out for too long against legal regulation. In this context, the loss of trust in the political institutions and the administration in Switzerland is understandable.

Switzerland's legislation was and is devised to be compatible with Europe wherever possible. However, this does not mean that foreign authorities are more trusted. Rather, pragmatism and well-considered state and economic political interests are determining factors.

Summary

The following four points summarise the characteristics of the debate on biotechnology in Switzerland:

- With the direct democratic instrument of the referendum, groups outside institutionalised politics were able to engage with biotechnology and to influence the debate for more than a decade.

- The strongly represented and influential biotechnology lobby in Switzerland, seat of several international corporations and renowned research institutes, long viewed legal regulation with scepticism and favoured instead policies of self-regulation.

- While biotechnology applications in humans had already been regulated comparatively early, at least on a constitutional level, the regulation of extra-human applications followed relatively late.

- Distinct differences exist between the three language and culture regions with respect to their perception and valuation of biotechnology.

Acknowledgement

Funding for the Swiss project was provided by the Federal Ministry of Education and Science, the Swiss National Science Foundation and Inter-pharma.

Figure 1. A regulatory profile for Switzerland

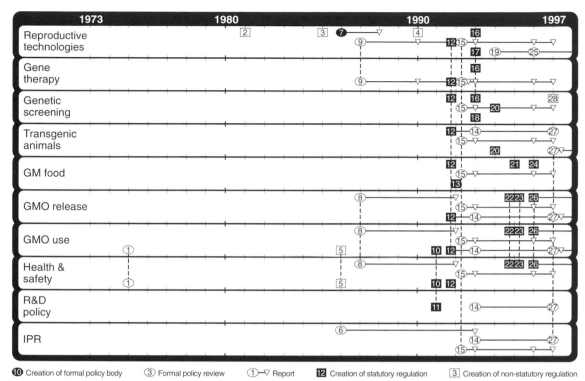

1 *1975.* NIH guidelines are adapted by the Arber Committee chaired by the Nobel-prize-winning microbiologist Prof. Werner Arber and established by the Swiss Academy for the Medical Sciences (SAMW).

2 *1981.* SAMW issues medical–ethical guidelines for artificial insemination (November 17) .

3 *1985.* SAMW issues recommendations for the use of *in-vitro* fertilisation and the transfer of embryos (23 May).

4 *1990.* SAMW issues medical–ethical guidelines for medically assisted reproduction (31 December).

5 *1986.* Swiss Interdisciplinary Committee for Bio-logical Safety (SKBS) established. It is an informal committee, successor to the Arber committee, established by the Swiss Society for Natural History, which became the Swiss Academy for the Natural Sciences (SANW), and the Swiss Academies for the Medical Sciences and the Technical Sciences. The SKBS is a self-regulated body for science and industry, in close cooperation with federal administration. It also registers use of GMOs. The guidelines are non-statutory but remain decisive for any use of GMOs.

6 *1986.* Parliamentary motion by MP Auer, a member of the Liberal Party representing the interests of the pharmaceutical industry. The motion requested an amendment of the existing regulation of Swiss Patent Law concerning biotechnology. Report on biotechn-ology and intellectual property rights, commissioned by the Federal Department for Justice, published in 1993.

7 *1986.* Federal Council commissioned a federal expert

committee on human genetics and reproductive medicine in September as a reaction to the Beobachter Initiative launched in 1985 (see note 9). This 'Amstad Committee', chaired by federal judge Eduard Amstad, submitted its report in August 1988 for the attention of the Federal Department of Internal Affairs and the Federal Department of Justice.

8 *1987.* Interdepartmental committee/body for the coordination of the procedures of approval for the use of rDNA organisms (KOBAGO) established by the Federal Council. In 1992 the report *Gene Technology Today and Future Perspectives* (on the implementa-tion of the constitutional amendment Art. 24novies) is submitted by KOBAGO for the attention of the Federal Council.

9 *1987.* The Beobachter Initiative was submitted. In 1985 the editor of the consumer magazine *Der Schweizerische Beobachter* launched a constitutional initiative 'against the misuse of reproductive medicine and genetic technology in humans'. In 1989 there was a counterproposal to the Beobachter Initiative by the Federal Council following the Amstad committee's report. In 1990 there was a second counterproposal to the Initiative by the State Council as reaction by parliament to the Federal Council's counterproposal which did not gain approval. Parliament's counterpro-posal (the amendment of the constitution by article 24novies 1-3) was finally put to the vote by the Swiss electorate in the referendum of 17 May 1993.

10 *1991.* A regulation in the case of incidents or hazards

in industrial plants and installations, including biotechnology explicitly (*Störfallverordnung*), came into force in reaction to the incident in *Schweizerhalle* (1986) and to the EU guideline 2020 (1990) and the forthcoming referendum on EEC membership (1992).

11 *1991.* Federal ordinance to establish so-called priority programmes in research, one of which was 'Priority Programme Biotechnology' starting in 1992 and funded by the Swiss National Science Foundation.

12 *1992.* Amendment of the Federal Constitution by article 24novies 1-3. This constitutional article regulates the use of gene technology mainly in the human domain. It set the framework for regulation on all levels and was accepted by the majority of the Swiss electorate (74%) in a poll (17 May). This referendum and consequently the new constitutional article were the final outcome of the Beobachter Initiative and its two counterproposals.

13 *1992.* New law regulating foodstuffs: article 9 dealt with GMOs (19 October).

14 *1993.* A constitutional initiative on the protection of living beings and the environment from genetic manipulation submitted on 25 October, later called the Gen-Schutz Initiative. It was launched by the Swiss Working Group on Gene Technology (SAG) in April 1992, shortly before the vote on the constitutional amendment 24novies (17 May). The initiative requested the banning of production, release and patenting of GMOs (animals and plants): it thus focuses on the extra-human domain.

15 *1993.* Report by the Interdepartmental Working Group on Gene Technology (IDAGEN), commissioned by the Federal Council in 1992 to evaluate the legislative process in biotechnology and to propose an adequate implementation strategy for article 24novies. Nine legislative projects were proposed. May 1996: Schweizer report on the proceedings of the implementation of the nine projects proposed in the IDAGEN report for the attention of the National Council's Committee for Science, Education and Culture.

16 *1993.* Rules for the registration of experiments on humans were issued by the Intercantonal Body for the Control and Registration of Medicines. These come into force on 1 January 1995.

17 *1993.* Federal ordinance on *in-vitro* diagnostics (use of GMOs) on 24 February.

18 *1993.* Medical–ethical guidelines for genetic screening on humans issued by SAMW.

19 *1994.* Constitutional initiative for humane reproduction, for the protection of humans from manipulative reproductive medicine, is submitted.

20 *1994.* Ethical principles and guidelines for scientific experiments on animals issued by the Swiss Academies for Medical Technologies and for Natural Sciences: non-statutory but decisive.

21 *1995.* Revision of the Federal Ordinance on Foodstuffs regulating registration by the Federal Office for Health and establishing the obligation for declaration for GMOs in food (1 March).

22 *1995.* Revision of the law for environmental protection to incorporate obligatory registration of and approval for the use of GMOs (in force, 1997).

23 *1995.* Revision of the law for epidemics.

24 *1996.* Federal Ordinance for the Procedure of Approval for Genetically Modified Food (in force, 1 December 1996).

25 *June 1996.* Federal bill for the Law for Human Reproduction Medicine: indirect counterproposal to the popular initiative against misuse of reproduction medicine (see note 19).

26 *November 1996.* Federal ordinance for the establishment of an official body for biological safety. This expert committee for biological safety replaces the SKBS. In force since January 1997.

27 *1997.* Gen-Lex Motion launched on 4 March as a reaction to and a consequence of the Gen-Schutz Initiative which was initially opposed by the Federal Council in 1995 without a counterproposal. In parliament too, both, State and National Council, opposed the initiative without a counterproposal. Instead, both chambers submitted the so-called Gen-Lex Motion which charges the Federal Council with accelerating the legislative process concerning biotechnology in the extra-human domain and closing the considerable gaps in the law especially concerning the 'dignity of all creatures' (Art. 24novies paragraph 3) and biodiversity. June 1997: Second Schweizer report submitted in reaction to the Gen-Lex Motion. The report was commissioned by the Federal Council and evaluated the status quo and the proceedings of implementation of the nine legislative projects of the IDAGEN report (see note 15). It identified large gaps in the regulation of biotechnology and made further proposals. December 1997: Proposals of the Federal Council on the regulation process in biotechnology in the extra-human domain – the so-called Gen-Lex Bill, answering the Gen-Lex Motion. The Bill was sent for a process of consultation which ended in March 1998.

28 *1997.* Non-statutory guidelines for the use of genetic screening for diagnostics/predictive medicine issued by SAMW.

Figure 2. Intensity of articles on biotechnology in reporting by the Swiss newsagency SDA

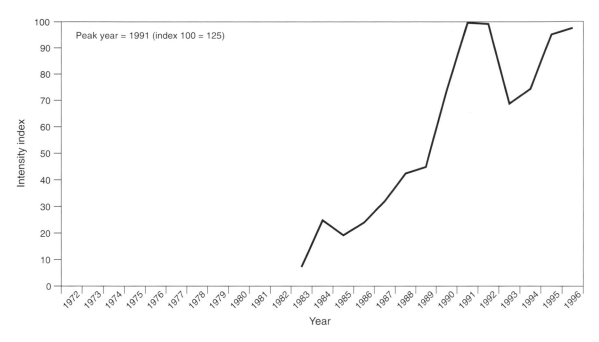

Figure 3. Attitudes to applications of biotechnology in Switzerland, 1997

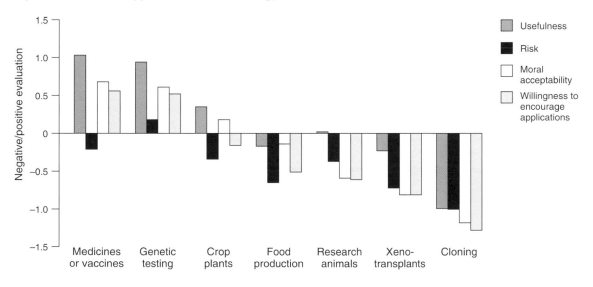

Table 1. Swiss media profile of reporting by the national news agency (for an explanation of terminology please see Appendix 5)

Phase	Freq.[a] (%)	Frame (%)		Theme (%)		Actor (%)		Benefit/risk (%)		Location (%)	
1. 1983–86	10	*Economic*[b,c]	*52*	**Economic**	32	**Industry**	58	**Neither**	67	**Switzerland**	68
		Political	23	Medical	25	Political	21	Benefit only	13	Belgium	10
		Scientific	8	Political	15	Scientific	13	Risk only	12	Japan	5
				Basic research	12			Both	8		
2. 1987–92	37	**Political**	**40**	**Regulations**	20	**Political**	32	**Neither**	45	*Switzerland*	80
		Economic	21	Medical	17	Industry	30	*Risk only*	30	USA	7
		Scientific	17	Political	16	NGOs	18	Both	14		
		Ethical	6	Economic	16	Scientific	13	Benefit only	11		
				Security & risk	13						
3. 1992–93	21	**Political**	28	*Regulations*	23	**Political**	30	**Neither**	63	**Switzerland**	77
		Scientific	26	Medical	19	Industry	27	Benefit only	17	USA	7
		Economic	20	Political	16	*Scientific*	20	Risk only	13	Belgium	4
		Ethical	6	Basic research	14	NGOs	16	Both	7	*Italy*	4
				Security & risk	10						
4. 1994–96	32	**Economic**	35	**Economic**	24	*Industry*	43	**Neither**	58	**Switzerland**	66
		Scientific	26	Medical	19	Political	20	*Benefit only*	22	USA	16
		Political	18	Basic research	13	Scientific	19	Risk only	11	Italy	3
				Political	12	NGOs	11	Both	9		
				Regulations	11						

a Percentage of corpus in the period; total $n = 606$.   b Bold indicates highest frequency within phase.
c Italics indicates highest frequency within category.

Table 2. Type of benefit and risk evaluations

	1983–86 (%)		1987–92 (%)		1992–93 (%)		1994–96 (%)	
Benefit only	Economic	15	Health	12	Health	9	Health	15
	Research	3	Economic	5	Economic	6	Economic	12
Risk only	Ecological	5	Moral	10	Moral	8	Health	5
			Ecological	9			Moral	4
			Health	6				

Table 3. Swiss media profile of reporting on biotechnology by Neue Zürcher Zeitung (for an explanation of terminology please see Appendix 5)

Phase	Freq.[a] (%)	Frame (%)		Theme (%)		Actor (%)		Benefit/risk (%)		Location (%)	
1. 1972–87	27	*Progress*[b,c]	49	*Medical*	38	*Scientific*	68	**Neither**	53	**Switzerland**	27
		Accountability	9	*Basic research*	37	Industry	13	*Benefit only*	30	Germany	11
		Ethical	7	*Animal*	7	Political	9	Both	10	USA	11
								Risk only	7	UK	10
2. 1988–96	73	*Accountability*	25	**Medical**	23	**Scientific**	33	**Neither**	37	*Switzerland*	56
		Progress	22	Regulations	19	Political	26	Benefit only	26	USA	13
		Ethical	10	Basic research	18	Industry	17	*Risk only*	23	Germany	6
		Economic	10	Security & risk	8	NGOs	13	Both	14		
		Globalisation	7	Political	7						

a Percentage of corpus in the period; total $n = 211$.   b Bold indicates highest frequency within phase.
c Italics indicates highest frequency within category.

Table 4. Type of benefit and risk evaluations

	1972–87 (%)		1988–96 (%)	
Benefit only	Health	8	Health	17
	Economic	7	Economic	9
			Research	4
Risk only	Moral	9	Moral	15
			Consumer	6
			Ecological	4

Table 5. Relevance of and attitude towards biotechnology

		Switzerland				Europe			
		German	French	Italian	Total	Total	Germany	France	Italy
Heard about bio-	yes	79	72	82	78	53	60	54	51
technology in media	no	20	27	18	21	47	40	46	49
Personal	important	48	64	71	53	54	47	54	57
relevance of	medium important	27	24	18	26	28	28	31	29
biotechnology	unimportant	25	12	11	21	18	25	14	14
Attitude towards	will improve	35	45	37	37	47	36	49	57
biotechnology	no effect	13	8	2	11	10	18	11	5
	make things worse	36	16	32	32	19	23	15	15

Table 6. Attitudes to applications of biotechnology (percentages of 'definitely agree' answers)

Application		*Useful*	*Risky*	*Morally acceptable*	*Should be encouraged*
Medicines or vaccines	Switzerland	41	18	26	25
	Germany	37	13	22	25
	EU[a]	**49**[b]	15	**31**	**34**
Genetic testing	Switzerland	40	14	27	27
	Germany	35	11	21	24
	EU	**53**	13	**34**	**39**
Production of foods	Switzerland	14	27	11	10
	Germany	18	25	11	12
	EU	**20**	25	**16**	**15**
Increase resistance of	Switzerland	23	27	15	13
crop plants to insects	Germany	24	25	16	16
	EU	**34**	25	**23**	**23**
Development of genetically	Switzerland	20	26	9	11
modified animals for	Germany	16	19	7	8
laboratory research	EU	**27**	23	**13**	**16**
Production of organs for	Switzerland	16	**36**	7	7
human transplants	Germany	16	25	8	9
	EU	**24**	30	**12**	**14**
Clone living creatures	Switzerland	7	51	4	5

a EU percentages are weighted by size of countries.
b Bold indicates highest frequency within attitude.

Table 7. Attitudes to benefits, risks and regulation of biotechnology

	Switzerland (%)		Europe (%)	
	Agree	Disagree	Agree	Disagree
It is not worth putting special labels on genetically modified foods	19	**78**[a]	18	**74**
The regulation of biotechnology should be left mainly to industry	15	**73**	20	**62**
Biotechnology is so complex that public consultation is a waste of time	19	**72**	29	**60**
I would buy genetically modified fruits if they tasted better	23	**67**	29	**56**
We have to accept some degree of risk, if it enhances competitiveness	24	**67**	28	**56**
Current regulations are sufficient to protect people from any risks	18	**66**	23	**53**
Biotechnologists will do whatever they like irrespective of regulations	**65**	25	**56**	31
Only traditional breeding methods should be used	**62**	26	**56**	29
Religious organisations need to have their say to on regulation	32	**58**	39	**47**
Traditional breeding methods can be as effective as biotechnology	**41**	34	**43**	30

a   Bold indicates highest frequency per country per statement.

Notes and references

1   Federal Office for Economic Policy, *Technology Policies of the Federation* (Bern, 1992).

2   SHIV Schweizerischer Handels- und Industrieverein/Vorort in cooperation with Bundesamt für Statistik BfS, *Research and Development in Swiss Private Industry 1996* (Zurich, 1998).

3   The Swiss National Science Foundation is the most important state institution for the advancement of research in Switzerland. It was founded in Berne in 1952 as a civil law foundation. The 53 Foundation Councillors come from universities, federal and cantonal authorities, and cultural and economic institutions. Seventy-five per cent of the funds are distributed among accomplished researchers. The Foundation also supports the next generation of scientists. Twenty per cent of the funds go to the arts and social sciences, 40% to mathematics and the natural and engineering sciences, and 40% to biology and medicine.

4   Binet, O, 'Gentechnologie in der Schweiz. Eine politischökonomische Analyse' in *WWZ-Beiträge*, Vol. 26 (Basle: Wirtschaftswissenschaftliches Zentrum der Universität Basel, 1996), pp31ff.

5   Binet, O, p32. A study from Interpharma, which represents the interests of the big Swiss pharmaceutical firms that do research, on the position and perspective of genetic technology in Switzerland (an economic analysis based on company data) from August 1996, estimated that 6.5% of the profit of pharmaceutical companies in Basle came from biotechnology. This study predicts a rise to 38% in the year 2005.

6   Switzerland's leading position in international scientific research was confirmed by the Citation Index in the February 1997 issue of *Science*.

7   Roche: Genentech (1990); Sandoz: Systemix (1991), Genetic Therapy (1995); Ciba: Chiron (1994).

8   Binet, O, p24.

9   The pharmaceutical company Ares-Serono in Geneva, for example, is especially successful. It has brought several recombinant medicines to the market within recent years.

10   Kobach, K W, *The Referendum: Direct Democracy in Switzerland* (Bern, 1993).

11   Since the beginnings of the federal state in 1848, a total of 121 initiatives have been submitted. Of these only 12 have been successful, plus six counterproposals.

12   The amendment includes three paragraphs: **1** Human beings and the environment are protected against misuse of biotechnology and reproductive medicine. **2** Federal government declares the protection of human dignity, personality, and the family when regulating the use of human genetic material. **3** Federal government is entitled to regulate the use of genetic material of animals, plants and other organisms with the obligation not only to acknowledge the dignity of creation and the security of human beings, animals, and the environment, but also to protect the genetic diversity of animals and plants.

13   *Der Schweizerische Beobachter* is the Swiss magazine with the largest circulation and readership. It appears twice weekly. An important section for readers' letters functions as their 'wailing wall'.

14   Binet, O (1996); Hieber, P, 'Zukunftstech-nologie mit Vergangenheit', in Busset, T, Rosenbusch, A and Simon, C (eds), *Chemie in der Schweiz: zur Geschichte der Forschung und der Industrie* (Basle: Christoph Merian Verlag, 1997), pp255–80.

15   The executive committee consists of 1400 single members, 20 collective members and the following 13 organisations: Ärzte und Ärztinnen für Umweltschutz, Basler Appell gegen Gentechnologie, EcoSolidar, Erklärung von Bern, Greenpeace Schweiz, Schweizerischer Bund für Naturschutz, Stiftung für Konsumentenschutz, Schweizer Tierschutz, Swissaid,

Vereinigung schweizerischer biologischer Landbau-Organisationen, Vereinigung zum Schutz der kleinen und mittleren Bauern, WWF Schweiz, and Zürcher Tierschutz.

16 VOX Analysis, Part 2, August 1992. GfS-Forschungsinstitut, Schweizerische Gesellschaft für Sozialforschung, Zurich, Switzerland.

17 Buchmann, M, 'The impact of resistance to biotechnology in Switzerland: a sociological view of the recent referendum', in Bauer, M (ed.), *Resistance and New Technologies: Nuclear Power, Information Technology, and Biotechnology* (Cambridge: Cambridge University Press, 1994).

18 Once the expert committees and government department are satisfied with their draft proposal for a law, it is unveiled before the public, and the formal process of consultation (*Vernehmlassungsverfahren*) begins. The department solicits written reactions from groups not included in the expert committee. Also, a short statement is published in the major newspapers asking for comments and suggestions. Thus, participation in the consultation process is unrestricted. The groups taking part in the pre-parliamentary process fall roughly into three categories: special interests or citizens specifically affected; relevant expert groups or individuals; groups or individuals that are considered so important that

they are often invited to participate in hearings on matters in which they have no direct stake and no particular expertise. Such participants have effectively come to possess an institutional mandate to take part in virtually all deliberations of prime political importance. Among them are the *Vorort* (the Swiss Organisation for Trade and Industry), the *Gewerbeverband* (trade organisation), and the *Gewerkschaftsbund* (united unions). The favoured status of these groups further widens the power gap between insider and outsider groups.

19 News reports have been available from an electronic database since 1983.

20 Pharmaceutical products and vaccines are also coded in the superordinate category 'medical issues'. In the Swiss media, pharmaceutical products and vaccines are mainly reported in an economic, rather than medical–scientific, context.

21 Reproductive and childbearing issues (e.g. *in-vitro* fertilisation) are coded in the superordinate category 'medical issues'. However, these issues are most often presented in a political-regulatory context in Swiss media reporting, especially when associated with the Beobachter Initiative.

22 NGO is a superordinate category which includes, besides environmental organisations, other actors such as trade unions, the Church and consumer groups.

Address for correspondence

Prof. Heinz Bonfadelli, IPMZ - Institut für Publizistikwissenschaft und Medienforschung, Universität Zürich, P.O. Box 507, CH-8035 Zürich. Tel. +411 6344664, fax: +411 6344934. E-mail bonfadel@ipmz.unizh.ch, Website ipmz.unizh.ch

United Kingdom

Martin W Bauer, John Durant, George Gaskell, Miltos Liakopoulos and Eleanor Bridgman

Introduction

The development of biotechnology in the UK can only be understood in the wider political and economic context of the country. The UK is a parliamentary democracy and a constitutional monarchy. There are two chambers of Parliament: the House of Commons, comprising over 600 members elected to represent individual constituencies in a 'first past the post' voting system; and the House of Lords, comprising an even larger number of unelected members including hereditary peers, life peers and bishops of the Church of England. Governments are elected for five-year terms of office, but elections may be called at any time. The UK acceded to the EU in 1973.

There are three principal political parties: the Labour Party (traditionally left of centre), the Liberal Democratic Party (traditionally centre), and the Conservative Party (traditionally right of centre). There are also smaller political parties representing the nationalist causes in Scotland and Wales, and both the Republican and Unionist causes in Northern Ireland. In general, however, the UK voting system makes it difficult for other single-issue parties (e.g. the Green Party) to gain a foothold in national politics. Labour held power between 1974 and 1979, the Conservatives then held power from May 1979 to May 1997, and Labour regained power under Prime Minister Tony Blair from May 1997.

The UK has a relatively strong science base, particularly in the field of molecular genetics. In postwar Cambridge, for example, first the Cavendish Laboratory and then the Medical Research Council Laboratory of Molecular Biology (LMB) became internationally significant foci of research on the molecular structure of nucleic acids and proteins. The great impetus given to the new discipline of molecular biology by Crick and Watson's discovery of the structure of DNA in 1953 persisted through the late 1950s and 1960s, with important work on the nature of the genetic code and the structure and synthesis of proteins. This was maintained in the 1970s and 1980s with the first flowering of modern biotechnology. In 1975, for example, Milstein and Kohler developed monoclonal antibody technology at the LMB in Cambridge; and in 1985, Jeffreys developed DNA fingerprinting technology at the University of Leicester.

In many ways, the UK was relatively well placed to exploit the economic potential of molecular genetics in the 1970s and 1980s. As we have seen, there was a large amount of 'home-grown' scientific and technological expertise. In addition, the UK had considerable industrial strength in the fields of agrochemicals and pharmaceuticals. This having been said, the general performance of UK industry in the 1970s and early 1980s gave successive governments cause for concern. In the case of biotechnology, there was anxiety about the extent to which industry would capitalise upon new discoveries in the development of successful new products. After 1979, therefore, the Conservative administration under Margaret Thatcher sought to create a supportive environment for industrial innovation based upon new biotechnologies.

Through the 1980s and 1990s, large UK companies such as Glaxo-Wellcome, Zeneca and Smithkline Beecham began to invest seriously in biotechnology. At the same time, an increasing number of small venture-capital companies were created, often involving academic–industry business partnerships. It is clear that the American example was influential here, particularly in the early 1980s, when considerable publicity surrounded the fact that many prominent US molecular biologists were becoming involved in the creation of start-up biotechnology companies. By the mid 1990s, the UK had more large, medium and small biotechnology companies than any other European country, and UK bioscience industry was the second largest in the world after the USA, with an estimated market value in 1995 of £310 million.[1]

Public policy

Overview

The UK government has been engaged in policy-making for biotechnology since the very beginnings of the field in the early 1970s.[2] The twin aims of successive UK governments in this area have been to promote the growth of the sector as a whole in the interests of the UK economy and to regulate flexibly particular areas of practice in response to actual or potential public concerns. In this, as in other areas of science and technology policy, the UK has acted as a member state of the EU that nonetheless has particular affinities with and sensitivities towards North America. From the outset, UK biotechnology policy has been influenced by events in the USA (the dominant player in the field of biotechnology on the world stage), but increasingly, the UK response to biotechnology has been shaped in the European context. In the late 1980s, the UK had a disproportionately great influence on the first phase of policy-making in the EU.[3] Thereafter, the UK approach to policy-making in this area may be summed up as an attempt to maintain a supportive US-style climate for the development of biotechnology within a more highly constrained European context.

In contrast to many other European countries, UK science and technology policy-making generally favours a pragmatic, case-by-case approach. The tendency is to avoid detailed regulation in the first instance and to opt instead for flexible arrangements – often involving voluntary codes of practice in preference to formal statutes – that are capable of responding rapidly to subsequent developments. Biotechnology policy initiatives in the UK over the past 25 years have largely conformed to this style. In general, the principal actors have cooperated in the introduction of a series of more-or-less ad hoc regulatory mechanisms operating through a complex network of departmental and interdepartmental advisory committees, agencies and commissions.

The UK government was one of the first to introduce regulations on controlling biotechnology and began with a voluntary approach with the establishment of the Genetic Manipulation Advisory Group (GMAG) in 1976. The first statutory requirements for the use of GMOs were introduced in the 1978 GM regulations (under the Health and Safety at Work Act 1974). The UK has never considered the possibility of introducing a cross-sectoral 'gene law', as is found for example in Germany, and has instead opted for a range of sector-specific policy-making. Biotechnology has only occasionally been the subject of explicit attention in the UK Parliament, and it has never become a party political matter of any great consequence.

UK biotechnology policy has gone through four principal phases. Phase 1 (*c.* 1973–80) was dominated by the concern to ensure the health and safety of laboratory workers involved in recombinant DNA technology; phase 2 (*c.* 1981–89) was dominated by the concern to support the economic potential of the new biotechnologies; phase 3 (*c.* 1990–96) was dominated by the increasing importance of policy-making at the European level and the need to deal effectively with a growing number of sector-specific applications; and so far phase 4 (*c.* 1996 onwards) has been dominated by consumer concerns about new products, particularly in the food sector.

Phase 1 (c. 1973–80): coping with risks to laboratory workers. The key event in the first phase of UK biotechnology policy-making was the publication of the 'Berg Letter' which provoked a wide-ranging debate internationally about the potential hazards associated with recombinant DNA (rDNA) technology.[4] The UK participated in the voluntary moratorium on rDNA experiments in 1974–75, and in 1976 (the year that the US National Institutes of Health published its safety guidelines for rDNA experiments) it established GMAG to oversee practice in this area. In 1978, UK guidelines for rDNA working practices were published, introducing the first statutory requirement for use of GMOs. From 1974 to 1976, the USA may be said to have set the policy agenda internationally; but after 1976, the UK became an agenda-setter in Europe for policy-making in relation to rDNA technology.

Phase 2 (c. 1981–89): maximising economic benefits. The second phase of UK biotechnology policy-making was heralded by the publication of a UK Government White Paper on biotechnology in 1981. This White Paper was a response to the Spinks Report in 1980 which highlighted the potential of the strong UK research base while recognising that the UK was in danger of losing its leading edge by failing adequately to support biotechnology R&D.[5] The report called for funding and coordination across government departments, and identified the need to promote technology transfer. The White Paper looked to British industry to provide the necessary resources to develop the technology. At this time, considerable attention was given in the UK to the success with which new dedicated biotechnology firms were being established in the US, often through partnerships between academia and industry. Spinks recommended establishing a government-supported company to exploit publicly funded research. This was realised in 1980 when the National Enterprise Board provided 44% financial backing to establish Celltech.

The UK Government Department of Trade and Industry (DTI) established a biotechnology unit in 1981 (later renamed the Biotechnology Directorate), with the aim of identifying and initiating programmes of research between academia and industry. Between 1982 and 1985 the DTI allocated £16 million to such programmes. Subsequent DTI schemes such as 'Biotechnology Means Business' helped in securing the largest biotechnology industry in Europe. By 1989 the USA and the UK had identified the need to regulate environmental release in order both to address growing environmental concerns and to facilitate commercialisation of gene technology. With the publication of parallel reports, a split emerged in the treatment of biotechnology across the Atlantic.

The National Research Council Report (1989) in the USA, opted for a product-based approach to regulation. By contrast, a report by the UK Royal Commission on Environmental Pollution (1981) suggested that the 'unpredictability of nature' justified process-based regulation.[6] This view resulted in horizontal process-based legislation to regulate biotechnology. Later, this distinction between product- and process-based regulation would result in conflict over the import into Europe and labelling of genetically modified soyabeans.

Phase 3 (c. 1990–96): towards a mature regulatory environment. In 1990, several events signalled the start of the third phase of UK biotechnology policy-making: the passage of EC directives 90/219 (Contained Use) and 90/220 (Deliberate Release) marked the increased importance of the European Commission in this area of policy-making; the passage of the Environmental Protection Act (1990) and the 1992 regulations on deliberate release to the environment represented the UK response to the implementation of Directive 90/220, the approval of three genetically engineered chymosins for use in 'vegetarian' cheese initiated the active phase in the UK debate about new food biotechnologies; and the passage of the Human Fertilisation and Embryology Act terminated the active phase of the UK debate about new reproductive technologies.

In 1993, the House of Lords published a report on regulation of the UK biotechnology industry and global competitiveness. Suggesting that earlier fears about the safety of genetic technologies had proved to be unfounded, the report recommended a more relaxed approach to regulation. The White Paper written in response to this report endorsed this approach, and emphasised that the UK was at the forefront of efforts to persuade the European Commission to adopt a product-based approach to regulation. At around this time, the policy community gave increasing attention to a variety of sector-specific issues. In 1991, the launch of the Nuffield Council for Bioethics signalled the increasing importance of ethical issues, particularly in the context of new medical biotechnologies. Since 1991, a plethora of new advisory committees to government has been created to deal with ethical and social issues in particular areas of medical biotechnology (e.g. gene therapy, xenotransplantation, genetic testing).

Phase 4 (c. 1996–present); consumer concerns. The fourth phase of the debate was marked by the introduction into the European market of Monsanto's GM 'Roundup Ready' soyabean. Interestingly, by 1996 European food companies had already successfully introduced several GM food products (so-called 'vegetarian' cheese and GM tomato paste) into the UK market. In the case of 'Roundup Ready' soyabeans, however, the issues of segregation and labelling quickly became contentious. Public debate of these issues in the UK was conducted in the context of the long-running BSE controversy, which had served to sensitise all the key stakeholders to the importance of maintaining public confidence in new food processes and products. Hard on the heels of GM soyabean came the announcement of the birth of Dolly the sheep. Once again, a specific application created a huge amount of public interest and generated calls for new sector-specific regulation. By the time of writing (July 1998), the debate about the growing of GM crops had become more heated than at any other time since 1973.

Public engagement with policy-making

In dealing with the question of public engagement with biotechnology policy-making in the UK, we should recall at the outset Jasanoff's judgment that in general the British style of policy making tends to be 'informal, cooperative, and closed to all but a select inner circle of participants'.[7] Biotechnology policy-making in the UK has generally conformed to this description. Having said this, it should be noted that the UK approach to policy-making does allow for particular kinds of public engagement. Obviously, the work of Parliament itself is open to public scrutiny but, in addition, advisory committees to government (which constitute key parts of the machinery of UK biotechnology policy-making) are able to respond to wider public interests and concerns both through their membership (which may include not only professional experts but also various representatives from different interest groups) and through their engagement in various forms of direct public consultation (e.g. calling for evidence on key issues, or publishing draft policy documents for comment).

Evidence of responsiveness to wider public concerns may be found in the memberships of some of the key biotechnology advisory committees. From the outset, GMAG was established with both expert (scientific, industrial, etc.) and non-expert (trade union and public interest) representation. Similarly, the Advisory Committee on Releases to the Environment (ACRE) was established with a broad range of experts including an 'environmentalist' member. In the 1970s, trade unions (then more politically powerful than they were to become after the election of Margaret Thatcher in 1979) were well represented in the policy process. After experiencing early difficulties with the introduction of approved novel food biotechnologies, the Advisory Committee on Novel Foods and Processes (ACNFP) chose to strengthen its membership with a consumer representative and to engage more closely with the public in the course of its work on novel food biotechnologies, for example by holding press conferences in association with its regular meetings and by placing minutes of its meetings in the public domain. Throughout their work, the biotechnology advisory committees have consulted publicly through the issue of consultative documents, calls for evidence, etc. However, none has chosen to meet in public, and none has yet become actively involved in more radical forms of public engagement such as citizens' juries or consensus conferences.

Non-governmental organisations (NGOs) have played an important part in public debate about biotechnology in the UK. In the late 1980s, environmental organisations (e.g. Friends of the Earth, Greenpeace, and the Green Alliance) took greater interest in biotechnology. While Greenpeace chose to campaign principally at the European level on the issue of patenting, the Green Alliance was influential in securing a provision for public access to information on GMO releases through a system of public registers in the 1990 Environmental Protection Act (EPA). While Britain has never had a unified alliance against biotechnology comparable with, for example, the Green movement in Germany or the referendum movement in Switzerland, biotechnological issues have served to bring certain NGOs together in less formal groupings. For example, in 1994 Greenpeace decided to fund the Genetics Forum in order to monitor the public register of GMO releases. Since 1991 the Green Alliance has organised regular meetings between stakeholders concerned with biotechnology issues, bringing together such groups as Greenpeace, WWF-UK, the Genetics Forum, the National Council of Women and regulators from the Department of the Environment in an attempt to coordinate key areas of concern. In the 1990s, there have been signs of the emergence of 'unholy alliances' amongst NGOs on both sides of the debate. For example, radical environmental organisations have found common cause with conservative 'pro-life' campaigners in opposing some forms of medical biotechnology, while patients' welfare organisations have joined with some sections of industry and academia in supporting research and novel treatments relating to human genetic disorders.

In the 1970s, there was considerable interest in public participation in science and technology policy-making. This declined sharply in the 1980s, under a Conservative administration that was more interested in the public as consumers than as citizens. More recently, however, the policy community has become more interested in public consultation. In 1994, the Science Museum, London, organised the first UK National Consensus Conference on Plant Biotechnology with funding from the Biotechnology and Biological Sciences Research Council.[8] In the spring of 1997, the Wellcome Trust organised a 'People's Parliament' on ethical issues in the new genetics; and in the autumn of 1997, a Citizens' Jury on the ethical and social issues associated with genetic testing in medicine was organised in Wales. Also in 1997, the UK Minister

for Science announced that he intended to hold a public consultation exercise on the wider (including ethical) issues raised by recent advances in the biosciences. In the context of a new Labour administration which appears more committed to public consultation than its predecessor, it is still far from clear how far the UK may yet decide to go in the direction of a more 'participatory' style of policy-making for biotechnology.

Key issues

The evolution of the policy debate in the UK is summarised in the UK policy profile (Figure 1). A series of key issue areas have dominated UK biotechnology policy-making in turn over the past quarter of a century. In rough order of their appearance in the policy debate, these are: health and safety at work; industrial and economic competitiveness; environmental protection; consumer protection; and ethics. In part, the successive emergence of these issue areas reflects the growth of biotechnology itself – starting in the 1970s with novel and largely untested techniques of genetic manipulation, progressing to early industrial applications around 1980, and finally starting to make a noticeable impact on consumers (particularly in the agricultural and medical sectors) in recent years; but in part, too, this trajectory reflects wider social and political pressures on the policy process. For example, the intense debate about new reproductive technologies that took place in the late 1980s was driven more by public and political concern about the ethics of experimenting on human embryos than by technical or commercial interests,[9] and the increasing attention paid to the regulation of intentional releases of genetically modified organisms reflected the growing public and political significance of environmental issues in the 1980s.

Biotechnology in the UK media

The aim of the UK media analysis is to chart the media discourse on biotechnology over the period 1973–96. The focus has been on the opinion leading elite press represented by *The Times* (1973–87) and the *Independent* (1988–96). A random sample of 539 articles was selected and analysed (see Appendix 4 for methodological details). *The Times* and *Independent* serve a general audience with higher education and social standing, the latter with a slightly younger readership than other quality papers. Some 58% of Britons read a daily newspaper regularly. British newspapers are segmented into quality and popular newspapers: 17% read a quality daily and 18% read a quality Sunday paper. The *Independent*, since 1988 the newspaper of record of the British Library, has a circulation of 250,000 and a readership of 870,000, or 2% of the population (National Readership Survey, 1997). As biotechnology is studied here as an emerging issue and newspapers are still the opinion leading source for radio and television programmes, we consider the elite press in aggregate as a valid proxy for the British media as a whole.

A total of 5471 relevant articles on biotechnology and genetics were identified over 24 years. Comparing the total coverage, the average reportage over 24 years and the year by year coverage, the British cultivation of biotechnology is consistently well above the EU average of comparable media outlets such as *Le Monde* in France, the *Frankfurter Allgemeine Zeitung* in Germany, or *Corriere della Sera* in Italy (see Appendix 6, Table 2). The significance of this high-intensity coverage is discussed below. The relative number of articles in *The Times/Independent* is shown in Figure 2. Coverage in *The Times/Independent* had three cycles: 1973–79, with a peak in 1978; 1980–86, with peak years 1983–85; and a continuous rise from 1987 to the present. It is unlikely that the upward trend had reached its peak in Britain by 1996 (878 articles); indeed our continued count of articles rose to just below 1000 articles in 1997.

Phase structure of media coverage

Taking into account the changing intensity of coverage and the discursive frames we distinguish five phases of press coverage of biotechnology: 1973–79, 1980–83, 1984–87, 1988–92, and 1993 onwards. Table 1 shows the phase structure characterised by the distribution of five content variables.

Phase 1 (1973–79): the dawn of a new technology. In the first phase, the intensity of media coverage was low (on average, less than one article per week). The coverage was dominated by the discourse of progress (69%), and it was mainly concerned with medical themes (42%) and basic research. Around half of all stories are utilitarian celebrations of the benefits of biotechnology. At the same time, basic research was given considerable prominence, especially in short reports of current work in *Nature*. Scientific researchers clearly dominated the action, which was mainly confined to the UK and the USA. Other countries were seldom mentioned.

The spirit of phase 1 was that of the cultivation of a new technology by scientists in the UK and the USA. The risks of rDNA research were reported and discussed in the aftermath of Asilomar (1975), but the promise of progress, particularly on the medical front, was uncontested. The only other frame in the period is nature/nurture. This is because there was much debate at this time about the inheritance of intelligence. The themes of

laboratory security and health and safety at work echo similar debates surrounding Asilomar in the USA. However, these themes were less prominent in the daily press than they were in the weekly specialist press such as *New Scientist* and the *Economist*. We note that the term 'biotechnology' was not used until the early 1980s; the terms of the day were 'recombinant DNA' and 'genetic manipulation'.

Phase 2 (1980–83): the economic prospect for the twenty-first century. In phase two, the intensity of coverage remained low (rising to an average of one or two articles per week). The economic frame emerged strongly in the early 1980s (37%), and this reduced the progress frame by half. Political and economic issues replaced medical and health and safety issues as the dominant themes of the period. The reporting of economic benefits from potential industrial applications (32%) dominated, and there was a new focus on UK home news. In this phase, there was a marked shift from stories about basic research to an overwhelming preoccupation with utilitarian stories about potential benefits (77%). For the most part, attention continued to focus on the UK (especially) and the USA.

The developments in phases 1 and 2 are similar to those that Goodell has identified for the USA:[10] from an initial debate about safety and ethical issues at Asilomar, through the curtailment of the controversy, to the celebration of the prospects of a new industry. In the UK, media coverage mirrors (and presumably helps to reinforce) growing governmental enthusiasm for effecting a marriage between basic scientific research and industrial application. The example of the USA, where molecular geneticists were joining with venture capitalists in the establishment of successful new companies, was much publicised. Public debate in the UK focused on the policies needed to encourage such partnerships at home. Increasingly, biotechnology came to be seen at this time as a key strategic technology – along with information technology – for the twenty-first century. These concerns are reflected in the increased presence of both political and industrial actors on the media stage.

Phase 3 (1984–87): nothing but progress. In phase 3, the intensity of media coverage rose considerably (reaching an average of one article per day by 1987). Media attention spread to a variety of themes such as new methods of plant and animal breeding, and new techniques of genetic identification. In particular, the discovery of DNA fingerprinting by British scientists in 1985 was widely reported, raising the theme of 'identification' to its highest level (21%). Such was the impact of DNA fingerprinting that by the early 1990s it had become the major anchor in the UK for both popular press coverage and public representations of human genetics.[11]

A key feature of this period is the overwhelmingly celebratory tone of the reporting. The discourse of benefit increased to 86%, and non-utilitarian and risk–danger stories continue to be significant by their absence. In this extremely positive environment, scientific researchers (60%) regain the position of relative prominence which they had enjoyed in phase 1. In short, the mid 1980s are the most triumphalist period in the history of biotechnology reporting in the UK. While critical concerns may have been voiced, they were given little or no opportunity to be heard in the mass media at this time.

Phase 4 (1988–92): NGOs, ethics, and public accountability. In phase 4, the upward trend in the intensity of reporting continued. At the same time, there were changes in the form and content of the coverage. Although progress continued to dominate, the ethical and economical frames reappeared. At the same time, public accountability emerged as a new frame sponsored by NGOs. Environmental, patient and consumer organisations emerged for the first time as significant actors. As frames diversify, so do themes. Issues related to the regulation of biotechnology were given a prominent space for the first time in this phase, while the theme of basic research continued to decline.

Many of these changes appear to reflect a new emphasis on regulation which reflected the growing maturity of the field. Significantly, this emphasis was accompanied by a shift towards a more ambivalent discourse of consequences: risk/benefit and exclusively risk stories become more prominent and so do stories which are neutral about the consequences of biotechnology. For the first time, the EU and the world as a whole appear as significant elements in this coverage.

Phase 5 (1993 onwards): re-emerging ambiguities: the global challenge to British industry. The final phase sees a remarkable increase in the intensity of coverage, to an average of two to three articles per day in a single newspaper by 1996. By now, the range of issues associated with biotechnology in the media was sufficiently large to confer news value upon the sector as a whole. There is even a sense that genetics, geneticists and some evolutionary biologists have achieved a certain amount of glamour.

There are further changes in the frame structure of the coverage at this time. The progress frame (34%) loses further ground relative to the economic and the ethical frames; the medical frame remains prominent, particularly in the context of the discovery of an increasing number of genes responsible for diseases such as cystic fibrosis, and, at the same time, the nature/nurture frame re-emerges as a result of a series of 'gene-for-*x*' stories about characteristics as varied as obesity, sexual orientation, female 'sensitivity', reading ability and schizophrenia.

Another interesting feature of phase 5 is the appearance in the media coverage for the first time of commentary on public opinion polls and media coverage of biotechnology. This media acknowledgement of the significance of both the public and the media themselves in the biotechnology debate reflects the increasingly contested nature of the debate at this time.

In summary, we observe a number of striking features of media coverage of biotechnology in the UK. Overall, an attentive reader of the elite press over nearly 25 years will have encountered biotechnology as a shining example of progress in which basic advances in molecular genetics are being harnessed to health and wealth creation in an increasingly large number of different ways. In the late 1990s, this positive image will have been tempered by a certain amount of concern about risk-related and ethical issues; but overall, the impression cultivated by the press is overwhelmingly positive, with a marked emphasis on the medical benefits that accompany new developments in the field.

Public perceptions

There have been a number of studies of public perceptions of biotechnology in the UK during the 1990s.[12] In many respects, the results paint a rather similar picture. First, the public do not reject biotechnology out of hand, but discriminate between classes of application on the criteria of perceived worth, necessity and hazard. Medical and environmental applications are supported, but there are concerns about GM foods and the use of animals in laboratory research. When thinking about biotechnology in general terms, the public express a sense of unease. This is based on concerns about gene transfers across species boundaries, together with feelings that government spokespersons and regulators are biased towards an industry driven by commercial interests and a certain amount of scepticism that the longer-term safety issues are given insufficient attention.

For many British people, the BSE crisis is seen as a case in point: it is a warning of the fallibility of expertise and of the inadequacy of existing regulations in matters of new technology. Grove-White and colleagues note how people may discuss the issues both as consumers and citizens, leading to apparently contradictory views.[13] While as consumers people may purchase GM foods, as citizens they may be concerned about the social implications of the technology. Many of the studies report that people want far more public consultation on biotechnology, a theme reinforced in the recommendations of the First UK National Consensus Conference on Plant Biotechnology in 1994.

These studies provide a context for assessing the

results of Eurobarometer 46.1. We have seen that the UK is a relatively prominent player in the field of biotechnology, that it has a long history of active policy-making and a pattern of relatively intense and largely positive elite media coverage in this field. Nevertheless, in 1996 biotechnology was not a topic of consuming interest to the majority of the British public. When asked, 'What comes to mind when you think about modern biotechnology?' some 30% of the British public replied, 'Don't know'. Prior to the survey interview, 51% said they had never talked to anyone about biotechnology and 45% said they had seen nothing about the subject in the media over the previous three months. However, while biotechnology is not an issue of mass awareness in the UK, there is clearly an attentive minority. Some 30% said both that they had talked about the issue before, and that they had read or seen something about it in the media in the previous three months. In answer to another question, 35% thought that biotechnology was important to them personally.

Biotechnology has emerged in the context of a UK culture that is generally optimistic about the contribution of science to society. The survey found that 72% of British people think that five selected technologies will improve our way of life. In relation to other European countries, this indicates a relatively positive culture towards science. The views about biotechnology are somewhat polarised, however, with 45% optimistic and 25% pessimistic. This polarisation may be due in part to the intensity of media coverage of the issue, which provides the opportunity for both positive and negative commentary as evidenced in the profile of the media coverage. Only Denmark and Switzerland show similarly high levels of polarisation.

Comparisons of results on the optimism measure with the results on similar measures in the two earlier Eurobarometer biotechnology surveys in 1991 and 1993 show no change in optimism about biotechnology for the period 1991–93, but a decline of 6% for the period 1993–96. This decline in optimism is not matched by declining optimism in scientific culture as a whole. Taking the change in science culture as a baseline, the decline in optimism about biotechnology is 8%. It may be significant that this decline has occurred at a time when a number of specific applications of biotechnology – the 'Flavrsavr' tomato, for example – have come to public attention.

The responses to the open ended question, 'What comes to mind when you think of biotechnology/ genetic engineering?' (split ballot) captures the semantics of biotechnology and genetic engineering in the public domain. The main associations in order of frequency of use are: 'Don't know', 'negative', 'messing', 'diseases', 'babies', 'improve', 'medicine', 'nature', 'animals', 'food', 'cures'.

For women the image of biotechnology/genetic engineering is much more negative than for males. For all respondents the frequency of 'don't know' responses is greater when the stimulus word is 'biotechnology' rather than 'genetic engineering', reflecting the media's preference for the latter term. Given the polarisation between optimistic and pessimistic expectations of biotechnology, it is interesting to note that there are different semantic structures for the two groups. For the optimists, the associations are generally positive: 'disease', 'improve', 'medicine', 'new', 'crops', 'genes', 'research', etc. In contrast, for the pessimists the associations are: 'negative', 'messing' and 'nature'.

The UK public has a relatively high level of biotechnological knowledge (see Appendix 2, Table 2k). At the same time, answers on a number of 'subjective' knowledge items designed to test the existence of menacing images of biotechnology is rather low, with only the Netherlands showing a lower score. About 22% of British people think that ordinary tomatoes don't have genes while genetically modified ones do, 15% believe that eating a genetically modified fruit could modify a person's genes, and 28% consider that genetically modified animals are always bigger than ordinary ones. There is an interesting relationship between attentiveness to biotechnology and knowledge. Non-attentives have significantly lower knowledge and higher scores for menacing images.

Higher levels of attentiveness, personal importance and knowledge are all associated with more polarised views about biotechnology, as reflected in the scores for optimism and pessimism. For example, at the lowest level of knowledge 37% are optimistic while 20% are pessimistic. At the highest level of knowledge the respective percentages are 46% and 25%. If attentiveness, personal importance and knowledge can be taken as overall indicators of involvement with the issue of biotechnology, then it may be said that greater involvement is associated with increasingly polarised views about biotechnology. A path analysis shows that attentiveness is associated with educational level, age and gender (with men being slightly more attentive than women).

We turn now to the key questions on attitudes towards six different biotechnological applications (see Figure 3). As is the case for many EU countries, there are different opinions about the various applications, which can be grouped into favourable (medical applications and genetically modified crops), neutral (GM foods) and unfavourable (transgenic animals and xenotransplantation). Of note is the fact that the UK public are more polarised in their attitudes than the average European. Where the European mean is positive, the UK public are more positive; where the European mean is negative, the UK public are more negative. The negativity towards animal applications perhaps reflects a special sensitivity in the UK towards animals in general and towards research on animals in particular. But it is also of note that the distribution of opinion is skewed towards the positive pole.

To clarify the foundations of support, we may enquire how far 'encouragement' is predicted within the data set by the measures of use, risk and moral acceptability. Multiple regressions of encouragement on use, risk and moral acceptability show high correlations, with a total of 63% explained variance. Essentially, this analysis suggests that, when judging an application of biotechnology, the question that comes to most people's minds is whether the application is useful for society and morally acceptable. Issues of risk and safety are of much less importance in determining the extent of public support, notwithstanding the centrality of these issues in regulatory debates.

Qualitative (focus group) research conducted in spring 1996 provides some indications of the nature of moral concerns about biotechnology in the UK. In discussion, many participants talked about 'good' and 'bad' genetics. Within good genetics come medical applications which are associated with treatment and good health. Bad genetics includes applications which might lead to eugenics, monsters and unnatural practices. The combination of a general sensitivity about the use of animals in scientific research and the much-publicised picture of a mouse with a transplanted human ear have served to confirm in many people's minds the view that animal applications are intrinsically immoral. Overall, good genetics is acceptable because it is seen to serve higher values. In contrast, bad genetics is viewed as transgressing the boundary between the natural and the unnatural. As such it is widely judged to be morally unacceptable.

Food is a special case which resists simple classification within the frame of good and bad genetics. While food in general is a necessity and part of good health, both the survey results and the qualitative research suggest that for many people in the UK genetically modified foods are associated with bad genetics. Thus, a substantial minority of respondents think that GM foods are larger (monsters), uniformly perfect (eugenics) and likely to affect people's genes (infection). Furthermore, GM foods are not seen as a necessary development; rather, they are viewed as being purely in the interests of producers and retailers. Underscoring this perceived lack of usefulness of GM foods is the fact that only 30% of respondents think that biotechnology will contribute to the solution of world hunger.

In the UK, as in many other European countries, there is little confidence in national authorities to regulate biotechnology effectively. Given the comparatively long history of advisory committees and other regulatory initiatives in the UK, the failure

to foster public awareness must rank as a serious shortcoming. The qualitative research evidenced almost no knowledge about the regulatory system, and the Eurobarometer shows that less than 10% of respondents select the British Parliament or other public bodies as the institutions best placed to regulate biotechnology. That nearly 40% selected international organisations may reflect either a lack of trust in national institutions or a perception that the regulation of biotechnology raises issues which go far beyond national boundaries.

The Eurobarometer data reveal that trust is issue-specific. For xenotransplantation, 41% of respondents select the medical profession and 17% select animal welfare organisations as the most trustworthy sources of information; whereas for GM crops, the most trusted sources of information are environmental organisations (27%) and farmers' organisations (20%). In both cases, national public bodies and industry attract very low levels of confidence. Overall these findings on regulation and trust appear to reflect a failure of a number of key institutions involved in biotechnology to secure public confidence.

The survey was designed to allow us to explore the relationships between a number of socio-demographic factors (age, gender, educational level) and perceptions of biotechnology. In general, we find rather poor relationships of this kind within the UK data: only age is a significant predictor of support (with younger people tending to be more supportive). Unlike the impressions we had obtained in the qualitative work, the survey data reveal no gender effect on attitudes to biotechnology. Further analysis indicates that scientific knowledge (which is positively correlated with support) is more important than educational level.

Overall, however, it is people's values and attitudes that are more strongly related to their views about the applications of biotechnology. Supporters tend to be more optimistic about the contribution of technology to everyday life, and to have more positive expectations about the impact of biotechnologies in the future. They are willing to accept some risk for economic progress, and they believe that current regulations are sufficient to protect people from risks. Supporters are also more likely to express trust in public authorities and national public bodies to tell the truth about biotechnology.

On the whole, then, in 1996 biotechnology was not a salient issue in Britain. While people were relatively supportive of science and technology and relatively knowledgeable about matters biotechnological, levels of attention to this particular technology were not high. In the UK there was a relatively positive attitude to all but animal applications of biotechnology, qualified by some moral unease. The most frequent association with the technology was 'don't know'. Beyond this we observe something of a bifurcation in associations with the term. Optimists were on the whole more likely to cite relevant applications and to be positive, while pessimists were concerned about the dangers of tampering with the natural order and tended to be negative.

Confronted by questions about specific applications, the main bases of judgment are utilitarian and ethical. Perhaps the public have anticipated the debates to come. What sort of society do we want to live in and bequeath to our children? How should we interact with the natural order and with non-human species? These seem to be political and moral rather than scientific and technical issues. For most of the UK public, our results suggest that determining whether the technology is acceptable on these non-technical grounds is more important than considerations of risk.

Integrative commentary

On all the measures used in this study, the UK emerges as one of the 'busiest' EU member states with respect to biotechnology: it has a strong science base, a large industrial sector, a long history of active policy-making, and a pattern of very intense elite media coverage of the subject. Yet the Euro-barometer results for the UK suggest that in 1996 biotechnology was not a particularly salient subject with the general public. Although there was an attentive minority whose attitudes towards biotechnology tended to be relatively polarised, a considerable proportion of the UK public appeared both untouched and untroubled by the subject. The reasons for this apparently paradoxical situation probably include the fact that media coverage of biotechnology in the UK is largely confined to the quality press which reaches only certain sections of the public, the fact that the UK political system mitigates against 'single-issue politics', and the fact that UK policy-makers have generally responded promptly to public concerns about particular issues.

Surveying 25 years of policy-making and media coverage in the UK, several trends emerge rather clearly. Although the detailed phase structures of the two chronologies are different, both reveal a broadly similar pattern: first, rDNA technology emerged in the 1970s as a novel area of scientific research with great potential medical benefits that nonetheless required regulation in the interests of professional (scientific) and public safety; second, towards the end of the 1970s both the policy community and the media awakened to the economic potential of biotechnology, not least as a result of commercial activities in the USA; third, in the mid 1980s the government continued to foster biotechnology programmes at the same time as the media became almost celebratory about the benefits of biotechnology; and fourth, in the late 1980s the policy community became increasingly caught up in a series of sector-specific regulatory issues as the media tempered its enthusiasm for biotechnology in the context of increasing calls for public accountability. In the 1990s, the close linkage between policy and media continues as both begin to show increasing interest in ethical issues; and the significance of this convergence is revealed in the results of the Euro-barometer survey which point to a public that is particularly sensitive to the moral dimensions of biotechnology.

This close coupling of policy-making and media coverage over 25 years strongly suggests that the elite media are an important part of the public debate about biotechnology in the UK. While the limits of accuracy of the two profiles do not allow us to make strong claims about cause and effect, it seems clear that the elite media have played a vital role in the cultivation of biotechnology within the public sphere: in general, we find rising levels and changing contents of media coverage in the periods leading up to key policy decisions (as for example, in the coverage of new reproductive technologies in the period before the passage of the Human Embryology and Fertilisation Act in 1990). This is a pattern we might expect to find if the public sphere, as expressed in the media, is influential in the political process.

The media are not the only important influence on the changing course of public debate about biotechnology in the UK. Structural changes in public attitudes are also of great importance. The late 1980s and early 1990s witnessed a general decline in the UK in public confidence in government and public authorities of all kinds. This general decline is consistent with our 1996 cross-sectional data on trust and confidence in relation to biotechnology. Among those actor groups involved directly in the conduct and regulation of biotechnology, only medical researchers and practitioners command high levels of public confidence. Others – both in government and in industry – appear to suffer from a lack of public credibility. This, coupled with increasing amounts of qualified or ambivalent media coverage, has almost certainly contributed to the creation of significant amounts of public unease in the UK in the mid 1990s about a number of specific biotechnologies.

A number of specific national sensitivities may also have contributed to the shaping of public attitudes towards particular biotechnologies in the 1990s. For example, public attitudes towards food biotechnology appear to have been affected by the UK's experience with BSE. This has served both to alert the public to the potential risks associated with the introduction of new technologies within the food industry and to undermine public confidence in the institutions involved in the regulation of food science and technology. Similarly, public attitudes towards animal biotechnologies appear to have been affected by long-standing public sensitivity in the UK towards issues involving animal welfare. Such sensitivity is reflected in much media coverage of animal biotechnology, and it would appear to have contributed to a relatively more negative UK public perception of transgenic animal research and research on xeno-transplantation.

A third national sensitivity would seem at first sight to be relevant to the shaping of UK public perceptions of biotechnology – namely, public attitudes towards the EU. For much of the period of this study, the UK government adopted a generally sceptical attitude towards many European political initiatives; indeed, so-called 'Euroscepticism' was an important factor in the 1997 general election. During this period, we observe an increasingly national focus to UK media reportage coupled with a continuing tendency to concentrate on the USA as the most important country on the world stage of biotechnology. The preoccupation with the USA as the chief standard of comparison for the UK has undoubted significance for biotechnology policy-making and practice, but the significance of the sceptical stance towards Europe is harder to assess. Our qualitative research suggests that most members of the public are largely ignorant of the policy-making institutions involved in the regulation of biotechnology at both the national and the European levels; and this makes it easier to understand the apparent mismatch between the increasing role of European institutions in the regulation of biotechnology and the increasing focus of the UK media on the UK scene.

It is not the purpose of social research of this kind to predict the future. However, it is interesting to speculate about some possible future trajectories of public perception of biotechnology in the UK. Within the patterns we have described, it is easy to detect the potential both for increasingly broad public support for biotechnology and for increasingly differentiated and 'fine-tuned' public attitudes towards specific biotechnological applications. The seeds of broad support appear to be contained within the UK public's generally optimistic view of science and technology, their generally very positive view of medical biotechnologies, and the apparent success of all specific food biotechnologies brought to the market thus far. The seeds of an increasingly differentiated view appear to be contained in the trend towards increasingly ambivalent media coverage, in the sharply different attitudes displayed towards the six specific applications, and in the tendency for more attentive members of the public to have more polarised views. Continuing national sensitivities about food and about the uses of animals in scientific research will probably remain important ingredients in the UK public debate.

As in many other European countries, it appears that moral and ethical considerations play a more important role in the formation of British public perceptions than do issues of risk and safety. This implies that public support for or resistance to biotechnology will depend more upon questions of value than upon questions of fact. The success with which scientists and regulators accommodate the moral sensitivities of the public may be a determining factor in the further development of particular applications of biotechnology. Of course, it is always possible that dramatic technical developments (such as Dolly the sheep) or major scandals or safety scares will alter the course of public debate. At the same time, we consider that underlying structural features of the debate – perceptions of the need for particular applications, changing levels of trust in key institutions, and the effectiveness with which policy-makers respond to moral concerns – will play a significant part in determining the future of biotechnology in the UK.

Acknowledgement

This report arises out of a project, 'Biotechnology and the British Public', funded by the Leverhulme Trust (F/4/bg).

Figure 1. A regulatory profile for the United Kingdom

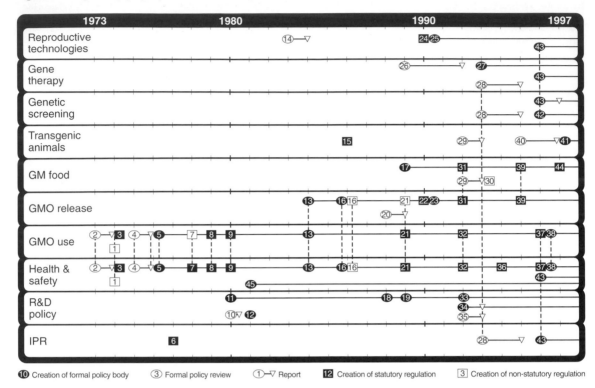

10 Creation of formal policy body   3 Formal policy review   1—▽ Report   12 Creation of statutory regulation   3 Creation of non-statutory regulation

1   *1974.* UK Department of Education and Science (DES) decided to implement a moratorium on recombinant DNA (rDNA) work until the UK government decides its own guidelines.

2   *1973.* Government set up the Ashby Working Party, to investigate the potential benefits and hazards of rDNA technology. On 13 December 1974 the Ashby report was published.

3   *1974.* The Health and Safety Executive (HSE) addressed health, safety and welfare at work in the Health and Safety at Work Act (HSWA 1974).

4   *1975.* Government established the Williams Working Party.

5   *1976.* DES established Genetic Manipulation Advisory Group (GMAG).

6   *1977.* The UK 1977 Patent Act came into force.

7   *1978.* The UK introduces Regulations on Genetic Manipulation (SI 1978 no. 752). These regulations required notification of experiments to the Health and Safety Executive (HSE) under the HSWA 1974. GMAG then provided safety advice; however, researchers were not required by law to follow recommendations made by GMAG.

8   *1979.* The UK introduced GMAG Assessment Scheme (which was revised in January 1980). The Scheme outlined four laboratory safety containment levels for categorising laboratories which perform genetic manipulation experiments.

9   *1980.* The GMAG Assessment Scheme was aban-

doned in favour of the GMAG Risk Assessment Scheme (January 1980).

10   *1980.* The Joint Working Party (members include Cabinet and Research Council advisors and members of the Royal Society) produced the Spinks report.

11   *1980.* The Heads of the Research Councils formed the Inter-Research Council Coordinating Committee on Biotechnology (IRCCCOB) to coordinate the work of the research councils.

12   *1981.* The Chairman of the Science and Engineering Research Council (SERC) responded to the Spinks report by establishing the Biotechnology Directorate to support strategic research and to establish university–industry links.

13   *1984.* GMAG was replaced by the Advisory Committee for Genetic Manipulation (ACGM), an advisory body to the HSE/C and other government departments. ACGM advise on safe working practices, categorisation of experiments, risk assessment, new techniques, health monitoring, training and controls.

14   *1984.* Following a year's review, the Warnock Committee produce a report.

15   *1986.* Laboratory animals, including those used for recombinant DNA experimentation, are protected under the Animals (Scientific Procedures) Act 1986.

16   *1986.* The ACGM working group, the Subcommittee on Planned Releases (soon known as the Intentional Introductions Subcommittee – IIS) was set up to develop voluntary guidelines on the release of GMOs.

HSE approved guidelines in April 1986.

17 *1989.* The Ministry of Agriculture Fisheries and Food (MAFF) set up the Advisory Committee on Novel Foods and Processes (ACNFP) .

18 *1988.* The Biotechnology Advisory Group (BAG) replaced the IRCCCOB. BAG coordinated the work of the research councils.

19 *1989.* The Advisory Board for Research Councils replaced BAG with the Morris Committee.

20 *1989.* After a year's deliberation, the report by the Royal Commission on Environmental Pollution (RCEP) is produced and debated in Parliament (6 July 1989). The report recommended statutory control of releases.

21 *1989.* The International Introductions Subcommittee (part of the HSC's ACGM) issued new statutory genetic manipulation regulations which specifically protected human health and safety, but also allowed for the consideration of environmental concerns. The regulations required 1 notification to the HSE in advance of a planned release of a GMO, and 2 compulsory local safety committees (prior to 1989 these were voluntary).

22 *1990.* In response to the RCEP report and the EC directive on deliberate release, regulations on the environmental release of GMOs were introduced in Part IV of the Environmental Protection Act (EPA). The regulations required safety assessments of the possible effects of GMO releases (submitted to the Department of the Environment). Some releases required consent. The EPA also required a public register detailing proposed release, risk assessment, ACRE's advice, and final decision from the Secretary of State including any conditions for release.

23 *1990.* The Intentional Introductions Subcommittee (under HSC) and the Interim Advisory Committee on Introductions are merged to become the Advisory Committee on Releases to the Environment (ACRE) to advise the Department of the Environment under the EPA.

24 *1990.* The Human Fertilisation and Embryology Act was passed. This enshrined in law many of the recommendations of the Warnock Report (1984), and represented something of a victory for the advocates of new reproductive technologies over those who had earlier supported Enoch Powell's 'private member's bill' (1985).

25 *1990.* The Human Fertilisation and Embryology Authority was established as part of the requirements of the HFE Act, whose role was to advise and report to government on developments in fertility treatments.

26 *1989.* The Committee on the Ethics of Gene Therapy (the Clothier Committee) was established).

27 *1993.* The Gene Therapy Advisory Committee (GTAC) was established under the Department of Health to look at ethical and safety issues.

28 *1995.* After two years of review, the House of Commons Select Committee produced a report, entitled *Human Genetics: The Science and its Consequences.* It called for a Human Genetics Commission to advise government on gene patenting, genetic medicines and the use of genetic information by employers and insurance companies, plus social and ethical issues. Government responded to this

proposal 11 months later in June 1996.

29 *1992.* In September, MAFF appointed the Committee on the Ethics of Genetic Modification and Food use (Polkinghorne Committee) to address the ethical concerns of production, marketing and consumption of transgenic animals. The report was published in 1993.

30 *1993.* The Food Advisory Committee (FAC) working with ACNFP, produced Guidelines for Labelling of Genetically Modified Foods. FAC stated that there was no need to label a GM food if the product was 'nature identical', but foods made from interspecific genetic modifications should be labelled using the words 'product of gene technology'.

31 *1992.* Under the 1990 EPA, the government published the Genetically Modified Organisms (Deliberate Release) Regulations 1992, which implemented EC Directive 90/220. The regulations were to be administered by the Department of the Environment.

32 *1992.* In response to EC Directive 90/219, the government updated the Health and Safety Executive Regulations (based on HSWA 1974), and introduced the Genetically Modified Organisms (Contained Use) Regulations 1992, for which the HSE is responsible.

33 *1992.* The Department of Trade and Industry (DTI) Chemicals and Biotechnology Division (CB) set up the Biotechnology Industry Government Regulatory Group.

34 *1992.* The government set up the Office of Science and Technology (OST) under the DTI. In 1993 the OST produced a White Paper on science and technology, entitled *Realizing our Potential.* This report recommended restructuring the research councils. Consequently the Biotechnology Directorate and the Biological Sciences of SERC were joined with AFRC to become the Biotechnology and Biological Research Council (BBSRC). The White Paper also included a strategy for raising public awareness.

35 *1993.* The Subcommittee of the House of Lords Select Committee on Science and Technology published a report entitled *Regulation of the UK Biotechnology Industry and Global Competitiveness,* which stated that the extensive approach to regulating biotechnology in the UK was threatening competitiveness.

36 *1994.* The Control of Substances Hazardous to Health (COSHH) Regulations required employers to assess the health risk of their work, adopt the appropriate safety measures, and inform the HSE of work with certain biological agents, in accordance with the EC Directive 90/679/EEC.

37 *1996.* The Genetically Modified Organisms (Contained Use) Regulations were introduced (SI 1996 no. 967) amending the 1992 regulations in accordance with Directive 94/51/EC.

38 *1996.* The ACGM was restructured and a technical subcommittee established to provide specialist technical advice to ACGM.

39 *1995.* Genetically Modified Organisms (Deliberate Release) Regulations 1995 came into force. The amended 1992 regulations implemented EC Directive 94/15/EC, which was an updated version of 90/220/EEC taking into account technical advances. Under these new regulations a 'fast-track' procedure was introduced requiring less information for release or

marketing of some GMOs and less amount of time for publicly advertising releases.

40 *1995.* The government established the Advisory Group on the Ethics of Xenotransplantation (AGEX). After a two-year review, a report by AGEX, *Animal Tissue to Humans* (1997), recommended delaying clinical trials, plus the need for a permanent advisory committee.

41 *1997.* The Department of Health established the UK Xenotransplantation Interim Regulatory Authority (UKXIRA) to advise on the use of animal tissues for transplants to humans.

42 *1996.* The government established the Advisory Committee on Genetic Testing (ACGT) to advise the Secretary of State for Health on the use of genetic

tests. The ACGT was chaired by Lord Polkinghorne (July 1996).

43 *1996.* The Human Genetic Advisory Commission (HGAC) was established (December) with a joint secretariat from the OST and Department of Health. Its remit was to review scientific developments and issues with social and ethical implications and to build public confidence in genetics.

44 *1997.* The Novel Foods and Novel Food Ingredients Regulations 1997 (SI 1997 no. 1335) came into force implementing EC Directive 258/97/EC.

45 *1981.* The government established the Advisory Committee on Dangerous Pathogens (ACDP) to advise on health risks and safe use of pathogens (including those which had been genetically modified).

Figure 2. Intensity of articles on biotechnology in the United Kingdom press

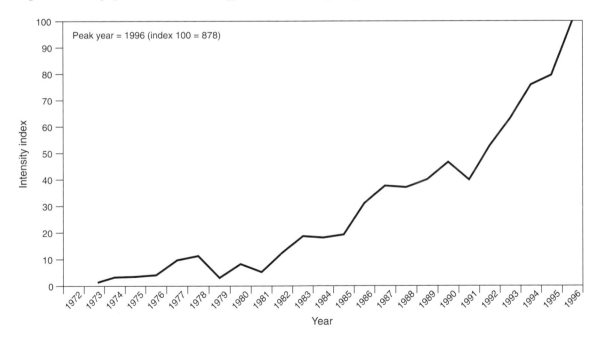

Figure 3. Attitudes to applications of biotechnology in the United Kingdom, 1996

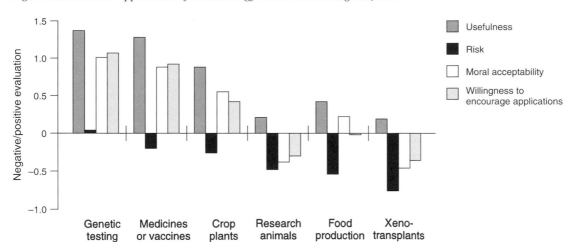

Table 1. United Kingdom media profile (for an explanation of terminology please see Appendix 5)

Phase	Freq.[a] (%)	Frame (%)		Theme (%)		Actor (%)		Benefit/risk (%)		Location (%)	
1. 1973–79	16	**Progress**[b]	**68**	**Medical**	**42**	**Scientific**	**78**	**Benefit only**	**48**	**USA**	**39**
		Nature/nurture	15	Basic research	26	Neither		Neither	24	UK	37
				Security & risk	9	Political	6	Both	20	Other Europe	13
				Indentification	6			Risk only	8	Other countries	10
2. 1980–83	15	**Progress**	**47**	**Economic**	**22**	**Scientific**	**49**	**Benefit only**	**77**	**UK**	**52**
		Economic[c]	37	Medical	22	Industry	32	Both	21	USA	30
		Ethical	5	Political	19	Political	11			Other Europe	12
		Accountability	5	Basic research	13						
				Identification	10						
3. 1984–87	15	**Progress**	**86**	**Medical**	**24**	**Scientific**	**60**	**Benefit only**	**86**	**UK**	**45**
				Indentification	21	Industry	24	Both	12	USA	31
				Basic research	19	Political	7			Other Europe	10
				Economical	10					Other countries	7
				Agriculture	7						
				Security & risk	6						
				Animal	5						
4. 1988–92	22	**Progress**	**45**	**Medical**	**23**	**Scientific**	**39**	**Benefit only**	**50**	**UK**	**50**
		Ethical	14	Regulations	13	Political	14	Both	23	USA	26
		Economic	12	Indentification	13	Industry	14	Neither	21	Other countries	8
		Accountability	8	Ethical	10	NGOs	11	Risk only	7	World	6
				Security & risk	6					Europe	6
				Economical	8						
				Basic research	7						
5. 1993–96	33	**Progress**	**34**	**Medical**	**29**	**Scientific**	**32**	**Benefit only**	**42**	**UK**	**63**
		Economic	21	Indentification	14	Industry	28	Both	28	USA	20
		Ethical	15	Economic	14	Political	18	Neither	20		
		Nature/nurture	8	Regulations	10	Media/public	8	Risk only	10		
		Accountability	7	Ethical	8	NGOs	7				
				Security & risk	8						

a Percentage of corpus in the period; total $n = 539$.
b Bold indicates highest frequency within phase.
c Italics indicates highest frequency within category.

Notes and references

1 Ernst & Young, *European Biotech '97: A New Economy* (Stuttgart: Ernst & Young International, 1997).

2 Yoxen, E, *The Gene Business: Who Should Control Biotechnology?* (Suffolk: Richard Clay/The Chaucer Press, 1983).

3 Cantley, M F, 'The regulation of modern biotechnology: a historical and European perspective', in Brauer, D, (ed), *Biotechnology – Legal, Economic and Ethical Dimensions* (Weinheim: VCH, 1995).

4 Berg, P, Baltimore, D, Boyer, H W, Cohen, S N, Davis, R W, Hogness, D S, Nathans, D, Roblin, R, Watson, J D, Weissman, S, Zinder, N D, 'Potential biohazards of recombinant DNA molecules', *Science* 185 (1974), p. 303.

5 Sharp, 'The management and coordination of biotechnology in the UK in 1980–88' in *Philosophical Transactions of the Royal Society*, 324 (1989), pp. 509–23.

6 Jasanoff, S, 'Product, process, or program: three cultures and the regulation of biotechnology', in Bauer, M (ed) *Resistance to New Technology: Nuclear Power, Information Technology and Biotechnology* (Cambridge: Cambridge University Press, 1995).

7 *Ibid.*

8 UK National Consensus Conference on Plant Biotechnology (London: Science Museum & BBSRC, 1994); Joss, S and Durant, J (eds), *Public Participation in Science: The Role of Consensus Conferences in Europe* (London: Science Museum, 1996).

9 Mulkay, M, 'Embryos in the News', *Public Under-standing of Science*, 3 (1994), pp. 33–45.

10 Goodell, R, 'How to kill a controversy: the case of recombinant DNA', in Freidman, S M, Dunwoody, S, and Rogers, C L (eds), *Scientists and Journalists: Reporting Science as News* (New York: The Free Press, 1986).

11 Durant J, Hansen, A and Bauer, M, 'The public representation of new human genetics', in Richards, M and Marteau, T (eds), *The Troubled Helix: Social and Psychological Implications of the New Human Genetics* (Cambridge: Cambridge University Press, 1996).

12 Martin, S and Tait, J, 'Attitudes of selected public groups in the UK to biotechnology', in Durant, J (ed), *Biotechnology in Public: A Review of Recent Research* (London: Science Museum, 1992); Frewer, L J, Shepherd, R and Sparks, P, 'The interrelationship between perceived knowledge, control and risk associated with a range of food-related hazards targeted at the individual, other people and society', *Journal of Food Safety* 14 (1994), pp. 19–40; Grove-White, R, Macnaghten, P, Mayer, S and Wynne, B, *Uncertain World: Genetically Modified Organisms, Food and Public Attitudes in Britain* (Lancaster: Lancaster University, Centre for the Study of Environmental Change, 1997).

13 Grove-White, R , Macnaghten, P, Mayer, S, and Wynne, B, *Uncertain World: Genetically Modified Organisms, Food and Public Attitudes in Britain* (Lancaster: Lancaster University, Centre for the Study of Environmental Change, 1997).

Address for correspondence

Dr George Gaskell, Methodology Institute, London School of Economics and Political Science, Houghton Street, London WC2A 2AE. E-mail gaskell@lse.ac.uk

Europe

Jean-Christophe Galloux, Hélène Gaumont Prat and Ester Stevers

Introduction

There can be few other times in history when the concept of 'Europe' has been as keenly contested as it is today. In the main, this is because the relationships between individual European nation states and the supranational political entity known as 'the European Union' (EU) are in a state of continual and active flux. Since the 1970s, the membership of the EU has steadily grown, and at the same time, its political aspirations have steadily expanded. Despite much doubt and hesitation – and not a little opposition from some member states – the general trend of the EU has been towards closer economic and political integration among the member states. As a result, decision-making at the European level has come to play a steadily larger and more prominent part in the economic and political lives of EU citizens.

Nowhere is this trend more obvious than in the field of biotechnology. In the 1970s, the EU played a smaller role in European science and technology policy-making than it does today. Indeed, even in 1978, when biotechnology was actively developing in Europe, most key initiatives (and all legal provisions) were located at the national level. With time, however, the key institutions of the EU – the European Commission, the Council of Ministers, and the European Parliament – came to play an increasingly important part in biotechnology policy-making. Today, it would be unthinkable for EU member states to consider biotechnology policy without taking full account of existing and planned arrangements at the European level.

For this reason, we have devoted a separate chapter to the history of European dealings with biotechnology over the past quarter century. This 'European profile' is concerned with policy-making only; for obvious reasons, it lacks sections devoted to media coverage (of which there has been very little at the European, as contrasted with the national, level) and public perceptions (which are dealt with separately in Part III of the book, which is devoted to the analysis of the 1996 Eurobarometer survey of biotechnology). Here, therefore, we seek to chart the growing involvement of the EU with biotechnology.

The biotechnology industry in Europe

European companies using modern biological techniques to develop commercial products in various field of biotechnology (health care, agriculture, environment and food processing) have tripled in number from 386 in 1993 to 1036 in 1996. The biotechnology sector employed 15,000 people in 1993 and 39,000 in 1996. A large number of these firms were founded after 1986. Europe hosts many companies operating internationally which have invested heavily in genetic engineering, including Astra, BASF, Bayer, Glaxo-Wellcome, Hoffman-Laroche, Limagrain, Novartis, Novonordisk, Rhône-Poulenc and Zeneca. Nearly half of biotechnology firms are located in northern European countries: 182 in the UK, 105 in Germany, 102 in France, 68 in Sweden and 48 in the Netherlands. Their activities may be summarised as follows: 320 (45%) in heath care and diagnosis; 184 (26%) in chemicals; 55 (8%) in agriculture; 37 (5%) in environment; 32 (4%) in food; and 88 (12%) in other sectors. The revenue generated by these industries amounted to 1.72 billion ecus in 1996.[1]

The investment in R&D for the same period was 1.5 billion ecus. European research is on the cutting edge in many areas of biotechnology such as plant molecular biology and genome research, however, the USA spends five or six times as much as Europe on research. The number of biotechnology patent applications in Europe increased from 2850 in 1990 to 4142 in 1996, but these lag behind American patent applications by a considerable amount.

Europe's biotechnological sector is expected to continue growing at the rate of 20 per cent a year. Nevertheless, the European biotechnology industry still suffers structural weaknesses such as insufficient collaboration between academia and industry, lack of coordination of research across member states, shared access to resources and infrastructure, and inadequate venture capital. Economic development in the biotechnology sector has been most successful in the USA compared with Europe and Japan.

Since the mid 1980s, the main task of European political structures concerning biotechnology, especially at the EU level, has been to create the conditions for the development of biotechnology comparable to the development observed in the USA and Japan, its two principal competitors. Priorities have focused on financing biotechnology-based research since the beginning of the 1970s (with, for example, the founding by 12 European governments of the European Molecular Biology Organisation – EMBO – in 1970). However, the first research programme dedicated to biotechnology was launched only in 1982 by the European

Commission: the Biomolecular Engineering Programme (BEP). Since then, all framework programmes for research and development have included biotech-nology aspects.

Other initiatives have been less successful. The coordination of the regulatory framework for research and production using genetic engineering techniques took place fairly quickly (with directives 219 and 220 adopted in April 1990), but debates on the coordination of patents on biotechnological inventions took more than 10 years. The resulting uncertainty has, according to industry, discouraged investors.

In 1998 the European regulatory framework for biotechnology is in place, and functioning, and is regularly updated. EU funding for basic research in biotechnology increases constantly. It remains to be seen whether Europe will improve its performance from a competitive point of view.

Policy-making bodies in Europe

Since the inception of Europe-wide institutions, their natures and the balance of power characterising their interactions have changed considerably. It is obvious that the public policies developed at the European level have been strongly influenced by this institutional evolution.

The European Union

The major changes undergone by the EU concern the inclusion of additional countries and formal modifications of the founding treaties. The UK, Ireland and Denmark joined the European Community in September 1973; Greece joined in 1981; Spain and Portugal in January 1986; and Austria, Finland and Sweden in January 1995. The European Parliament was directly elected for the first time in June 1979. In 1986, the ratification of the European Single Act allowed some important changes in the functioning of the European institutions. The Maastricht Treaty was signed on 7 February 1992 and the single market came into force on 4 January 1993.

Functions and power are divided among several bodies.

- The European Council consists of heads of state and government. It determines the main orientation of the EU and takes major political decisions.

- The Council of Ministers consists of the ministers of the member states, and is the principal decision-making body. The Council is composed of the ministers responsible for the subject matter dealt with during a particular session (e.g. agriculture, biotechnology). The Council is assisted by the Committee of Permanent Representatives (COREPER) in Brussels.

- The European Commission has the right to initiate proposals for legislation. It is responsible for the implementation of decisions, and sees to it that Community laws are respected.

- The European Parliament is directly elected every five years. It participates in the decision-making process and, more specifically, in the detailing of the budget. Depending on the relevant procedure, it either comments on proposals for legislation sent from the Commission to the Council or it has formal power to change or reject proposals. The Parliament also has the right of petition.

- The Court of Justice (CJCE) decides on litigations and interprets Community law.

- The Social and Economic Committee represents the opinions of relevant interest groups.

- The Committee of the Regions represents the opinions of the territorial communities.

The main players in the decision-making process are the Council, the Commission and the Parliament. Since the early 1980s, however, the role of the European Council has been steadily increasing. It plays a major role in priority-setting, and its support has been essential for the proposal of and voting on numerous legislative projects. The influence of the Court of Justice has also been growing since the late 1980s, in particular through its interpretation of Community law.

The administrative structure of the EU, put into place during the 1950s, is based on the French model. The Commission is headed by 10 Commissioners who are chosen by the governments of the member states, after consultation with the European Parliament. The Commissioners direct about 15,000 civil servants divided into 26 Directorate Generals (DGs) and 15 specialised services. Responsible for a specific policy area, each Commissioner is assisted by a Cabinet whose members are appointed at personal title. The Commissioners decide together on the projects which will be sent to Council and Parliament for legislation. The functioning of the Commission is thus based on collegiality and shared responsibility. The small number of European civil servants compared to the diversity and complexity of the issues treated by the Commission incites it to an extensive use of external expertise. This is especially the case in the areas of high technology. The increase of regulatory activities by the Commission since the mid 1980s augmented the use of consulting agencies and the experts who work inside the Commission for a given period. It is interesting to note that, although each country has a quota of civil servants working at the Commission, experts are hired on personal title independent of nationality. As a result, services depending to a high degree on external expertise tend to have an over-representation of certain nationalities. Often these experts are recruited from the UK and northern European countries.

The decision-making procedures are long and complicated owing to the many different bodies participating in the process. Procedures depend on the content and the legal form of the decision to be made. Accordingly, the number of the bodies implicated and the nature of their formal role in the process are in a constant state of flux. Consensus and compromise are the rule, and competition for power characterises the functioning of the different institutions. Competition takes place between the main bodies (the Council, the Commission and the Parliament), as well as within each of these bodies along the main lines of division – nationality, ideology and administrative service (DG). In the Council, each Minister defends national position and interests. In the Commission, strong rivalries exist between the different DGs, and nationality and ideology reinforce potential conflict among its agents. At the level of the European Parliament, the policy-making process is dominated by ideological differences among its parties but the influence of nationality is not completely neutralised. No decision can be taken, and no policy can be developed if the majority of member states or institutions do not agree.

Other European institutions

Although the EU is the major arena for public policies at the European level, it is not the only player. The Council of Europe also has an important political role in the field of biotechnology. On more technical points, institutions such as the European Molecular Biology Organisation (EMBO), the European Patent Office (EPO) and the European Science Foundation (ESF) deserve mention.

Council of Europe. The Council of Europe, created in 1949, was originally an association of Western Europe democracies. It includes EU member states and the European Free Trade countries. With the fall of the Iron Curtain, most Central European countries joined the Council of Europe. The stated aim of the Council is 'to achieve greater unity between its members for purpose of safeguarding and realising the ideals and principles which are its common heritage and facilitating their economic and social progress'. It is composed of a Committee of Ministers which is the decision-making body, and the Consultative Assembly, composed of members of national parliaments. It acts through Conventions (multilateral treaties representing a Europe-wide consensus) and Recommendations (guidelines addressed to governments of member states).

The Council of Europe is active in many fields including law, human rights, culture, education, communication and environment. It has acted as an international clearing house for reflection and action on the social, legal and ethical implications of the developments in biomedical sciences since the beginning of the 1980s. The Committee of Ministers set up a multidisciplinary ad-hoc Committee of Experts on the Progress of Biomedical Science (CAHBI) in 1989. Besides technical experts, its participants include representatives from the Parliamentary Assembly, the EEC, the World Health Organisation, UNESCO and OECD, in order to carry out all the activities of the Council of Bioethics. However, the initiative comes from governments of member states. The Council of Europe avoids imposing rigid and uniform solutions but aims to stimulate member states to reform their existing laws in the direction of greater harmonisation. The results of its work can be seen in the main conventions dealing with broad fundamental principles such as the conventions for the protection of animals and the Biomedical Convention (November 1996), and in a series of Recommendations issued since 1979 in the field of biotechnology (for example, genetic engineering, 1982; rDNA, 1984; use of human embryos and fetuses in scientific research, 1989).

European Science Foundation. The ESF was founded in 1974 as a meeting place for the research councils of Europe (not only those from the EU – in 1990 it included delegates from 20 countries). It has the mission of setting objectives and stimulating basic research in Europe. Its annual meeting functions as a kind of mini-parliament of 70–80 scholars to review its activities. Since its inception and during the first regulatory phase in Europe, it had an important role in articulating the opinions of the European scientific community (through its official national councils) concerning the regulation of rDNA. It worked closely with the European Community. The main result of its work was to promote a convergence of scientific principles for rDNA safety in the late 1970s and early 1980s.

European Molecular Biology Organisation. EMBO was founded by a group of molecular biologists in 1963, and was formally recognised in 1970 by European governments with the setting up of the European Molecular Biology Conference. Unlike the ESF, EMBO is composed of individual scientists in the discipline. Together with the ESF, it was an active participant in the early stages of the regulation of rDNA.

European Patent Organisation. The European Patent Organisation includes the European Patent Office (EPO) and the Administrative Council. Established by the Munich Convention on European Patents (5 October 1973), it is an international organisation to which 18 states belong (15 member states of the EU plus Monaco, Liechtenstein and Switzerland). The aim of EPO is to deliver patents through a unique and centralised procedure. The Administrative Council is responsible for the administrative supervision of the Organisation and is composed of a delegate from each member state. Its financial autonomy is based on the fees paid by patent applicants and it has close relationships with industry and the EU. The EU Commission and EPO have worked together since about 1986 on the Directive dedicated to biotechnological inventions.

Public policy

The public policies on biotechnology in Europe cannot be summarised in the limited space allowed here. Our aim is to show some trends, essentially at the EU level (Figure 1). It must be kept in mind that the debates about biotechnology have from the start been international in character. The policies initiated and implemented at the European level are the result of highly complex networks of interactions between EU bodies themselves (between DGs, Commission, Parliament and Council), between the EU and international actors (competitors, e.g. the USA, or governmental organisations such as OECD), and between the EU and its member states.

The debate on biotechnology regulation began with the Asilomar Conference held in California in February 1975, although the first policy event at the EC level appeared three years later (1978) with the 'Proposal for a Council Directive Establishing Safety Measures against the Conjectural Risks Associated with rDNA'. It should also be noted that the situation in the USA often influenced the development of European policies in the field of biotechnology: during the first stage (1978–83) towards non-legislation; during the 'intermediary period' (1983–86) by pointing out the relative weakness of European industrial performance vis a vis the USA; in the mid 1980s on the publication of the 'Coordinated Framework for the Regulation of Biotechnology' by the President's Office of Science and Technology', which reoriented the regulatory framework; and during the second regulation phase (1991 onwards) towards more flexibility in the regulatory framework. The situation in the USA has been carefully watched for the past decades and the technical standards adopted there often followed. The discussions at OECD were identically influential, especially in the debate on biosafety.

Regulatory solutions adopted at the national level have been influential in several cases on European policy, e.g. the setting up by Denmark in the mid 1980s of the first legislation specifically on biotechnology, and discussions (since 1988) on the German 'Gene Law' (passed in May 1990). These kinds of initiatives made clear the threat to a unified common market presented by divergent national legislations. The influence of national regulatory models are not always of a competitive nature: their adoption at the EU level corresponds more to a need than a necessity. This is certainly the case in the establishment of the Group of Advisers on Ethical Implications of Biotechnology (GAEIB), created in 1991 as a result of institutionalised ethics commonly observed in France and Southern European countries.

The initiation of regulations at the EU level was also a function of the Commission's responsibilities. For example, up to the implementation (in 1989) of the Single European Act, there was no specific legal basis for R&D programmes. Authorities had to use the very general provisions of article 235 which required a unanimous vote of the member states.

The Council of Europe developed its own regulatory framework in parallel to the EU. It reached the same consensus for non-legislation at the beginning of the 1980s (Recommendation 934 (1982) on genetic engineering). As opposed to the EU, the Council of Europe focused on fundamental rights at stake in the development of biosciences and especially biomedicine. The two sets of legislation are coherent. The two institutions maintain close relationships in order to harmonise their work as far as possible.

Several phases can be distinguished in the evolution of public policies concerning biotechnology in Europe.

Phase 1 (1975–83). Non-legislation. Expertise on genetic engineering matters within the Commission services were located in DG XII (the Directorate-General for Science, Research and Development). It was placed in charge of biotechnology. The main aim of DG XII was to set up research programmes. There was no global view of the sector nor a perception of the need for special legislation. DG XII withdrew its proposal for a directive on 'safety measures against the conjectural risks associated with rDNA' in 1980 and replaced it with a non-binding Council Regulation. DG XII also willingly withdrew from such matters in the beginning of the 1980s. Owing to its scientific culture, DG XII had a broad scientific network and narrow links with ESF and EMBO, increasing the influence of the scientific community in the definition of the policies. As the Economic and Social Committee (ESC) stated in 1979: 'The conjectural risks associated with the work involving the production or utilisation of rDNA are probably non existent or small.'

Phase 2 (1983–86). Reorganisation. This phase took place when the Commission recognised the need for a global approach to problems posed by biotechnology: biosafety, consumer issues and the bioindustry, regulation of products and their distribution, etc. It implied a coordinated and integrated response to these wide-ranging but interconnected challenges through administrative structures which were essentially vertically divided. Opinion became deeply divided on the control of rDNA processes. DG VI (Agriculture), DG III (Internal Market and Industrial Affairs) and DG XI (Environment and, until 1991, Consumer Protection) entered the field and took control from DG XII. In 1983 the Biotechnology Steering Committee was created with a special secretariat, the Concertation Unit for Biotechnology in Europe (CUBE), followed by the Biotechnology Regulation Interservice Committee (1985), a forum for policy debates between DGs. DG XI became chef de file for biotechnology legislation, but not for technical standards. This was a consequence of the entry of environmental interests into the policy debates on biotechnology. DG XI brought to these issues its extensive regulatory experience. Legislation was also essential for DG III to create a common market for food products, pharmaceuticals, etc.

Phase 3 (1986–91). First legislation. This phase led to the elaboration of the main texts regulating biotechnology in Europe. Directives 219 and 220 related to contained use and deliberate release into the environment of GMOs were adopted in April 1990, just before Directive 679 on the protection of workers from the risks related to the exposure to biological agents at work. The draft directive on the legal protection of biotechnological inventions was published in October 1988. At the same time, DG VI proposed a community system of plant variety rights. In 1987, Directive 87/22 addressed pharmaceuticals produced by biotechnological or other high technology processes. This set of legislation, considered restrictive compared to US legislation, was not always coherent and was difficult to implement. Particularly, contradictions between horizontal and sectoral regulation, and between process and product regulation, were not resolved. In addition, the whole process of implementation, at the EU level and at the national level, was expected to take many years.

Phase 4 (1991 onwards). Second legislation. It became obvious that industry and the scientific community had lost the political battle regarding legislation, regulation and control of biotechnological processes and products. The role played by public opinion became increasingly important. Major companies started to invest heavily in biotechnology at the end of the 1980s, and industry had organised itself more efficiently (e.g. by the creation of the Senior Advisory Group for Biotechnology – SAGB – in 1989). Biotechnology featured prominently as one of the three technologies that were most promising for sustainable development in the next century.[2] Biotechnology came in for renewed scrutiny from the Community's authorities. At the administrative level, the conflicts occurring between DGs over

biotechnology, remaining largely unsolved by the Biotechnology Steering Committee (BSC) and the Biotechnology Regulation Interservice Committee (BRIC) for six years because of a lack of authority, were dealt with by the establishment of the Biotechnology Coordination Committee (BCC) in March 1991. This institution became more powerful than the previous ones. Chaired by the Commission's Secretary-General, the BCC covered all sectors and activities of the Commission in the field of biotechnology, with the participation of all relevant services. BCC was essential to the development of a well-balanced Community policy in biotechnology. As a result, the regulatory framework began to be revised so that the legislation could be more easily adapted to technical progress by regulatory Committee procedures and so that there was an easing of administrative requirements.

Actors and arenas

Directorate Generals (DGs)

DGs play an essential role in the Commission. Their goals differ, as do their domains of competence, and thus they defend different interests and strategies in the field of biotechnology.

DG III is the natural interlocutor of bioindustry associations. It is aware of international industrial competitiveness and defends the sectoral legislation, especially in foods and pharmaceuticals. DG III has been very concerned with a progressive development of coordinated Community-wide legislation. It has been influential in the drafting of biotechnology directives, and can be said to be the leader (since the mid 1980s) of Commission policy in biotechnology.

DG VI, largely preoccupied with the management of the common agricultural policy, has become increasingly involved in biotechnology policy.

DG XI (created in 1973), in charge of environmental questions and responsible for environmental protection, acted as 'the advocate and guardian' of directives 219 and 220 of 1990.[3] It defends the 'horizontality' of the regulatory framework. The experience of DG XI with chemical legislation has become a paradigm for regulating the products of biotechnology. DG XI has been particularly influential in the field of biotechnology.

DG XII's main interest has been in winning resources for larger R&D programmes and managing them efficiently. It has little political weight in the conflicts over regulation. It is often viewed as the voice of the scientific community.

Other DGs have been involved but with less importance: DG I (External Relations) was responsible for the international consequences of the policy adopted vis a vis the GATT negotiations in particular. DG V (Employment, Industrial Relations and

Social Affairs) took charge of worker safety regulation, for which it advocated a 'horizontal' approach.

Industry

Industry did not constitute a lobby before the end of the 1980s. It has had a role since 1984 when it was invited to meet the Commission, to discuss the Community actions in biotechnology and mechanisms for liaison between industry and the Commission. To coordinate the action of industry, the European Biotechnology Coordination Group was established in 1985, regrouping the main federations of bioindustries in Europe. This structure was finally abandoned by industry as ineffectual. SAGB, created in 1989, assumed the job in part. Associations such as EUROPABIO (which has more than 600 member companies throughout Europe) developed a more accessible discourse on biotechnology. EUROPABIO founded the European Bioethics Committee in 1997 to address moral questions raised by the industrial activity in the field of biotechnology. The negative vote on the Patent Directive in 1995 taught the industry how to use bioethical concerns. Industry has traditionally had close relations with DG XI, and it has been able to exert influence more effectively on the Council than elsewhere.

Scientists

Scientists have been present at different stages of decision-making. They benefit by their position of knowledge and expertise (see the role of EMBO and ESF in phase 1 of public policy). They act also as experts for the Commission. Their influence, still important, has declined since the mid 1980s when policy entered a legislative stage.

Political parties

The role of the European Parliament has evolved in accordance with the institutional reforms of the EU. Although its role is rarely decisive, it represents a part of the consensual system on which the decision-making process relies. The 1995 vote on the Patent Directive proposal remains the only example of opposition to the Commission. This was largely due to the intense lobbying of ecologists and to a lack of perception (in the proposal then discussed) of the moral consequences of patenting. The European Parliament is also the only arena in the EU where ideological debates take place. The political parties represented in the Parliament have had no real doctrine on biotechnology, with the exception of the Green Party which also acts as a pressure group. Some Members of the European Parliament specialise in biotechnology and lead debates.

Consumers

A consumer policy was established in the mid 1980s at the Community level. Consumer interests have been taken into account in policy debates on biotechnology products, especially at BRIC meetings, and these interests are quoted in most official reports. The European Consumers' Organisation (BEUC) is active in the field together with new actors such as the Consumer and Biotechnology Foundation. Their role increased with the marketing of products issued from biotechnological processes. Patients Associations (e.g. European Haemophilia Consortium, European Alliance of Muscular Dystrophy Associations, European Alliance of Genetic Support Groups) have been very active in lobbying for regulations linked to pharmaceuticals and patenting. Their position (in favour of a directive on patenting) harmed the position of Greenpeace (anti patenting).

Non-governmental organisations (NGOs)

NGOs are well represented in Brussels, but it is difficult to measure their real influence. Environmental associations, such as Greenpeace and Friends of the Earth, have been fairly successful in the decision-making process: the cases for directives on patents and labelling remain the best examples. These kinds of lobbyists may be more influential with the European Parliament than with the Council.

The public

The reactions and participation in decisions of the public were not a matter for concern during the first phase. However, the public were regularly mentioned in reports on the subject, primarily concerning information and education.[3] The Commission, at the end of the 1980s, granted support to a growing number of publications and initiatives for public information, for example, the Bridge Programme, consumer dialogue workshops and related publications. The Commission gave support to the newly constituted Task Group on the Public Perception of Biotechnology of the European Federation of Biotechnology. The first Eurobarometer survey took place in 1991 and has been repeated twice (1993 and 1996). Public opinion remains a priority issue for EU authorities. The Commission has continued to augment R&D programmes with studies on social and ethical dimensions.

Conclusion

Clearly, European dealings with biotechnology to date have been complicated and multivalent. The technical and industrial complexities of modern biotechnology (with its obvious cross-sectoral applications and implications) have combined with the political complexities of the EU itself to create situations that are hardly conducive to straightforward and speedy policy-making. Once the decision was taken to opt for process- rather than product-based regulation at the European level, it was inevitable that biotechnology policy-makers would have to deal with a very large number of individually difficult and delicate issues.

In this context, the passage of the directives on Contained Use and Deliberate Release must be regarded as considerable achievements. Other directives – such as those on patenting and labelling – have not been so easy to secure; and

the uncertainties surrounding them have caused significant economic, industrial and political difficulties, not least in the context of Europe's trade relationship with North America. In recent years, particular member states have experienced difficulty in implementing EU law on various aspects of biotechnology, to the point where this subject has come to be viewed as a test-case of loyalty to the EU itself.

Clearly, the European debate about biotechnology has a long way still to go. It is hard to see that the decisions which lie in the future will be any easier than those which lie in the past.

Figure 1. A regulatory profile for Europe

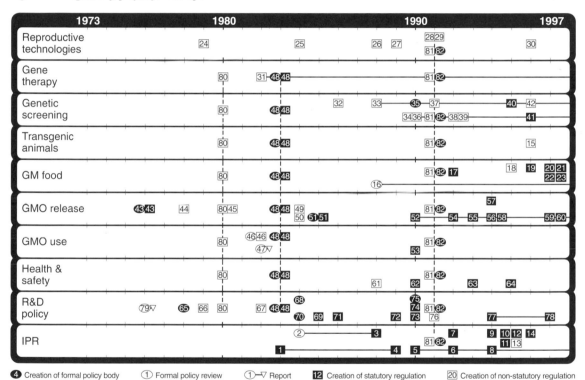

1   *1983*. Ciba-Geigy case (T 49/83) in which the European Patent Office (EPO) accepted the patentability of chemically treated seed.
2   *1984*. The European Council recognised the importance to development of better legal protection for biotechnology inventions.
3   *1988*. Publication of the first draft of the directive on the legal protection of biotechnological inventions (21 October).
4   *1989*. Oncomouse case, in which the examinator of the EPO refused to grant a patent for a transgenic animal (14 July).
5   *1990*. Oncomouse case, in which the Examining

Division of the EPO reversed the decision of the examinator and authorised the granting of the patent (3 October).
6   *1992*. Opinion of the Group of Advisers on Ethical Implications of Biotechnology (GAEIB) on ethical questions arising from the Commission proposal for the 'patent directive' (12 March).
7   *1992*. EPO granted a patent for the oncomouse (3 April).
8   *1994*. Regulation no. 2100/94 setting up protection for plant varieties at the Community level (27 July).
9   *1994*. Relaxine case, in which the EPO stated that an 'invention related to a human gene' was not an invention

as it was against morality and public order (8 December).

10 *1995.* European Parliament rejected the directive on legal protection of biotechnological inventions (1 March).

11 *1995.* Regulation no. 1238/95 concerning the functioning of the Community Office for Plant Varieties (31 May).

12 *1995.* Regulation no. 1768/95 (application of article 14 alinea 3 of Regulation no. 2100/94, 24 July)

13 *1995.* New draft of the directive on legal protection of biotechnology inventions published (December).

14 *1996.* Opinion no. 8 (GAEIB), stating that inventions related to human body elements are not amoral (25 September).

15 *1996.* Opinion no. 7 (GAEIB) on the genetic modifications of animals (21 May).

16 *1988.* Proposal to the Council (COM/88/351 C2.107/88) for a decision to adopt a research and development programme on food science and technology (23 April).

17 *1992.* First proposition of the Commission concerning novel foods and ingredients (7 July).

18 *1995.* Opinion no. 5 (GAEIB) on labelling of foods derived from modern biotechnology (5 May).

19 *1996.* Commission authorised the marketing of transgenic Colza (6 February).

20 *1997.* Commission authorised the marketing of Ciba-Geigy's transgenic maize (23 January).

21 *1997.* Regulation no. 258/97/CEE of the European Parliament and the Council concerning novel foods (27 January).

22 *1997.* Recommendation of the Commission concerning the marketing and labelling of novel food and ingredients (29 July).

23 *1997.* Regulation no. 1813/97/CE of the Commission concerning the labelling of foodstuffs based on GMOs (19 September).

24 *1979.* Council of Europe's recommendation on the human artificial insemination project.

25 *1984.* Mr Lizin presented a proposal for a resolution concerning Community regulation of artificial insemination.

26 *1988.* European Parliament Resolution no. 372/88 concerning *in-vivo* and *in-vitro* artificial fertilisation.

27 *1989.* Report on human reproductive technologies, Council of Europe.

28 *1991.* The Commission established a working group on Human Embryos and Research (HER).

29 *1991.* Council Recommendation no. 1160 concerning the draft of the European Bioethics Convention (28 June).

30 *1996.* Council of Europe 'Convention for the protection of human rights and dignity of the human being with regard to the application of biology and medicine' (November).

31 *1982.* Recommendation no. 934 of the Council of Europe concerning genetic engineering. (26 January)

32 *1986.* Recommendation no. 1046 of the Parliamentary Assembly of the Council of Europe, on the use of human embryos and foetuses for diagnostic, therapeutic, scientific, industrial and commercial purposes.

33 *1988.* Resolution of the European Parliament on ethical and legal problems of genetic manipulation.

34 *1990.* Proposition of the Commission to the Council with a view to a decision adopting a specific research and technological development programme on health and human genome analysis.

35 *1990.* Resolution no. 3 on bioethics, concerning the mission of the CDBi (Steering Committee on Bioethics) which was established to prepare a draft of the Convention on Human Rights and Biomedicine.

36 *1990.* Committee of Ministers of the Council of Europe: Recommendation no. R(90)13 on prenatal genetic screening, prenatal genetic diagnosis and associated genetic counselling.

37 *1991.* European Council Recommendation (no. 1160) on Genetic Engineering.

38 *1992.* Committee of Ministers of the Council of Europe: Recommendation no. R.92.1 of 10 February 1992 on analysis of DNA for penal process purposes.

39 *1992.* Committee of Ministers of the Council of Europe: Recommendation no. R(92)3 of 10 February 1992 on genetic testing and screening for health purposes.

40 *1995.* Directive of the European Parliament and of the Council with regard to the processing of personal data and on the free movement of such data (24 October).

41 *1996.* Committee of Ministers of the Council of Europe: Convention on Human Rights and Biomedicine (19 November).

42 *1996.* Opinion no. 6 (GAEIB) on the ethical aspects of prenatal diagnosis (20 February).

43 *1976.* The European Science Foundation (ESF) creates the Ad Hoc Committee on DNA Research.

44 *1978.* European Commission, proposal for a Council Directive establishing safety measures against the conjunctural risks associated with recombinant DNA work (5 December).

45 *1980.* European Commission Draft Council Recommendation concerning the registration of Recombinant DNA work (July) .

46 *1982.* European Council: Council Recommendation (no. 934) of 30 June 1982 of Parliamentary Assembly concerning the registration of work involving recombinant DNA.

47 *1982.* European Science Foundation: Report on the Current State of Regulations concerning Recombinant DNA Work (result of an inquiry conducted among the members of the former Liaison Committee for Recombinant DNA Research).

48 *1983.* Creation of the Biotechnology Steering Committee and its secretarial arm: the Concertation Unit for Biotechnology in Europe (CUBE).

49 *1984.* Committee of Ministers of the Council of Europe: Recommendation no.16 on research with recombinant DNA.

50 *1984.* Council of Europe: Recommendation no. R(84) concerning the registration of work involving DNA (15 September).

51 *1985.* Creation of the Biotechnology Regulation Inter Service Committee (BRIC, July).

52 *1990.* Council Directive no. 90/219 (23 April) on the contained use of GMOs (JOCE no. L117, 8 May).

53 *1990.* Council Directive no. 90/220 on the deliberate release into the environment of GMOs (23 April).

54 *1992.* Decision of the Commission (92/3215) authorising the marketing of recombinant vaccine (18 December).

55 *1993.* Commission Decision no. 93/572 concerning marketing of products which contain GMOs, application of the article 13 dir. no. 92/220 (19 October).

56 *1994.* Commission Directive no. 94/15/EEC adapting Council Directive 90/219/EEC in the light of technical progress (15 April).

57 *1994.* First authorisation of the Commission (no. 94.385) to release genetically modified tobacco seeds (8 June).

58 *1994.* Commission Directive no. 94/15/EEC adapting Council Directive 90/219/EEC in the light of technical progress.

59 *1997.* Decisions no. 97/392 and 97/393 of the Commission concerning the marketing of a genetically modified colza (6 June).

60 *1997.* Directive no. 97/35/EC adopted by the Commission adapting Council Directive no. 90/220/EEC in the light of technical progress (18 June).

61 *1988.* Proposition of the Commission concerning Council Directive on the protection of workers from the risks related to exposure to biological agents at work (June).

62 *1990.* Council Directive no. 90/679/CEE (26 November).

63 *1993.* Directive no. 93/88/CEE of the Council modifying Directive no. 90/679/CEE (12 October).

64 *1995.* Directive no. 95/30/CEE of the Commission adapting Directive no. 90/679/CEE in the light of technical progress (30 June).

65 *1978.* Creation of the European Federation of Biotechnology, a forum for professional associations and academic institutions.

66 *1979.* Proposal of the Commission to the Council concerning a programme of research and development in the biomolecular engineering sector (11 February).

67 *1982.* The first research programme on biotechnology is launched at the Community level: Biomolecular Engineering Programme (BEP, 1982–86).

68 *1984.* The start of research policy in the field of biotechnology; European Biotechnology Coordination Group is set up in the early 1985 as a main partner in the preparation of biotechnology research and development programmes (December).

69 *1985.* Biotechnology Active Plan (BAP) launched (1985–89) and revised in 1987.

70 *1984.* Creation of the Biotechnology Steering Committee (February) later named Concertation Unit for Biotechnology in Europe.

71 *1986.* Creation of European Laboratories Without Walls (ELWW).

72 *1989.* Council approves a programme of research and development in the agro–industrial sector (1988–93) based on biotechnology (23 February): European Collaborative Linkage of Agriculture and Industry through Research (ECLAIR).

73 *1990.* Adoption by the Council on 23 April of the Third Framework Programme for research and technological development (1990–94).

74 *1990.* Programme on Biotechnology Research for Innovation, Development and Growth in Europe (BRIDGE, 1990–93).

75 *1990.* Technical Committee on Biotechnology established to assess the requirements in the areas of biotechnological research, industrial production, agricultural and environmental applications (September).

76 *1991.* European Information Service on Biotechnology established.

77 *1994.* Fourth Framework Programme (1994–98); 552 million ecu dedicated to biotechnology programmes and 336 million ecu to biomedical programmes.

78 *1997–98.* Negotiation of the Fifth Framework programme (1999–2003): included in the three thematical programmes are discovering living resources and ecosystems.

79 *1976.* First prepared report on possible initiatives in the promotion of a European biotechnology industry.

80 *1980.* Withdrawal of the previous proposal and replacement by a Council Recommendation (non-binding) allowing member states to adopt laws, regulation and administrative provisions requiring notification, not authorisation, of recombinant DNA work.

81 *1991.* Communication of the Commission to the Council of Ministers and the European Parliament (17 April): 'Promoting the Competitive Environment for the Industrial Activities based on Biotechnology within the Community'. This document considers biotechnology to be a key technology for economic development in the Community.

82 *1991.* Creation of the Biotechnology Coordination Committee (BCC). It covers all sectors of Commission activities in the realm of biotechnology with the participation of all relevant services and assists the Commission in preparing decisions.

Notes and references

1 Ernst & Young, *Biotechnology in Europe '96* (London: Ernst & Young, 1997).

2 Delors, J, *Our Europe: the community and national development*, trans. B. Pearce (London: Verso, 1992). (French title: *La France par L'Europe*).

3 Cantley, M F, 'The regulation of modern biotechnology: a historical and European perspective', in Brauer, D (ed), Biotechnology - Legal, Economic and Ethical Dimensions (Weinheim: VCH, 1995).

4 In particular, the FAST report. (Luxembourg: Commision of the European Communities in asscociation with *Futures*, 1984).

Address for correspondence

Professor Jean-Christophe Galloux, Avocat à la Cour, 52 Avenue des Champs-Elysées, 75008 Paris. Tel. +331 53 83 6068, fax, +331 53 83 6060, E-mail gallouxjc@paris.coudert.com

Part III
Public perceptions

Public perceptions of biotechnology in 1996: Eurobarometer 46.1

George Gaskell, Martin W Bauer and John Durant

Introduction

In the autumn of 1996, the European Commission conducted for the third time a 'Eurobarometer' survey on public perceptions of biotechnology. Eurobarometer 46.1 included a number of trend items from the earlier surveys, together with several new questions devised to extend the investigation to cover more recent developments and issues. Eurobarometer 46.1 was fielded in all 15 member states of the EU. In addition, our colleagues in Norway and Switzerland obtained national funding for the fielding of the same questionnaire. Thus, the survey provides comparable data for 17 European countries. It explores the relatively informal representations of biotechnology amongst representative samples of the public in all 17 countries.

This report provides an overview of the principal findings of Eurobarometer 46.1. It is divided into two main sections. Section 1 reviews the key results from the questions that provide background information. This includes attitudes of a general nature, engagement with biotechnology, knowledge, views about regulation, and trust in different sources of information. Section 2 covers the most important set of questions in the survey concerning attitudes to six selected applications of biotechnology.

The report complements the portrait of public perceptions presented in the country profiles in Part II of this book. The reporting of the survey results is not in the same order as the questions in the questionnaire. This is because the methodological requirements of the ordering of questions for data collection are not necessarily congruent with substantive content and interpretation. Readers with an interest in a particular question or questions should refer to Appendices 1 and 2, which contain the questionnaire and detailed results on all items for the 17 participating countries. All tables are weighted to $N = 1000$ for each country. The maximum error attaching to all national figures is +/–3%.

Section 1. General findings

Impact of technology on our way of life

For each of six technologies (solar energy, computers and information technology, telecommunications, new materials and substances, space exploration and biotechnology) people were asked, 'Do you think it will improve our way of life in the next 20 years, it will have no effect, or it will make things worse?' (Q6). For modern recombinant DNA technologies a split ballot was used in all countries except Switzerland: half the sample in each country were asked about 'biotechnology' and half were asked about 'genetic engineering' or its nearest linguistic equivalent for each country. This question came at the beginning of the survey to avoid any priming or context effects from other questions.

In Appendix 2 (Table 1), the raw percentages for each technology by country are shown. Here we will look at some of the general patterns in the results. Figure 1 shows that in general Europeans are more optimistic about other technologies than they are about biotechnology/genetic engineering. At the same time, there is more pessimism about biotechnology/genetic engineering, with one in five Europeans believing that it will 'make things worse'.

The results of the split ballot on biotechnology/ genetic engineering are shown at the bottom of Figure 1 (see page 202). These results reveal that the terms 'biotechnology' and 'genetic engineering' appear to have different connotative meanings to respondents. Thus, biotechnology is associated with more optimism (50.1%) and less pessimism (15.1%) than genetic engineering, where the comparable figures are 39.4% and 26.8%.

Figure 2 tells us a little more about national differences in the connotations of the terms 'biotechnology' and 'genetic engineering'. In most countries the term biotechnology has more positive connotations than genetic engineering. In the extreme case of Finland, more than twice as many respondents were optimistic about biotechnology as were optimistic about genetic engineering. On the other hand, in Spain, Portugal, Italy and France genetic engineering is approximately as positively regarded as biotechnology.

It is tempting to speculate that this pattern of results arises from a combination of positive and negative associations of the two terms in the different countries. For example, in countries such as Austria and Finland it may be that many respondents associate the prefix 'bio' with nature and health. It is possible than in countries such as Portugal and Spain the term genetic engineering is associated for many respondents with engineering and technology, which are highly regarded activities in these countries. In contrast, it is possible that for the majority of Europeans the differences between biotechnology and genetic engineering are reflective of the negative connotations of engineering in relation to life, sometimes linked in the media with references to 'manipulating' or 'tampering' with living things.

Table 1c in Appendix 2 shows the extent of 'don't know' responses on this item in the different countries. It is clear that familiarity with biotechnology and genetic engineering varies considerably. In Greece, for example, some 47% say 'don't know' in response to both of these terms; this contrasts with around 15% in Denmark and the Netherlands. These varying percentages of 'don't know' responses need to be borne in mind when interpreting the split ballot results. In a country such as Greece different assessments may reflect more general associations with the terms than in a country such as Denmark, where more people are familiar with modern biotechnology.

In all countries, there is a modest tendency for people who are optimistic about the other five technologies also to be optimistic about biotechnology/genetic engineering (for biotechnology, $r = 0.34$; for genetic engineering, $r = 0.25$). Taken together with the higher levels of 'don't know' responses for biotechnology/genetic engineering (about one-quarter of all respondents), this suggests that many Europeans are uncertain about biotechnology and tend to respond to it in ways that reflect their general views about 'bio' and 'technology'.

These questions on the impact of different technologies on our way of life may be used as indicators of the extent to which particular countries have generally positive or supportive cultures towards technology in general and biotechnology in particular. For purposes of description these will be called 'technology culture' and 'biotechnology culture'.

Figure 3 shows the percentage of optimists for biotechnology/genetic engineering and the mean percentage of optimists for the other five technologies. In every country there is a relatively positive technology culture: the majority of respondents are optimistic about the contribution of the five technologies to everyday life. For biotechnology, in contrast, the culture is generally less favourable: in only four countries (Italy, Spain, Portugal and Belgium) is there a majority of optimists. These results suggest that any scepticism towards biotechnology is not simply determined by a general syndrome of 'technophobia'. However, that there are links between technology culture and biotechnology culture is shown by the fact that as the one declines so does the other (and at a relatively faster rate). This is consistent with the point made earlier about the probable influence of the wider connotations of key terms such as engineering and technology.

Figure 4 looks more closely at expectations for biotechnology, showing the percentages of optimists and pessimists in each country. In every country

apart from Austria, there is a greater percentage of optimists than pessimists. The ratio of optimists to pessimists can be taken as an index of what might be termed bipolarisation of opinions. At one extreme is Portugal, with a ratio of optimists to pessimists of 6.5, closely followed by Spain at 5.7. Here are countries in which the vast majority of the public are optimistic. At the other extreme and showing greatest bipolarisation is Austria, with a ratio of optimists to pessimists of 0.8, followed by Norway (1.1), and Switzerland (1.2). These are countries in which roughly equal numbers of the public are optimistic and pessimistic. Denmark, Greece and Germany are also relatively polarised. High levels of bipolarity are likely to be associated with high levels of controversy about biotechnology/genetic engineering. Interestingly, two of the most polarised countries, Austria and Switzerland, have recently held popular initiatives on aspects of genetic engineering and in both cases, these exercises have been associated with intense media coverage and public debate.

The questions about the impact of technologies on way of life were also asked in the earlier Eurobarometer surveys on biotechnology in 1991 and 1993. The results from all three surveys are shown in Table 1 (see page 207). Since 1991, general optimism about technology has shown a statistically insignificant increase from 67.8% to 69.4%, while specific optimism about biotechnology/genetic engineering has shown a slight but statistically significant decrease from 50.5% (1991) to 47.5% (1996). However, with biotechnology itself moving rather rapidly in Europe since 1996 – not least in the marketplace – it will be important to see what happens to this trend in further surveys.

At the national level there are a number of possible trajectories for optimism. There may be an increase, a decrease, or no change in one or both of general technological optimism and biotechnological optimism. Taking into account sampling error, in Ireland and the UK we see a bifurcation effect: an increase in technological optimism but a decrease in biotechnological optimism. In Germany, Greece, France and Luxembourg, technological optimism is stable but there is a decline in biotechnological optimism. Portugal shows increases in both, while in all other countries for which there is data there is a stable pattern.

For the comparison of the connotative meanings of the two terms from 1993 to 1996, we see overall stability in the percentages of optimists for both biotechnology and genetic engineering, but increases in the percentages of pessimists: for biotechnology, there is an increase from 11% to 15%; and for genetic engineering, there is an increase from 21% to 27%. There are relatively few country trends which reach statistical significance. Among these, Belgium and the Netherlands show increasing optimism for genetic engineering; while Greece and the UK show decreasing optimism for genetic engineering. In the Netherlands there is increasing optimism for biotechnology, while in Greece, Spain and France such optimism declines.

Terminology has always been significant in this subject area. For a new technology with no obvious precedents for the public, the choice of labels is seldom neutral. In the English-speaking world, for example, the terms 'recombinant DNA technology', 'genetic engineering', 'genetic manipulation', 'genetic enhancement', 'genetic modification' and 'life sciences' have all been taken up (and many of them subsequently abandoned) over the past 25 years. Although the term biotechnology itself has a long 'pre-history',[1] it appears to have been adopted specifically in connection with genetic technologies in the late 1970s. The Eurobarometer data appear to indicate that this term has been more successful than genetic engineering in winning European public support; however, whether this will continue to be the case in the future is an open question. This potentially important issue will be dealt with in more detail in a second volume of research findings coming out of this project.

Engagement with biotechnology

From the outset, we have been concerned to develop robust measures of the extent to which people engage with biotechnology in their daily lives. Given the limitations of space, it was decided to include just two questions in this area. Respondents were asked whether they had heard about issues involving modern biotechnology over the previous three months, and whether they had ever talked with anyone else about modern biotechnology. Before turning to results on these items, it is worth recalling that the survey was conducted in November 1996; that is, at around the time that the European media were taking up the issue of the import of genetically modified soyabeans into Europe from North America but before the media explosion which followed the announcement of the cloning of Dolly the sheep. Our results, therefore, provide a snapshot of public engagement at a crucial point of transition in European debates about biotechnology.

For Europe as a whole some 55% of respondents reported that they had heard about modern biotechnology over the previous three months, and 50% stated that they had talked to someone about it prior to the interview. There are large cross-country differences. At one extreme, 79% of Swiss respondents reported that they had heard about modern biotechnology over the past three months; while at the other extreme, the comparable figure for Greece is 30%. Again, 78% of Swiss respondents but only

28% of Greek respondents reported that they had talked to someone about modern biotechnology prior to the interview. While these contrasts are not in themselves surprising (consider for example, that by 1996 the Swiss had already had one referendum on modern biotechnology and were preparing for another in 1997) they do point to widely differing levels of public engagement which must reflect widely differing forms of public debate about modern biotechnology in the 17 countries represented in the study. (See Appendix 2, Table 15.)

Figure 5 shows two indices of engagement. The strong version combines *heard about and talked about* biotechnology, whereas the weaker version comprises *either heard about or talked about* biotechnology. In 1996, there appears to have been three groups of European countries: a highly engaged group – Switzerland, Denmark, Austria, Finland, Norway, Germany and Sweden; an intermediate group – Luxembourg, the UK, France, the Netherlands, Italy and Belgium; and a less engaged group – Greece, Portugal, Ireland and Spain. The less engaged countries are notable for having relatively small biotechnology science and industry bases and relatively low levels of policy-making and media coverage. This contrasts with many of the highly engaged countries, such as Switzerland, Denmark, the Netherlands and Germany, in which a relatively large biotechnology science and industry base is combined with a relatively early entry into policy-making and relatively high levels of media coverage and public debate. Austria and Finland are exceptions to this rule, for reasons described in the country profiles in Part II of this volume.

Genes in the popular imagination

The Human Genome Project and related biomedical research activities have received wide publicity in recent years. In popular science the gene is one of the icons of the 1990s, at the frontiers of research and of the understanding of life.[2] Press coverage of new discoveries regularly highlights the identification of genes associated with particular human characteristics. Along with physical traits and disabilities, genes have been linked with mental and social characteristics such as intelligence, personality and even criminality.

For the public, images of genes and inheritance may come from many sources – education, the mass media, talk with family and friends, etc. As a result, people may differ in the extent to which they come to think of human characteristics as products of inheritance or living conditions. When we look at the views of a nation we may talk of widespread beliefs or representations regarding genetic or environmental determinism, nature and nurture respectively.

Whether such beliefs would be associated with views about modern biotechnology is an open question. To tap such beliefs, the following question was posed:

There are differing views about whether people inherit particular characteristics, that is whether they are born with these characteristics, or whether they acquire them mainly from their upbringing, or the conditions under which they lived. Please tell me whether you think each of the following characteristics is mainly inherited or mainly the result of upbringing and living conditions: body size, intelligence, homosexual tendencies, eye colour, tendency to be happy, criminal tendencies, attitude to work, athletic ability, susceptibility to mental illness and musical ability.

The raw data are shown in Appendix 2 (Table 3). The first and most striking feature of the data is the extent of national variation. Thus, 76% of Finns but only 39% of the French consider that intelligence is mainly inherited. Similarly, 32% of the Irish but only 9% of the Swiss believe that attitude to work is mainly inherited. A total of 24% of Italians and 23% of Austrians but only 9% of Swedes and 10% of Danes state that criminality is mainly inherited. To make sense of such variation, the results for all countries were subjected to a multivariate (factor) analysis. This revealed that the responses have a two dimensional structure. The first factor groups together musical abilities, athletic abilities, intelligence, mental illness and body size. This means that respondents who attribute musical abilities to inheritance are more likely also to attribute intelligence, mental illness and body size to inheritance. The second factor groups together criminal tendencies, attitude to work, happiness, and homosexuality. (Eye colour was excluded as almost everyone attributed this characteristic to inheritance.)

The factor 1 characteristics are widely believed to be genetically determined; indeed, some 21% of respondents attributed all five of them to inheritance. By contrast, the factor 2 characteristics – about all of which there has been considerable controversy amongst researchers themselves – were less widely viewed as genetically based. Thus, less than 3% of respondents attributed all of the factor 2 characteristics to inheritance.

The factor analysis was used to explore country differences. For each factor, a score was created for each country and the countries were classified as below or above the median score. Countries above the median for both factors (and thus above the median in their tendency to attribute all the listed characteristics to inheritance) are: Ireland, the Netherlands, Finland and the UK. Countries below the median on both factors (and thus less likely to attribute all the listed characteristics to inheritance) are: Portugal, Spain, France, Belgium, Luxembourg, Greece and Switzerland. Germany, Denmark and Sweden are above the median on cluster

1 but low on cluster 2. Italy, Norway and Austria are high on cluster 2 and low on cluster 1.

It is tempting to commence a detailed interpretation of the significance of these results. One interpretation might attribute stronger beliefs in genetic determinism in the northern countries to the cultural inheritance of the Protestant religion. Against this stereotypical view are the cases of Norway (high on cluster 2 and low on cluster 1) and Ireland (high on both clusters). Furthermore, on closer examination it emerges that the data on genetic determinism do not correlate well with data in other parts of the survey. While the data deserve further analysis (particularly in relation to media coverage of genetics), for the moment they remain interesting – in some cases, intriguing – findings.

Knowledge

There has been a long interest in the role of scientific knowledge, so-called 'scientific literacy', in the formation of public attitudes towards science and technology. A belief among some decision-makers is that a scientifically literate public will be less swayed by media misreporting and more inclined to show support for developments in science and technology. Miller argues that scientific literacy is essential for an informed public debate on scientific issues in democracies.[3] In a recent cross-national study, he finds that, in the USA, scientific literacy is a strong predictor of support for basic scientific research, but interestingly in Europe the association is much weaker.[4] Other research on European data finds differential relations between knowledge of and attitudes towards science depending on the level of national industrialisation.[5]

Eurobarometer 46.1 included a set of ten knowledge questions (a quiz) about aspects of basic biology and genetics. Respondents were asked to say 'true' or 'false' in answer to each of ten items (the 'don't know' option was also available). The questions and aggregate results are shown in Table 2 (page 208). (see also Appendix 2, Table 2.)

The questions are of two types. The first type concerns what might be deemed 'textbook knowledge': the sort of facts that a person might learn at school or from a television documentary (see Table 2, items 1, 3, 5–7, 9 and 10). These items were originally developed by Heijs, Midden and Drabbe.[6] The second type, while factual in nature, were designed to tap into images and impressions that people had offered in focus-group discussions conducted in the preparatory phase of the research (see Table 2, items 2, 4 and 8). For the textbook items, an incorrect answer is presumed to reflect a lack of formal scientific knowledge, whereas for the image items, an incorrect answer is presumed both to reflect lack of formal scientific knowledge and the presence of menacing images of biotechnology. For respondents who assent to the propositions contained in the image questions, it would seem that modern biotechnology involves the threatening possibilities of adulteration, monstrosities and infection.

Turning first to the textbook items, we deal first with five trend items adopted from the 1993 survey (Table 2, items 1, 3 and 5–7). These have been aggregated in a five-item scale. While five items are hardly an adequate index of scientific knowledge, the resulting scale possesses satisfactory scaling characteristics (Cronbach alpha = 0.59), and associations between the score on this scale and socio-demographic characteristics show plausible concurrent validity. Thus, as might have been expected, the youngest age group (15–39) has the highest knowledge level, men tend to be more knowledgeable than women, knowledge increases with level of education, and knowledge is higher among those who are more closely engaged with biotechnology. Similarly, the three image items were aggregated to form an 'image' scale, where higher scores reflect more menacing images of biotechnology.

Figure 6 shows the mean scores of the European countries on both the trend textbook items and the image items. Dealing with textbook knowledge first, we see wide variation in the country averages from 3.5 in Denmark and Switzerland to 2.3 in Portugal. There is a broader context to these results. In general, the countries with higher levels of knowledge have greater involvement in biotechnology: they have a better developed biotechnology science and industry, a longer history of regulation, and a more active public debate (as reflected, for example, in higher levels of engagement). By contrast, most of the countries with lower levels of knowledge (Portugal, Greece and Spain) have less involvement in biotechnology. A partial exception to this rule is Austria, whose combination of a more active debate and a relatively low level of knowledge makes it rather unusual. As the country profiles reveal, Austria is a relative latecomer to the European debate about modern biotechnology. Its combination of a high level of engagement, a low level of knowledge and a high level of public concern give it a unique position in Europe.

Turning next to the image scale, we find that, in general, as textbook knowledge decreases so menacing images increase. In itself, this is not surprising since the image items are also factual in nature. It is, however, interesting to observe that in countries with lower levels of textbook knowledge there are generally higher levels of 'menacing' imagery surrounding biotechnology. In addition, we note that in some countries with very similar levels of textbook knowledge, levels of menacing imagery are quite different. Thus, Germany and Italy have identical levels of textbook knowledge but threatening images are much more prevalent in Germany than in Italy. A similar contrast exists between Sweden and the Netherlands and Greece and Spain. Once again, we note that Austria has a very high level of menacing images.

A comparison of the results in 1993 and 1996 on the five trend textbook knowledge items shows that in every EU member state there has been a slight (statistically insignificant) increase in average knowledge levels. The significance of knowledge in relation to attitudes is discussed further in the second part of this report.

The regulation of biotechnology

The 1996 Eurobarometer included for the first time an item on the regulation of biotechnology. Clearly, regulation is a complex matter embracing a wide variety of issues many of which lie outside the scope of a single survey. In the qualitative research conducted to inform the development of the survey, the issue of the appropriate level at which regulation should apply had been raised. As a result, respondents were asked, 'Which of the following bodies do you think is best placed to regulate modern biotechnology?' The alternatives offered and the results for Europe as a whole are shown in Figure 7.

As can be seen, more Europeans voted for international organisations such as the United Nations and the World Health Organization (35%) than for either their own national institutions (national public bodies + national parliament = 17%) or the EU (6%). Interestingly, however, scientific organisations seem also to win something of a vote of confidence by being the second highest choice (22%).

Appendix 2 (Table 10) gives the results on this item for each country. In every country except Ireland international organisations are seen as best placed to regulate biotechnology, with the Dutch public giving most support (50%). While scientific organisations are the second choice in most countries there are considerable differences. For example, 32% of Swedes vote for them in contrast to 11% of Norwegians. National public bodies and parliaments receive the highest vote of confidence in Ireland, Norway and Austria, but across Europe as a whole they fare rather badly in public confidence regarding the regulation of biotechnology.

At first sight, these results are surprising. It might have been expected that respondents would turn first to their own national institutions as the most obvious source of regulation. That this is not generally the case may reflect either a lack of trust in national institutions and/or a recognition of the international nature of many of the regulatory issues. Other studies have reported a general decline of public confidence in national political institutions.[7]

On the other hand, we should not ignore the possibility that the public are increasingly aware of the global impacts of new technological developments and, as such, acknowledge the need for regulation of biotechnology at an appropriate international level. For this reason, we should be cautious in assuming that the results represent a vote of confidence in either of the particular international organisations (the United Nations and the World Health Organization) mentioned in the question.

Trust in sources of information

The issue of who is best placed to regulate biotechnology is linked to issues of trust and confidence. In recent years, increasing attention has been paid to the issue of trust as a crucial element in the public perception of science and technology.[8] The 1993 Eurobarometer contained an item on confidence in sources of information. Respondents were asked to say how confident they were in a series of different organisations to tell the truth about modern biotechnology. This question was repeated in the 1996 survey, and the results for both surveys are shown in Table 3.

It can be seen the pattern of results in 1993 is largely preserved in 1996. Once again, consumer organisations and environmental organisations are most widely trusted, followed by school and university and (well down the list) public authorities. Taken with the previous findings regarding the suitability of national public authorities to regulate biotechnology, this result underscores the relative unwillingness of the European public to rely upon their national institutions. At the same time, they reinforce the role of NGOs in the public debate. The modest increase in confidence in consumer organisations may be a trend to monitor as food biotechnology becomes more widely debated in Europe. At the same time, the slight decline in confidence in environmental organisations also deserves continued monitoring.

Figure 8 shows the results on trust in public authorities by country for 1993 and 1996. In 1993, Denmark displayed the highest level of trust in public authorities. In 1996, the situation was rather different. Trust in public authorities has fallen dramatically in Denmark and increased significantly in Greece and Spain. Given that the results are for first choice only, they may reflect the changing visibility of any among the various institutions offered (including, for example, NGOs). Finally, we note that the new data in 1996 for Norway and Finland reveal relatively high levels of trust in public authorities.

Given the obvious importance of the 1993 findings concerning the relatively high levels of trust in NGOs, it was decided to extend this part of the survey in 1996 by including a second split-ballot item designed to explore the extent to which trust is issue-specific. Respondents were asked to select from a list of 12 different institutions which one they had most confidence in to tell the truth about two different areas of modern biotechnology: new genetically modified food crops grown in fields; and introducing human genes into animals to produce organs for human transplants. The results are shown in Figure 9. (See also Appendix 2, Table 13.)

For new genetically modified food crops, environmental organisations were trusted most (22%), with consumer and farming organisations joint second (16%). For xenotransplantation, by contrast, the medical profession was chosen by 44% of the sample, followed by animal welfare organisations (11%). In terms of truth and trust, therefore, Europeans do discriminate between sources of information; and trust is strongly issue-specific. Environmental organisations and the medical profession are generally the most widely trusted institutions within their own spheres of competence.

It is important to note a structural difference between environmental organisations and the medical profession so far as the two chosen examples of biotechnology are concerned. Environmental organisations are not responsible for delivering new food crops to the public, but rather act as watchdogs. The medical profession, however, does carry direct responsibility for the provision of new medical technologies to the public. The fact that providers of agricultural biotechnologies are far less widely trusted than providers of medical biotechnologies is surely significant in the shaping of public attitudes in the two areas.

General attitudes

The 1996 Eurobarometer contains two batteries of attitude questions: a set of ten general items, and a set of questions on six specific applications. In this section, we summarise the results on the general attitude items. (See Appendix 2, Table 9.)

Table 4 provides aggregate results for the 17 countries on the ten general attitude items. It will be seen that a majority of Europeans doubt the sufficiency of current biotechnology regulations (question 1). Other issues about which Europeans appear to feel strongly are: that regulation should not be left to industry alone (question 4); that genetically modified foods should be labelled (question 6); and that public consultation about biotechnology is desirable, notwithstanding the complexity of the subject-matter (question 5). On balance, the responses on these items would appear to reflect a cautious approach to modern biotechnology, with concerns about risk and regulation tempering enthusiasm for the introduction of new applications particularly in the area of food biotechnology.

To investigate the attitudinal structures underlying these responses more closely, the ten items were subjected to multivariate (factor) analysis. Three clusters of attitudes were identified. These segment the public along the following dimensions:

- those relaxed about risk and regulation versus those anxious about risk and regulation (items 1, 4 and 9);

- those with a preference for traditional methods versus those accepting modern biotechnology (items 3 and 10);

- those believing in the need for public involvement (including consumer choice) versus those rejecting the need for such involvement (items 5 and 6).

Looking at the socio-demographic profiles of those with high and low scores on each of these factors, we find that: those relaxed about risk and regulation are more likely to be men and to think that biotechnology will improve our way of life; those with a preference for traditional methods are more likely to be women, to be older, to have more menacing images of biotechnology, and to think that biotechnology will make our way of life worse; and those believing in the need for public involvement are more likely to be younger, more knowledgeable, and, interestingly, to be polarised as to whether biotechnology will improve our way of life or make it worse.

Expectations

Another group of questions in the survey dealt with public perceptions of the future expectations of biotechnology. Respondents were invited to forecast the likelihood of certain things happening as a result of developments in modern biotechnology. Table 5 provides aggregate results on ten items, divided into two groups: a first block dealing with positive scenarios and a second block dealing with negative scenarios.

Only two of the ten scenarios offered produced agreement amongst 70% of respondents: one is positive (solving crimes) and the other is negative (creating new diseases). This striking result illustrates a more general feature of the results, which is that significant numbers of respondents appear to possess both positive and negative expectations of modern biotechnology. Such expectations can only be expected to create ambivalent attitudes. At the same time, we observe that more respondents have positive expectations of medical and diagnostic biotechnologies than do so of agricultural and food biotechnologies. Indeed, only a minority of respondents thought it likely that biotechnology will substantially reduce world hunger over the next 20 years. On the other hand, almost half of all respondents did think it likely that biotechnology will make a contribution to the reduction of environmental pollution.

Factor analysis shows that the two groups of items form two separate scales with high reliability (Cronbach alpha = 0.68). Figure 10 shows the mean scores of the European countries on the two scales of positive and negative expectations. As will be seen, most countries have more positive than negative expectations of biotechnology. Consistent with its generally positive attitudes towards modern biotechnology, Finland shows the biggest excess of positive over negative expectations. An interesting exception is Greece, where negative expectations slightly outnumber positive ones. The interrelationships between public expectations, public attitudes and media coverage in relation to biotechnology will be considered in the concluding chapter of the book.

Section 2. Attitudes to applications of biotechnology

At the core of the survey was a group of questions concerning specific attitudes to applications of biotechnology. Respondents were asked whether they thought each of six biotechnologies was useful, risky, morally acceptable, and should be encouraged. The six technologies were described to respondents as follows (headings in brackets refer to labels used in figures and tables):

1 Using genetic testing to detect inheritable diseases such as cystic fibrosis (genetic testing).

2 Introducing human genes into bacteria to produce medicines or vaccines, for example to produce insulin for diabetics (medicines).

3 Taking genes from plant species and transferring them into crop plants to make them more resistant to insect pests (crop plants).

4 Using modern biotechnology in the production of foods, for example to make them higher in protein, keep longer or change in taste (food production).

5 Developing genetically modified animals for laboratory research studies, such as a mouse that has genes which causes it to develop cancer (research animals).

6 Introducing human genes into animals to produce organs for human transplants such as into pigs for human heart transplants (xenotransplants).

Responses were recorded on the following scale: strongly agree = 2, moderately agree = 1, moderately disagree = -1, and strongly disagree = -2. A 'don't know' category was coded as 0 for the purposes of analysis. Mean scores across Europe are shown in Figure 11. The raw data are in Appendix 2, Tables 4–8.

For the public in the 17 European countries, we see that there is considerable differentiation in the evaluations of usefulness, risk, moral acceptability and support across the six applications. All the six biotechnological applications are seen as potentially useful, with genetic testing, medicines and crop plants the most useful. Crop plants, food production, the use of transgenic animals for research and xenotransplantation are seen to involve risks. Only the use of transgenic animals for research and xenotransplantation are thought of as morally unacceptable. Overall, the European public believe that applications involving genetic testing, medicines and crop plants should be encouraged, whereas transgenic animals for research, food applications and xenotransplantation should not be encouraged.

This pattern of results strongly suggests that the European public are not fairly described as 'antibiotechnology'. Rather, people appear to judge specific applications on their individual merits.

The overall pattern of encouragement for the six applications hides some striking differences between the European countries. In Table 6, the country scores for the level of encouragement for each of the six applications are depicted. The table is divided into a grid where each cell represents 0.25 of a unit within the five-point response scale. Here, increasingly positive scores represent greater support, increasingly negative scores represent greater opposition, and scores around 0 represent neither support nor opposition. Perhaps the clearest result is the widespread support for biotechnology in the areas of genetic testing and the production of new medicines and vaccines: in all countries, these applications appear on the positive side of the table. But the table also shows considerable variation across Europe in terms of the general level of support for the applications, and also the relative level of support for the different applications.

Looking at the top and the bottom of the table, we see that in Spain and Portugal all the applications are supported, whereas in Austria only medicines and genetic testing are supported – and even these get luke-warm support. The countries showing the least support are Austria, Switzerland, Norway, Germany, Sweden and Denmark. Interestingly, these are among the most economically developed countries in Europe. Italy, Finland, Greece, Spain and Portugal are the most supportive, and in an intermediate position we see Luxembourg, Ireland, the Netherlands, the UK, France and Belgium.

We also see different sensitivities across the countries, as judged by the application which receives least support. In Austria, Sweden, Denmark, Luxembourg, France and Greece food biotechnologies are not supported and have the most negative judgement of the six applications. Xenotransplants cause most concern in Switzerland,

Norway, Ireland, the Netherlands, the UK, Finland and Portugal. Transgenic animals are of most concern in Germany and Belgium.

Finally, with respect to food applications of biotechnology there are nine countries where the public is not supportive (Austria, Switzerland, Norway, Germany, Sweden, Denmark, Luxembourg, the UK and Greece) and six where the public is supportive (Ireland, the Netherlands, Belgium, Finland, Portugal and Spain). This survey result from late 1996 seems to anticipate the controversy and conflict over GM foods that we see across Europe in 1998.

How do these results for 1996 compare with those of the Eurobarometers conducted in 1991 and 1993? The 1996 section on applications involved new questions designed to reflect developments in the field. However, the previous surveys asked questions of a similar kind about some of the same areas of application: medicines, genetic testing, crop plants and GM foods. In both 1991 and 1993 all four applications were considered 'worthwhile and should be encouraged', with food production gaining the least support. Indeed, the evaluation of GM foods declined from 1991 to 1993. In the 1996 survey, we see indications of a continuation of this trend, with foods now receiving a slightly negative score (−0.15).

Logics of support

For each of the six applications of biotechnology, respondents were asked to judge usefulness, risk, moral acceptability and level of encouragement. A number of hypotheses can be made as to how people combine these four dimensions of assessment. A person may, for example, focus on the risk issue. If the application in question is seen as risky, then it may be judged that it cannot be either useful or morally acceptable and should not to be encouraged. Alternatively, the key issue for respondents may be moral acceptability. Again, it is possible that in this rather unfamiliar area people may have a general view about whether something ought to be encouraged, and then make their judgements on the other dimensions in ways consistent with this initial reaction. Across a population there may be a variety of such patterns of response, or 'logics'.

For the purposes of our analysis we shall assume, following the order of the questions, that the final question concerning the level of support is based on some combination of use, risk and moral acceptability. With this in mind, a multiple regression analysis was undertaken in order to determine the best predictors of encouragement. This analysis indicates that the model provides a good statistical explanation of encouragement. The average variance accounted for over the six applications is 65%, and in no case does the figure fall below 60%. Overall, moral acceptability is the best predictor of encouragement (average $\beta = 0.52$), followed by usefulness (average $\beta = 0.39$). Surprisingly, risk has very low predictive value (average $\beta = -0.06$).

The absence of a relationship between risk and encouragement is remarkable, particularly in light of the importance attached to issues of risk and safety in scientific debate and public policy-making.[9] This suggests that there is a disjunction between expert reasoning (focusing on risk) and lay reasoning (focusing on moral and ethical issues), a finding that clearly requires further exploration. Another issue of relevance here is the fact that 'use' and 'moral acceptability' are co-correlated predictors of 'encouragement'. This makes it difficult to undertake further interpretation of the data using multiple regression analysis.

In light of this, it was decided to adopt a different approach to the interpretation of the attitudinal data. The structure of the question set involves asking a sequence of four questions about each of six applications. We may hypothesise that respondents implicitly or explicitly adopt a stance which involves a coherent relationship among the four dimensions: usefulness, risk, moral acceptability and encouragement. To explore this hypothesis, it is necessary to identify patterns of response ('logics') across the four dimensions. If the questions are treated as dichotomous choices (i.e., if we exclude 'don't know' responses), then theoretically respondents can choose any one of 16 different combinations or logics, as follows:

useful	yes or no
risky	yes or no
morally acceptable	yes or no
encourage	yes or no

Since we are interested in the more commonly used logics, we select only those that are used by at least 5% of the population. Across the six applications, three logics always dominate. These dominant logics of support, risk-tolerant support and opposition are shown in Table 7.

It is interesting to note that the supporters and opponents take opposite positions on all the four dimensions. Before looking at the distribution of these logics across the six applications, two caveats are of note. First, this analysis is not based on all respondents, but rather on the subset of respondents who expressed a definite view on each of the four dimensions (for animals, 50.6% of the respondents are included; for genetic testing 65.6% – see Table 8 for details). Second, we have imposed a threshold value for inclusion of 5% using a particular logic. For genetically modified animals and xenotransplantation we will see that two additional logics meet this threshold value and are included. With these caveats

in mind, Table 8 shows the distribution of the logics for the six applications.

Taking medicines and genetic testing first, we see that the dominant logics are those of the supporters and the risk-tolerant supporters. For medicines, only 8% of all the respondents with an opinion fall into the opponent category; for genetic testing the comparable figure is 7%. For both applications there are more supporters than risk-tolerant supporters, but for each application there is a substantial percentage who think that there are risks but notwithstanding express support. For genetically modified crops, while the supporters and risk-tolerant supporters are the largest groups, there is a larger percentage of opponents: some 18%. Turning to food applications, the situation is reversed. The most common logic is that of the opponents: 30% view this application as not useful, risky, morally unacceptable and not to be encouraged. While this group is smaller than the total of supporters and risk-tolerant supporters, it shows that genetically modified foods generate a sizeable number of opponents.

For transgenic animals and xenotransplantation two additional logics meet the 5% threshold. In terms of use, risk, moral acceptability and encouragement these new logics are NNNN and YYNN. These two logics are forms of opposition: the first represents the view that an application is neither useful nor morally acceptable; and the second represents the view that an application is useful but not morally acceptable. The common feature of these two oppositional logics is the absence of moral acceptability, implying that this characteristic overrides perceptions of usefulness and/or absence of risk.

Another interesting feature of this analysis is the level of consensus on the logics used for the different applications. Medicines and genetic testing have the highest levels of consensus: here, the supporters and risk-tolerant supporters are in the overwhelming majority. (Interestingly, these two applications also have the highest percentage of people expressing a view on all four dimensions.) These two applications are followed by genetically modified crops and genetically modified foods, where again the two supporting logics are in the majority, albeit to a lesser extent. For transgenic animals and xenotransplantation there is much less agreement as to how these are to be evaluated. Here, there are roughly equal splits between opponents and supporters for transgenic animals, and a clear majority of opponents for xenotransplantation.

The pattern of the logics across the six applications in Table 8 suggests that perceptions of usefulness, risk and moral acceptability are rather consistently combined to shape overall support in the following way. First, usefulness is a precondition of support. In none of the logics do we see 'not useful' associated with support. Second, people seem prepared to accept some risk as long as there is a perception of usefulness and no moral concern; but third, and crucially, moral doubts act as a veto irrespective of people's views on use and risk. This can be seen in the cases of transgenic animals and xenotransplantation, with the logic YYNN.

We reported above that multiple regression analysis showed that risk plays only a small role in the determination of levels of support. This is confirmed by the inspection of the logics. As a final check, we conducted separate regression analyses on the three most common logics. The pattern of the results was unchanged: moral acceptability is the best predictor of support, followed by usefulness, with risk contributing almost nothing.

The finding that risk is less significant than moral acceptability in shaping public perceptions of biotechnology holds true in each EU country and across all six specific applications. This has important implications for policy-making. In general, policy debates about biotechnology have been couched in terms of the technical and scientific evaluation of risks to the environment and/or human health. If, however, people are more swayed by moral considerations, public concern is unlikely to be alleviated by technically based reassurances and/or regulatory initiatives that deal exclusively with the avoidance of known risks.

Table 9 show the distribution of the common logics by country. If we are correct in assuming that use, risk, moral acceptability and support are the public's basic dimensions of technology assessment, then it is of note that between three and five common logics hold across all the European countries. We see that the prevalence of particular logics differs between the countries, and across the six applications there are specific patterns. Thus, on all applications Finland has the highest proportion of supporters, while the Netherlands has the highest proportion of risk-tolerant supporters. Similarly, Austria and Norway have the highest proportions of opponents on most applications. In some cases, particular national sensitivities may be observed. Thus, the UK generally displays relatively low proportions of opponents; but in the cases of transgenic animals and xenotransplantation it has relatively high proportions of opponents.

Socio-demographic and attitudinal correlates of different logics

It is of interest to inquire whether the three common logics are related to background socio-demographic characteristics (e.g. age, sex and educational level), to other personal characteristics (e.g. knowledge and religiosity) and/or to attitudes towards biotechnology. For this analysis, we have selected food biotechnologies for the following reasons. First, the

patterns of association between the logics and these respondent characteristics are fairly consistent across the six applications. Second, through purchasing, preparation and consumption, food is the 'closest' application of biotechnology to members of the public in their everyday lives. Finally, food is more than an instrumental good: it is part of the social and moral fabric of a society, linked to values and other fundamental concerns.

Table 10 presents the relevant findings. The table introduces a new variable based on Inglehart's distinction between 'materialists' and 'post-material-ists'.[10] Inglehart argued that a process of inter-generational value change is gradually transforming the politics and cultural norms of advanced societies, with a shift from materialist to post-materialist value priorities. The emphasis on economic growth and security is replaced by concerns for the environment and for self expression. The Eurobarometer survey did not have room to include Inglehart's scale. However, a proxy was created out of question 17, which asked respondents to prioritise issues that would influence their vote in a forthcoming election. Materialists were defined as those who chose both 'crime' and 'unemployment', while post-materialists were defined as those who chose both 'education' and 'environment'.

The table shows the bivariate cross-tabulations, but these can be misleading as many of the variables are intercorrelated. To control for this we used a logistic regression analysis on all the variables with the exception of the final four rows, that is the assessments of genetically modified foods on the criteria of use, risk, moral acceptability and encour-agement. These were excluded on the grounds that these were the very basis for the allocation of people to one of the three groups. It can be clearly seen that the supporters and risk-tolerant supporters share many similarities but are strikingly different from the opponents. Thus, we adopted a two-stage analytic procedure: first, we contrasted the supporters with the opponents; and second, we contrasted the supporters with the risk-tolerant supporters.

The first contrast, between the supporters and opponents, produced a reasonably well-fitting model with 77% of cases correctly classified on a cut criterion of 0.5. Opponents were somewhat better predicted at 84%. In contrast with opponents, supporters are significantly more likely to be men, younger, and materialist; they are more likely to have fewer menacing images of biotechnology, to be optimistic about the contribution of biotechnology to everyday life, to believe that current regulations are sufficient, and to be willing to accept some risks for the sake of economic progress. Supporters are more likely to say that they would buy genetically modified foods, and to expect biotechnology to substantially reduce world hunger; they are more likely to trust public authorities to regulate biotechnology, and less likely to believe that biotechnology will create dangerous new diseases.

The second contrast between supporters and risk-tolerant supporters is less clear. Only 56% of cases were correctly classified using the same cut-off criterion. To this extent the differences reported below, although statistically significant, should be interpreted with caution. In contrast with supporters, however, risk-tolerant supporters are less likely to be materialist, optimistic about the contribution of biotechnology to life, to trust public authorities, to believe that current regulations are sufficient, and to say they will buy genetically modified fruits. They are more likely to believe that biotechnology will lead to dangerous new diseases and to agree that religious authorities should have a say in the regulation of biotechnology.

It is worth noting that some of the variables in Table 10 which appear at first sight to distinguish between the three groups (e.g. education and participation in the debate) are not predictors of group membership when all the other variables are taken into account. Equally, it is notable that the level of textbook knowledge does not distinguish between the groups, a point to which we will return in the next section.

As reported above, statistically significant differences between risk-tolerant supporters and supporters are not strong. On certain value and attitudinal items the two groups differ in a manner that is open to a plausible interpretation. We also find some further indications of different representations among the two groups when we look at the specific attitudes to food biotechnology, the final four rows in Table 10. Here, not only do risk-tolerant supporters differ greatly from supporters in their assessment of risk (the defining difference between the groups) but also they differ to a lesser extent in their estimation of usefulness, moral acceptability and encouragement. In other words, while risk-tolerant supporters are prepared to discount the risks associated with genetically modified foods, they do not appear to discount them entirely. Overall, their level of support for genetically modified foods is slightly but significantly lower than that of supporters. The perception of risk, therefore, leads to a more cautious enthusiasm for biotechnology.

The finding that the opponents of food biotechnology are more likely to be women, to be older, to have a lower educational level, to be highly religious, and to be people holding less materialist values suggests that the opponents are not a single grouping but a combination of post-materialists ('green' opponents) and those holding more traditional values ('blue' opponents).[11]

Overall, taking account of all the socio-demographic characteristics and the attitudinal variables, this analysis of the social segmentation of the logics points to a consistent and coherent pattern. Many of the characteristics of supporters and opponents, as depicted by the survey responses, are what might have been expected. However, it must be said that few of the segmenting variables strongly distinguish between the three groups. This is not surprising, perhaps, given that the analysis is based on an aggregation across 17 countries with different cultures and traditional patterns of social segmentation. We might expect, for example, that religion or environmentalism would have differential significance in different countries, suggesting that a country-by-country analysis is a necessary next step. However, as indicated in many of the national profiles featured in this volume, the segmentation of attitudes towards biotechnology is not particularly strong. It may be that because biotechnology is a relatively novel issue with relatively low salience for many people (at least in 1996), we are only just seeing the emergence of segmented attitudes.

Finally, we looked to see how those who have more developed opinions on food biotechnology differed from those with less developed opinions. For this, we compared all those respondents who held one of the three logics for food biotechnology; that is, all those who gave a positive or negative response to the four questions, with those who were excluded from the analysis of logics on account of one or more 'don't know' responses. The contrast between the two groups in rather striking. Those who employ a logic, implying the existence of more developed opinions, have higher textbook knowledge, are better educated, have greater engagement with biotechnology, and have more positive and negative expectations for biotechnology. These characteristics may be seen as the resources from which more developed attitudinal positions are formed.

Knowledge and attitudes

It is often argued that knowledge is an important determinant of support for science and technology: the more informed the public, the more likely they are to be supportive. For a new (and, to many members of the public, relatively unfamiliar) science like biotechnology, this knowledge–support hypothesis might, in part, explain some of the ambivalence observed amongst members of the public. Yet the results from the Eurobarometer show that while knowledge is clearly related to the formation of attitudes, it is not a simple relationship. For example, the correlation between support for each of the six applications of biotechnology and the five-item textbook knowledge scale is very modest. The largest correlation is with medical applications ($r = 0.13$). Similarly low correlations are observed for a nine-item knowledge scale and an aggregate measure of support over the six applications. Furthermore, whether a person is optimistic or pessimistic about the contribution of biotechnology to life is not correlated with knowledge.

By contrast, in the Eurobarometer report by DG XII higher correlations between an index of knowledge and an index of support are reported ($r = 0.22$),[12] leading to the conclusion that knowledge influences support. This analysis is based on aggregate country-level data and with only 15 data points the resulting correlation is statistically insignificant. The given interpretation amounts to the ecological fallacy of explaining aggregate relations, and in this case statistically insignificant relations, at the level of individuals.

Our analyses appear to confirm the finding from a number of other studies that knowledge plays at best a modest part in determining attitudes towards science and technology. However, the situation is probably both more complicated and more interesting than this. Our results suggest, for example, that people with higher knowledge levels are more likely to express a definite opinion about biotechnology, in our terms to have a logic; but, as we have seen, knowledge does not distinguish between holders of the three logics. Furthermore, looking beyond the survey we note that experts, presumably with a sophisticated knowledge of biotechnology, do not always agree. Knowledge is one of the resources that contributes to the formation of opinions, but those opinions may be positive or negative. This in itself is not particularly surprising. In general, we would expect those who know more about a subject to have more definite opinions about it, but while the motive for some to learn more may be to inform their support, the motive for others may be to inform their concern. In this way, the existence of opposing attitudes amongst those with higher knowledge lowers the overall correlation between knowledge and attitudes.

Conclusion

This analysis and interpretation of the Eurobarometer survey on biotechnology shows a complex pattern of attitudes both within and between the European countries. It is apparent that the publics of Europe are discriminating in their views about different applications of biotechnology. Some applications, particularly in the medical arena, are widely supported, while others, for example transgenic animals, raise widespread concerns. The results of the Eurobarometer survey constitutes a valuable and informative feedback mechanism from the public to the many actors and groups involved in different ways in the development of biotechnology. As we argued in the first chapter of this volume, biotechnology is emerging in an environment which

includes public perceptions. These perceptions may at times constitute a challenge or irritation to the biotechnology complex. How scientists, industrialists and regulators respond to these challenges, in terms of symbolic or material changes to the biotechnology project, will determine in part whether there is an accentuation or attenuation of public concerns. We speculate that public opinion, objectified as perceptions, may be increasingly influential – directly, through consumer choice (exit decisions) and indirectly, through the voice that the public has in the political arena as public policy is increasingly informed by social research – not least the focus group.

One of the most significant results is the finding of consistent patterns of assessments or 'logics' concerning the applications of biotechnology. The majority of respondents with more developed opinions view applications of biotechnology through the perspective of one of three logics. These we labelled the logics of support, risk tolerant support and opposition. Although risk is one of the four dimensions of the logics, it appears to play only a marginal role in the expressed level of support. Perceived use and moral acceptability are the more influential dimensions. Thus, unlike earlier debates about new technologies, morality rather than risk appears to be driving public perceptions. The survey does not tell us what people understand by moral acceptability and risks as general constructs in relation to particular applications of biotechnology. Because the terms were provided within the survey questions, the nature of the underlying concerns to which they give voice is necessarily uncertain.

The finding suggests that for the public the language of moral concern is preferred to the language of risk. In the three following hypotheses we speculate on the reasons for this preference. The 'either/or' hypothesis suggests that lay people tend to think about questions of danger and safety in absolute rather than relative terms. On this view, the notion of risk as a measure of dangerousness may be alien: like the outcome of a referendum or a court of law, there is a decision to be made one way or another. A second 'maturity' hypothesis suggests that moral acceptability reflects a general sense of unease among people who have had relatively little exposure to the issues. On this view, with experience and greater knowledge a discourse of generalised moral concern may give way to a discourse of particular concern about known risks. A third 'disqualification' hypothesis suggests that risk is no longer credible in public debate. People do not see it as a way to assess or to compare technological developments because of the perceived failure of the objective risk approach. Expert assurances on the absence of scientific risks in a variety of contexts have been overturned in the light of further experience, and it is possible that this has fostered public scepticism about risk assessment as a whole. Clearly, other hypotheses are possible, but it is of note that the European public appear to incline to the precautionary principle when judging applications of biotechnology. In this sense the public are implying that the onus is on the innovators and regulators to demonstrate safety and, in particular, the absence of longer-term dangers.

For different applications of biotechnology and in different countries, the relative prevalence of the three logics varies. While data from a single survey cannot be expected to explain the origins of these logics, there is suggestive evidence linking the logics to other attitudes and value orientations. As argued above, further research is required to understand how people construct images of usefulness, risk and moral acceptability, and how different combinations of these attributes are combined. Whatever the basis of the logics, however, they may be usefully employed as a systematic and comparative device for the assessment of new applications of biotechnology in future research.

Another key finding concerns the issue of trust. The survey documents low levels of trust in many of the bodies responsible for the regulation of biotechnology. This may be a consequence of either a generalised distrust in social and political institutions of all kinds or a specific distrust in the institutions regulating new technologies. Distrust may arise from ignorance (as people simply fail to recognise the existence of regulations or regulatory agencies) or from knowledge (as people perceive that the regulation of biotechnology has been or is likely to be mishandled; or as they become sceptical about the effectiveness of regulation because of perceived 'loopholes' or lack of transparency). Whatever the reasons, the finding from the Eurobarometer survey that the World Health Organization and the United Nations are seen as best placed to regulate biotechnology should probably be understood as a token of concern rather than as a well-thought out proposal.

It has been suggested that, in an increasingly complex world where everyone is dependent upon the special expertise of many others, trust is a functional substitute for knowledge.[13] We need to be confident that 'someone who knows' is there to do an effective job of regulation.[14] Particularly in

situations of high uncertainty and potentially great public concern, we might expect trust to become a feature of the way issues are viewed; in the absence of trust, the perceived risks and moral dangers are likely to appear greater. We find that there is a significant tendency for those people who express trust in public authorities to tell the truth about biotechnology to have a systematically more positive view: they are more likely to say that biotechnology should be encouraged, they are more likely to regard it as morally acceptable, and they tend to view it as less risky. Significantly, the largest effect of trust is found in the area of agricultural and food biotechnologies. Here, in an area that was the subject of considerable public debate across Europe when the survey was in the field, we find that trust is a crucial mediating factor in shaping public perceptions of biotechnology. Is the widespread absence of trust at the root of calls for more public consultation in the development of biotechnology?

Finally, we look forward from 1996 to the possibility of another Eurobarometer survey. The present results provide a snapshot of public perceptions for 1996. At this time, almost half of the sample said that they had little or no engagement with the topic and there was little or no media debate in a number of European countries. The finding that public perceptions had not changed much since 1993 is perhaps indicative of a relatively quiet period in the development of public representations of biotechnology. However, since 1996 there has been a much greater intensity in media coverage, public interest and policy debates. Dramatic discoveries and a number of high-profile applications have given biotechnology an increased significance for everyday life. It is now a subject that can hardly be ignored. But this is not to say that the 1996 Eurobarometer survey on biotechnology is outdated. While the percentage responses to particular questions would be expected to change even over a short duration, the key structural features and issues will retain their relevance. The evolution of the public's perceptions of usefulness, risk, moral acceptability and trust; the logics of support and opposition; and the relations of these specific attitudes to knowledge, general attitudes and social background: all of these need be charted over a longer time-frame by means of further survey research.

Figure 1. Impact of technologies on way of life

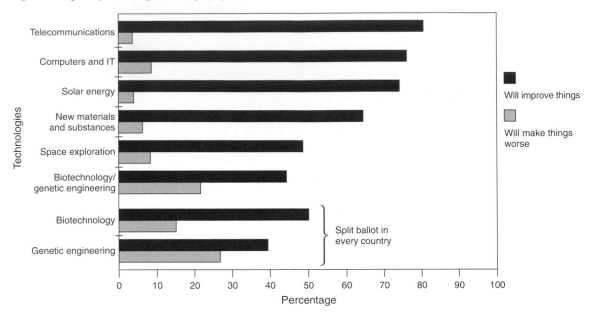

Figure 2. Optimism about impact of biotechnology/genetic engineering on way of life

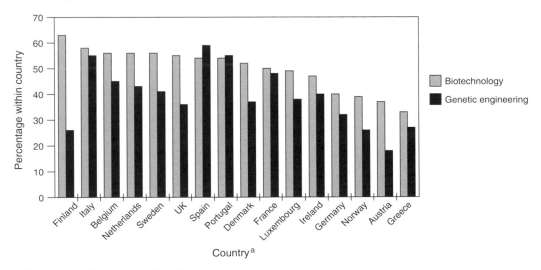

a   Switzerland did not run a split ballot.

Figure 3. Optimism about impact of technology on way of life

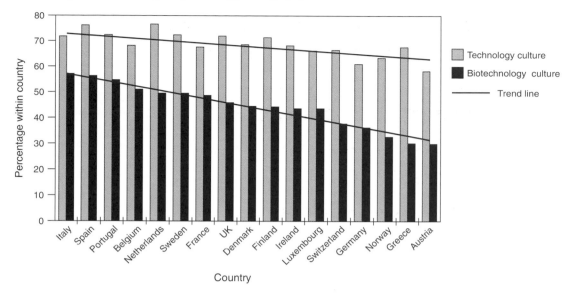

Figure 4. Impact of biotechnology on way of life: optimists and pessimists by country

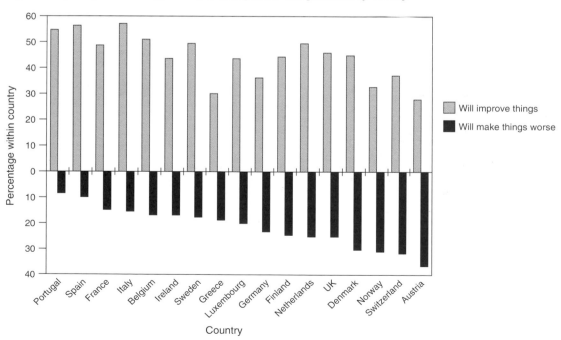

Figure 5. Engagement: 'Have heard of it' and/or 'talked about it'

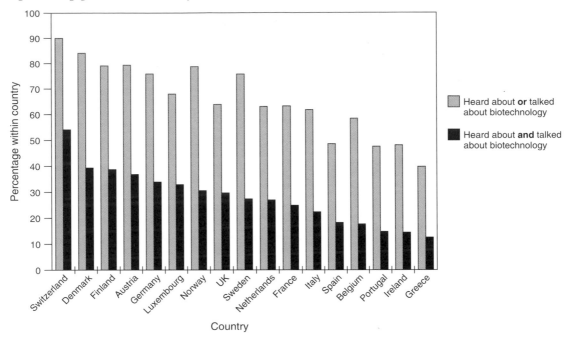

Figure 6. Textbook knowledge (5 trend items) and image scale

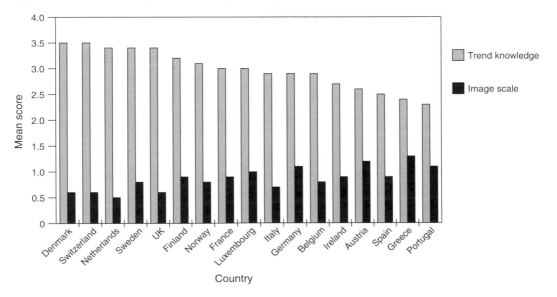

Figure 7. Who should regulate biotechnology?

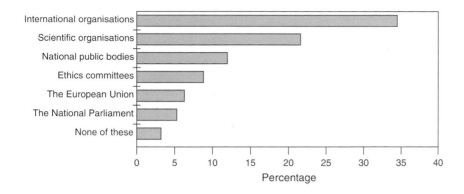

Figure 8. Trust in public authorities 1993–96

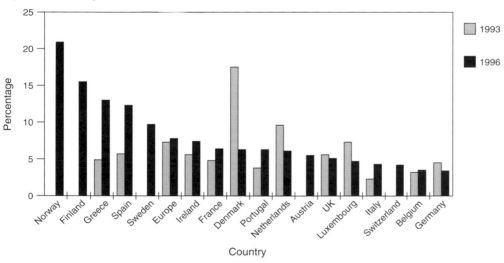

Figure 9. *Who can be trusted to tell the truth about genetically modified foods and xenotransplants?*

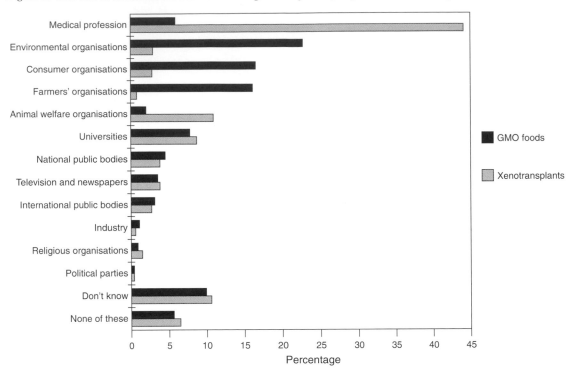

Figure 10. *Future expectations of biotechnology*

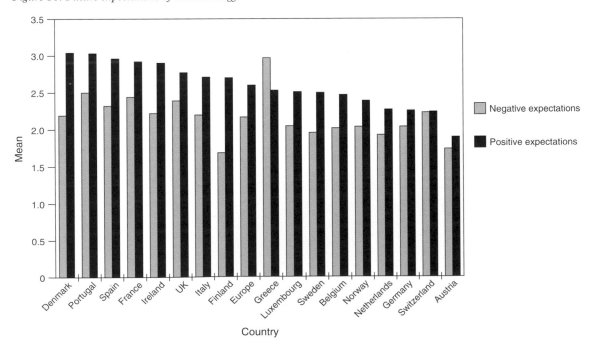

Figure 11. European attitudes to six applications of biotechnology

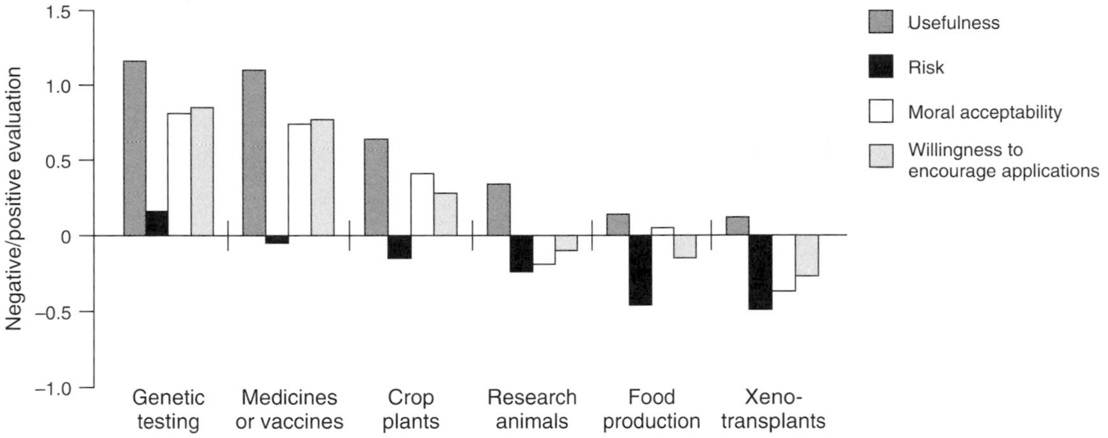

Table 1. *Trajectories of optimism in technoculture and biotechnology*

Country	Culture	1991 (%)	1993 (%)	1996 (%)	Country	Culture	1991 (%)	1993 (%)	1996 (%)
Belgium	Technoculture	68.2	66.2	68.2	Austria	Technoculture			58.2
	Biotechnology	49.0	46.5	51.0		Biotechnology			27.9
Denmark	Technoculture	69.0	70.4	68.6	Portugal	Technoculture	62.2	63.8	72.4
	Biotechnology	43.5	45.0	44.5		Biotechnology	46.5	53.5	54.7
Germany	Technoculture	59.4	57.2	61.0	Finland	Technoculture			71.4
	Biotechnology	43.5	33.5	36.2		Biotechnology			44.3
Greece	Technoculture	69.2	66.6	67.6	Sweden	Technoculture			72.4
	Biotechnology	39.0	37.0	30.1		Biotechnology			49.5
Spain	Technoculture	73.0	75.4	76.2	UK	Technoculture	68.7	70.0	72.0
	Biotechnology	58.0	60.5	56.3		Biotechnology	51.5	51.5	45.9
France	Technoculture	69.8	68.6	67.6	Norway	Technoculture			63.4
	Biotechnology	54.5	53.0	48.7		Biotechnology			32.6
Ireland	Technoculture	63.6	63.8	68.2	Switzerland	Technoculture			66.5
	Biotechnology	48.0	46.0	43.6		Biotechnology			37.7
Italy	Technoculture	69.4	70.6	71.8					
	Biotechnology	56.0	55.5	57.1					
Luxembourg	Technoculture	64.4	63.2	66.2	European	Technoculture	67.8	67.0	69.4
	Biotechnology	48.5	46.5	43.6	Community	Biotechnology	50.5	48.0	47.5
Netherlands	Technoculture	73.8	72.8	76.6	European	Technoculture			68.6
	Biotechnology	48.5	40.5	49.5	Union	Biotechnology			46.5

Table 2. The knowledge questions (means across all countries)

Here are some statements. For each of them, please tell whether you think it is true or false. If you don't know, say so and we will skip to the next statement.	True (%)	False (%)	Don't know (%)
1. There are bacteria which live from waste water.	83.9[a]	3.6	12.6
2. Ordinary tomatoes do not contain genes while genetically modified tomatoes do.	30.6	35.8[a]	33.6
3. The cloning of living things produces exactly identical offspring.	45.7[a]	20.3	33.9
4. By eating a genetically modified fruit a person's genes could become modified.	23.2	48.6[a]	28.3
5. Viruses can be contaminated by bacteria.	46.3	21.6[a]	32.1
6. Yeast for brewing beer consists of living organisms.	68.2[a]	12.5	19.3
7. It is possible to find out in the first few months of pregnancy whether a child will have Down's Syndrome.	79.6[a]	7.2	13.2
8. Genetically modified animals are always bigger than the ordinary ones.	34.6	36.1[a]	29.3
9. More than half of human genes are identical to those of chimpanzees.[b]	50.0[a]	15.6	34.4
10. It is possible to transfer animal genes into plants.	31.2	27.7[a]	41.1

a Denotes the correct answer
b This item was excluded from the analysis for technical reasons

Table 3. Trust in sources of information

Now I would like to know which of the following sources of information you have confidence in, to tell you the truth about modern biotechnology	1993 (%)	1996 (%)
Consumer organisations	24.8	30.4
Environmental organisations	26.3	22.3
Animal welfare organisations	5.7	5.4
Political organisations	0.7	1.4
Trade unions	1.1	1.5
Religious organisations	2.5	3.2
Public authorities	7.3	7.8
Industry	0.9	1.6
School or university	17.7	16.0
Don't know	12.9	10.0

Table 4. General attitudes to biotechnology

People have different views about the benefits and risks of modern biotechnology, and about how they should be regulated and controlled. I am going to read you a number of statements, For each one, please say whether you tend to agree or disagree	Tend to agree (%)	Tend to disagree (%)	Don't know (%)
1. Current regulations are sufficient to protect people from any risks linked to modern biotechnology.	23.9	52.2	23.9
2. Irrespective of the regulations, biotechnologists will do whatever they like.	54.1	31.6	14.3
3. Only traditional breeding methods should be used, rather than changing the hereditary characteristics of plants and animals through modern biotechnology.	56.7	28.3	15.0
4. The regulation of modern biotechnology should be left mainly to industry.	18.5	63.4	18.1
5. Modern biotechnology is so complex that public consultation about it is a waste of time.	27.3	60.7	12.0
6. It is not worth putting labels on genetically modified foods.	16.4	74.9	8.7
7. It would be better to buy genetically modified fruits if they tasted better.	26.9	58.1	14.9
8. Religious organisations need to have their say in how modern biotechnology is regulated.	35.8	49.2	15.0
9. We have to accept some degree of risk from modern biotechnology if it enhances competitiveness in Europe.	28.6	55.6	15.8
10. Traditional breeding methods can be as effective as modern biotechnology in changing hereditary characteristics of plants and animals.	43.9	29.7	26.4

Table 5. Expectations for the future

I am going to read you a list of ten possible things that might happen within the next 20 years as a result of developments in modern biotechnology. For each one, please tell me whether you think it is likely to happen within the next 20 years	Likely (%)	Unlikely (%)	Don't know (%)
1. Substantially reducing environmental pollution.	47.2	45.4	7.4
2. Allowing insurance companies to ask for a genetic test before they set a person's premium.	40.8	44.5	14.7
3. Substantially reducing world hunger.	34.2	58.4	7.4
4. Creating dangerous new diseases.	69.8	19.6	10.5
5. Solving more crimes through genetic fingerprinting.	69.8	19.2	10.9
6. Reducing the range of fruit and vegetables we can eat.	27.7	59.3	13.0
7. Curing most genetic diseases.	54.4	33.4	12.2
8. Getting more out of natural resources in Third World countries.	54.6	31.0	14.4
9. Producing designer babies	36.5	52.5	10.9
10. Replacing most existing food products with new varieties.	42.6	45.8	11.6

Table 6. Level of encouragement for six applications of biotechnology

	← Negative				Positive →					
	−1.0		**−0.5**		**0**		**0.5**		**1.0**	**1.5**
Austria	F	C/T	A		M/G					
Switzerland	T	A	F	C		G	M			
Norway		T	F	C/A		G/M				
Germany		A	T	F	C			G/M		
Sweden		F	A/T		C			G/M		
Denmark		F	T	A/C				G/M		
Europe			T	F/A		C		M/G		
Luxembourg				F/T/A	C			G/M		
Ireland			T	A	F	C	M	G		
Netherlands			T	A	F	C		M/G		
UK			T/A	F		C		M	G	
France			F	T	A	C		M	G	
Belgium				A/T	F		C	G/M		
Greece				F/T		A/C		M		G
Italy				T/F	A		C	M	G	
Finland			T		A	F		C/M	G	
Spain						A/F/T/C		M/G		
Portugal					T	F	A	C/M	G	

A = Research animals, F = Food, C = Crop plants, G = Genetic testing, M = Medicines, T = Transplants

Table 7. Dominant logics across six applications

Logic	Useful	Risk	Moral	Encourage
1. Supporters	Yes	No	Yes	Yes
2. Risk-tolerant supporters	Yes	Yes	Yes	Yes
3. Opponents	No	Yes	No	No

Table 8. The logic of judgements for six applications

Application	Useful	Risk	Moral	Encourage	(%)[b]
Food production	No	Yes	No	No	29.7
(53.2%)[a]	Yes	No	Yes	Yes	22.0
	Yes	Yes	Yes	Yes	21.0
Crop plants	Yes	No	Yes	Yes	34.6
(57.6%)[a]	Yes	Yes	Yes	Yes	26.2
	No	Yes	No	No	17.9
Medicines and	Yes	No	Yes	Yes	41.2
vaccines	Yes	Yes	Yes	Yes	37.0
(63.2%)[a]	No	Yes	No	No	8.1
Genetic testing	Yes	No	Yes	Yes	49.7
(65.6%)[a]	Yes	Yes	Yes	Yes	32.7
	No	Yes	No	No	7.0
Research animals	No	Yes	No	No	25.3
(50.6%)[a]	Yes	No	Yes	Yes	23.7
	Yes	Yes	Yes	Yes	20.1
	No	No	No	No	10.0
	Yes	Yes	No	No	8.6
Xenotransplants	No	Yes	No	No	32.9
(51.0%)[a]	Yes	Yes	Yes	Yes	20.4
	Yes	No	Yes	Yes	16.4
	No	No	No	No	10.4
	Yes	Yes	No	No	9.3

a The percentage of the total sample employing a logic.
b Since other logics were employed, but at less than 5% prevalence, these do not sum to 100%.

Table 9. Logics by country and applications

Crop plants				Medicines and vaccines			
Country	YNYY (%)	YYYY (%)	NYNN (%)	Country	YNYY (%)	YYYY (%)	NYNN (%)
Finland	65	16	11	Finland	65	15	10
Spain	33	29	10	Spain	36	37	5
Greece	50	17	20	Greece	57	19	10
Belgium	41	27	8	Belgium	43	40	4
Ireland	37	29	13	Ireland	41	40	8
Germany	37	19	21	Germany	46	29	11
UK	38	37	14	UK	42	52	6
Portugal	37	41	8	Portugal	31	54	5
Italy	44	31	13	Italy	41	36	7
Switzerland	28	18	26	Switzerland	39	34	13
France	34	30	17	France	35	48	6
Norway	24	18	34	Norway	30	30	12
Denmark	24	25	23	Denmark	53	33	6
Netherlands	23	49	11	Netherlands	20	67	6
Luxembourg	29	23	22	Luxembourg	53	31	7
Sweden	29	26	20	Sweden	45	38	7
Austria	12	11	36	Austria	28	21	14

Table 9. Continued

Genetic testing			
Country	YNYY (%)	YYYY (%)	NYNN (%)
Finland	73	13	5
Spain	43	37	3
Greece	83	13	3
Belgium	45	34	4
Ireland	46	42	4
Germany	52	26	8
UK	56	49	4
Portugal	38	55	3
Italy	53	40	3
Switzerland	50	22	13
France	53	42	4
Norway	37	22	17
Denmark	51	25	8
Netherlands	31	59	7
Luxembourg	55	27	8
Sweden	43	34	6
Austria	29	21	18

Food production			
Country	YNYY (%)	YYYY (%)	NYNN (%)
Finland	46	14	18
Spain	28	27	14
Greece	26	15	42
Belgium	26	21	18
Ireland	28	22	19
Germany	23	20	34
UK	24	28	26
Portugal	21	35	22
Italy	22	23	29
Switzerland	18	13	39
France	20	19	33
Norway	19	16	45
Denmark	15	14	38
Netherlands	14	43	16
Luxembourg	27	14	31
Sweden	13	17	42
Austria	9	8	39

Transgenic animals					
Country	YNYY (%)	YYYY (%)	NYNN (%)	YYNN (%)	NNNN (%)
Finland	38	10	22	8	8
Spain	21	28	12	7	7
Greece	38	14	23	5	4
Belgium	21	18	25	10	9
Ireland	24	21	24	8	9
Germany	18	13	35	9	16
UK	20	21	35	16	10
Portugal	23	43	12	7	3
Italy	21	26	24	10	7
Switzerland	22	12	39	11	11
France	25	24	22	10	5
Norway	27	18	30	5	7
Denmark	28	17	17	14	7
Netherlands	13	35	26	6	15
Luxembourg	30	14	19	7	15
Sweden	18	17	33	6	15
Austria	17	10	28	5	21

Table 9. Continued

Xenotransplants

Country	YNYY (%)	YYYY (%)	NYNN (%)	YYNN (%)	NNNN (%)
Finland	26	8	38	9	9
Spain	20	32	11	7	5
Greece	26	12	28	9	4
Belgium	14	25	28	10	9
Ireland	11	16	33	10	13
Germany	18	14	37	10	13
UK	12	26	38	17	8
Portugal	11	36	23	8	5
Italy	13	23	32	11	10
Switzerland	12	13	48	11	13
France	15	26	28	12	7
Norway	18	13	50	6	8
Denmark	21	17	31	14	10
Netherlands	9	37	34	6	15
Luxembourg	28	15	26	7	15
Sweden	19	20	35	6	12
Austria	12	11	33	5	23

Table 10. Segmentation of the logics of food biotechnology

	Supporters	*Risk tolerant supporters*	*Opponents*
Gender	M 56%: F 44%	M 55%: F 45%	M 45%: F 55%
Age (mean:mode)	41: 24 39	42: 25 39	45: 55+
Education (mode)	19+	16–19	<16
Post materialist	44.8%	46.9%	50.5%
Religious (mode)	No pattern	Low	High
Book knowledge	3.2	3.2	3.1
Menacing images	0.8	0.8	1.0
Engagement in debate	42%	40%	45%
Improve way of life	60%	63%	30%
Make life worse	11%	12%	38%
Current regulations are sufficient	33%	31%	17%
Need to accept risk for progress	42%	40%	17%
Would buy GM foods	48%	43%	9%
Labelling necessary	76%	77%	82%
Involvement of religious organisations	32%	38%	39%
Solve world hunger	44%	43%	24%
Create new diseases	62%	69%	79%
Trust in authorities	20.5%	17.5%	14.0%
Useful food	1.47	1.37	−1.58
Risky food	−1.34	1.22	1.62
Moral acceptable food	1.42	1.31	−1.56
Encourage food	1.43	1.31	−1.69

Notes and references

1 Bud, R, *The Uses of Life: a History of Biotechnology* (Cambridge: Cambridge University Press, 1993).

2 Nelkin, D, 'Forms of intrusion:comparing resistance to information technology and biotechnology', in Bauer, M (ed.) *Resistance to New Technology: Nuclear Power, Information Technology and Biotechnology* (Cambridge: Cambridge University Press, 1995).

3 Miller, J D, 'Scientific literacy: a conceptual and empirical review', *Daedalus* 112 (1983) pp29–48.

4 Miller, J D, Pardo, R and Niwa, F, *Public Perceptions of Science and Technology* (Fundacion BBV and Chicago Academy of Science, 1998).

5 Durant, J, Bauer, M, Midden, C J M, Gaskell, G and Liakopoulos, M, 'Industrial and post-industrial public understanding of science', in Dierkes, M and von Grote, C (eds), *Between Understanding and Trust: The Public, Science and Technology* (Reading: Harwood Academic, in press).

6 Heijs, W J, Midden, C J M and Drabbe, R A, *Biotechnology: Attitudes and Influencing Factors* (Eindhoven: University of Technology, 1993).

7 Renn, O and Levine, D, 'Credibility and trust in risk communications', in R E Kasperson and P J Stallen (eds), *Communicating Risks to the Public* (Dordrecht: Kluwer, 1991).

8 Renn, O and Levine, D, 'Credibility and trust in risk communications', in R E Kasperson and P J Stallen (eds), *Communicating Risks to the Public* (Dordrecht: Kluwer, 1991).

9 European Commission, *Europeans and Modern Opinions about Biotechnology: Eurobarometer 46.1.* (Brussels: Directorate General XII, Science, Research and Development, 1997).

10 Inglehart, R, *Culture Shift in Advanced Industrial Society* (Princeton: Princeton University Press, 1990).

11 Nielson, T H, 'Behind the color code of "no"', *Nature*, 15 (1997), pp1320–1.

12 European Commission (1997), p48.

13 Luhmann, N, *Trust and Power* (Chichester: John Wiley and Sons, 1979).

14 Freudenburg, W R, 'Risk and recreancy: Weber, the division of labour, and the rationality of risk perception', *Social Forces*, 71 (1993), pp909–31.

Address for correspondence

George Gaskell, Department of Social Psychology, London School of Economics and Political Science, London WC2A 2AE.

Part IV
Conclusion

Biology in the public sphere: a comparative review

Martin W Bauer, John Durant and George Gaskell

Introduction

This volume considers the place of modern biotechnology in the public sphere in Europe. Its aim is to
map the public dimensions of a new technology in
ways that permit meaningful comparisons to be
drawn between different aspects of the technology
and different parts of the public sphere. There is no
doubt that modern biotechnology (like other
strategic technologies before it) is – and always has
been – a very public technology. Since the 1970s it
has been the subject of regular public debate on the
part of scientists and technologists, economists and
investors, policy-makers and pundits, and commentators and critics. In the course of this debate,
biotechnology itself has been successively defined
and redefined, negotiated and renegotiated, as
different professional and political interests have
sought to shape the technology according to one set
of priorities or another.

The starting point for the analysis presented here
is that no neat separation can be made between
modern biotechnology itself and those public
representations of it that are the principal subject
matter of this study. Obviously, scientific and
technological developments have played vital parts
in the rise of modern biotechnology, but economic,
financial and legal decisions have also been extremely important in giving direction to these
developments; and so too have wider public and
political processes, without which it must be
presumed that biotechnology would have assumed a
form very different from that which it has today. If
modern biotechnology has become a system, then
this system must be regarded as having scientific,
technological, industrial, financial, ethical, legal and
political aspects, all of which are necessary to a
proper account of its functioning and trajectory.

Even the names given to this area of science and
industry have been heavily influenced by debate
within the public sphere. In the early days, the term
biotechnology itself was hardly used. Instead, the
English-speaking world commonly referred either to
'genetic engineering' or – in more technical discourse – to 'recombinant DNA (rDNA) technology'.
With time, however, what came to be perceived as
the negative connotations of 'genetic engineering' led
to the introduction of two new terms: first, 'genetic
manipulation', and then (as this term, too, came to
be viewed with suspicion) 'genetic modification'
(GM). Recently, in what may be a borrowing from

the German-speaking world, there has been a
noticeable increase in the use of the term 'gene
technology' (Gentechnologie).

The term 'biotechnology' itself long predates the
discovery of rDNA technology, and even the
discovery of DNA itself.[1] In its modern usage,
however, the term first came to prominence in the
late 1970s, just as the economic potential of rDNA
technology was first being widely trumpeted on both
sides of the Atlantic. At this stage, it seems clear that
'biotechnology' presented a number of attractive
features to industrialists and policy-makers alike. A
generic term embracing all of the different applications of rDNA technology, 'biotechnology' was a
useful way of signalling to industrialists, financiers
and policy-makers the need to support what was
judged to be an emerging and strategically significant sector of technology. With time, however, both
the diversification of 'biotechnology' and the
increasing social and political sensitivity of particular
'biotechnologies' led some supporters of the sector
to doubt the wisdom of continuing to use this
generic term at all. Recently, there have been signs
of a move away from the term, and new euphemisms
for modern biotechnology such as 'life sciences' have
made their first appearance on the public stage.[2]

The appearance and disappearance of labels in
this way is symptomatic of the rhetorical work that is
being done by terminology in public debates about
this most sensitive of modern technologies. Do
people wish to play up the technology, emphasising
the radical novelty of rDNA technology by portraying it as a revolutionary break with the past? Then it
may be useful to make use of 'biotechnology'. Or do
people wish, instead, to play down the technology,
emphasising the modesty of rDNA technology by
portraying it as an evolutionary step from traditional
plant and animal breeding? Then it may be useful to
ignore 'biotechnology' altogether, in favour of
'genetic modification', 'life science', or even no
particular term at all. Where the former option may
suit the needs of much economic and financial
debate, the latter may be preferred when dealing
with difficult legislative or regulatory matters.

However modern biotechnology may have been
labelled and defined, it has consistently presented
great functional differentiation within European
culture. Thus, scientists have tended to represent
biotechnology differently from entrepreneurs and
investment bankers; lawyers have tended to represent it differently from educators, environmentalists

or journalists; and so on. In a different context, Luhmann has distinguished between different areas of society (in his parlance, different subsystems) by distinguishing the particular publics associated with each subsystem together with the main symbolic foci of their attention,[3] as shown in Table 1.

Table 1. Biotechnology as a social system (based on Luhmann, 1990, 1995)

Subsystem	Actors	Arenas	Main symbolic focus
Economy	Producers Consumers	Market	Money Profit
Education	Students Teachers	School University	Learning Knowledge Skill
Politics	Voters Politicians	Parliament Street	Power Trust
Science	Laity Experts	Laboratory	Truth(s) Rationality
Medicine	Patients Doctors	Surgery Hospital	Disease Helath/illness
Religion	Laity Priests	Church	Absolute Truth God
Culture	Audiences Artists Creators	Museum Theatre	Beauty
Law	Judges Lawyers	Courts	Justice Legality

The system of modern biotechnology intersects with all of Luhmann's subsystems. As a result, it is actively present in all of the public arenas Luhmann describes. This multiplicity presents social scientists with the obligation to make choices about which arenas to study. In this study, we are mainly concerned with the top four subsystems in Table 1 (the economy, education, politics and science), and we dealt chiefly with the public as producers/consumers, students/teachers, voters/politicians and laypeople/scientists. We have chosen to gain access to these arenas by recording the evolving representation(s) of modern biotechnology within the public sphere, as these have been instantiated by policy processes, media coverage and public perceptions.

When tracking these public representations, we should remember that modern biotechnology is as culturally diverse in contemporary Europe as are the policy-making institutions, the mass media and the general publics that have to deal with it. If public representations of modern biotechnology in Europe today are many and varied, this is only partly because modern biotechnology assumes many different forms and functions; in part, also, it is because the public sphere itself is differently structured in different parts of Europe. Thus,

Scandinavian and Germanic traditions of science and technology policy-making are very different from one another, as are Anglo-Saxon and Mediterranean traditions of science journalism. The ways in which successive Danish or Greek governments have dealt with modern biotechnology reflect both the character of Danish and Greek political culture and the nature of Danish and Greek experience of biotechnology itself. Similarly, the ways in which the British media have represented modern biotechnology tell us as much about the British media as they do about British public attitudes towards gene technology.

Thus, in setting about a European comparative review of biotechnology in the public sphere at the national level, we are obliged to deal with a series of complex interactions between general (non-biotechnology related) and specific (biotechnology-related) factors. The national profiles in Part II of this volume provide the starting point for this review. These profiles provide synoptic overviews which at least begin to place particular public representations of modern biotechnology in the wider cultural contexts that are necessary for their proper interpretation. The main task of this final chapter is to stand back from these synoptic overviews in order to identify obvious patterns of similarity and difference. We begin with comparative reviews of the three principal arenas of the study: public policy, media coverage and public perceptions. Following these, we attempt an integrated comparative assessment of all three aspects of biotechnology in the public sphere.

In undertaking this review, we privilege no single arena of the public sphere (policy, media or public perception) and no single feature of public perception (awareness, knowledge, attitudes, etc.) Rather, we seek to use all of the empirical resources of the project in an attempt to identify significant similarities and differences between different European countries' public responses to biotechnology. In doing this, we shall make extensive use of the indicators provided in the national profiles. To simplify the task of country comparison, we shall group countries into three categories – low, medium and high – on many indicators. Where this is done, we have not employed a rigid criterion such as the tertile split; rather, we have looked across the scores of the different countries and identified significant steps that distinguish between groups of countries. For this reason, among others, the tables of comparative indicators provided below must be treated with a certain amount of caution. At best, they are higher-order abstractions from already abstracted data. They will have fulfilled their purpose if they serve to bring to our attention important patterns within the national profile data.

Policy debates across Europe

There have been many different policy responses to biotechnology in Europe since the 1970s. In part, these responses reflect the national diversity of Europe; but in part, also, they reflect the complexity of the interrelationships between national responses and responses at the European (EU) level. Considering first the national level, it is clear that the different European countries have responded to modern biotechnology in very different ways: some (such as Sweden) responded from as early as the mid 1970s, while others (such as Finland) responded much later on; some (such as Denmark) responded by passing general 'gene laws', while others (such as the Netherlands) opted for a sector-specific approach; and some (such as Germany) chose to implement relatively restrictive regulations, while others (such as France) adopted a stance that was rather more relaxed.

Not only formal but also informal policy responses to biotechnology have varied between countries. Thus, in some European countries (such as Italy) worries about the ethical implications of human reproductive technologies appear to have been the first major cause of public debate about modern biotechnology, while in others (such as the UK) the issue of genetically modified food has attracted most public attention; and in some countries (such as Austria) there has been considerable popular protest against most forms of modern biotechnology, while in others (such as Greece) there has been very little public reaction of any kind to the technology. Clearly, both the nature and the extent of public concern about modern biotechnology are very different in different parts of Europe.

In Table 2, we bring together a number of basic features of the national policy debates of the European countries represented in the study. The first variable shown is the date of entry of each country into the policy debate. Countries are distinguished into three groups: early entrants (which were active in policy-making in the 1970s); middle entrants (which became active in policy-making in the 1980s); and late entrants (which became active in policy-making in the 1990s). Alongside this, we show a measure of the average intensity of policy-making activity in each country over the period of the study. A third column then indicates whether or not each country has chosen to regulate modern biotechnology by means of some sort of cross-sectoral gene law. Finally, two other relevant indicators are included in the table: first, a measure of industrial activity in biotechnology in 1996[4]; and second, an indicator of activity in the field of agricultural biotechnology, in the form of a relative measure of the number of field releases of genetically modified organisms up to 1997[5].

Table 2. Policy debate: crossnational comparisons

Country	Entry policy[a]	Intensity policy[b]	Gene law	Size of biotechnology industry[b]	Number of releases of GMOs[b]
UK	1	2	No	1	1
Sweden	1	1	Yes	1	2
Switzerland	2	2	(Yes)[c]	1	3
France	2	2	No	1	1
Germany	2	1	Yes	1	2
Netherlands	2	2	No	1	2
Denmark	2	1	Yes	2	2
Finland	3	3	Yes	2	3
Austria	3	2	Yes	3	3
Greece	3	3	No	3	3
Poland	3	3	No	3	3
Italy	3	3	No	2	1

a 1 = early, 3 = late
b 1 = high, 3 = low
c Switzerland's distinctive constitution makes it difficult to compare with other countries. While Switzerland has not passed a gene law in the strict sense of the term, it has passed one constitutional amendment in 1992 relating to biotechnology by means of a national referendum. The 1998 amendment was rejected.

Reviewing the table, we note that among the countries under consideration the UK and Sweden are early entrants into the field of biotechnology policy-making, while Austria, Greece, Poland and Italy are all late entrants. On the whole, it is clear that earlier entrants tend to be countries with more intense policy debates. These countries tend to have a larger industrial activity in biotechnology, and they tend also to have a larger involvement in agricultural biotechnology, as measured by the relative number of authorised releases by 1997. None of these features is particularly surprising. On the whole, we would expect countries that are more active in modern biotechnology to perceive a greater need for policy-making in this area.

When it comes to the form taken by national policy-making for modern biotechnology, the situation is rather less clear. For example, there is no obvious correlation between the existence of national gene laws and any of the other variables in Table 2. This may be because the existence or non-existence of gene laws depends more upon national styles and traditions of law-making than it does upon national experiences with modern biotechnology per se. Alternatively, it may be because other factors outside the policy domain are influencing the stance adopted by policy-makers. We return to the second of these possibilities below. Clearly, if we wish to understand more about the particular character of national responses to biotechnology we need to go beyond these relatively crude indicators.

A striking factor in policy debates about biotechnology since 1970 has been the steadily growing influence of policy-making at the European (EU) level. In the beginning, the EU played only a relatively minor role in biotechnology policy-making. Increasingly after 1980, however, the significance of EU decision-making increased rapidly. With the passage of the two EU directives on contained use and deliberate release of genetically modified organisms in 1990, all member states of the EU were obliged to take note of (and, indeed, to implement) EU regulations for modern biotechnology. Some EU member states (e.g. Denmark, Germany, the Netherlands and the UK) implemented the directives relatively quickly, but others (e.g. Italy and Luxembourg) did so much more slowly. It is not clear that any EU member states regarded the directives as excessively restrictive; but it is clear that some member states regarded them as excessively liberal.

In Austria, the EU directives appear to have been the immediate occasion for an intense public and political debate about modern agricultural biotechnology. To an extent far greater than anywhere else, the public debate about modern biotechnology in Austria appears to have been conducted in the immediate context of Austria's accession to the EU in 1995. As certain sections of the Austrian public have viewed with alarm the obligation to implement the EU directives, so certain European institutions (not least, the European Commission itself) have viewed with dismay the possibility that Austria might actually refuse to accept EU law.

Ironically, while the EU directives were widely resented in Austria (and one or two other EU member states such as Luxembourg), they appear to have been accepted by a number of non-member states (such as Poland) as the occasion for a major reorientation of national biotechnology policy. As the Polish national profile makes clear, Poland had favoured a generally North American approach to the regulation of biotechnology until the political revolution that brought about the collapse of the old communist government. After this, Poland's desire to join the EU at the first available opportunity led it to implement the very different European approach represented by the two directives of 1990.

The growing influence of EU policy-making has complicated national policy debates about modern biotechnology in Europe. Not only has it provided a second important locus for debate and deliberation beyond the level of the nation state, but also it has introduced many of the newer EU member states to what are (or were) relatively unfamiliar and untried forms of policy-making. The delicate and at times rather strained relationships between key institutions such as the European Commission, the European Council and the European Parliament have no parallels in the national politics of EU member states; and it is clear that a considerable amount of 'institutional learning' has gone on in the course of fashioning workable European policies for modern biotechnology since the 1970s. Not only official policy-making institutions but also unofficial members of the policy community (including, for example, consumer and environmental agencies) have had to learn to do business at the European level.

Elite media coverage across Europe

It is worth starting our review of media coverage by reflecting on the growth of the coverage of biotechnology across Europe. The aggregate estimates of reportage on biotechnology and genetic engineering in 19 European newspapers during 1973–96 shows the ever-increasing cultivation of biotechnology: from a very low level during the 1970s, to a take-off after 1982, to a plateau between 1987 and 1991, and a veritable explosion after 1992 which is sustained to the present. Needless to say, these aggregates mask considerable national variation. In the following we will therefore only look closely at the years immediately prior to the Eurobarometer survey (i.e. 1995 and 1996). In these years, media coverage of biotechnology varied considerably across the participating countries. At one extreme, there were very few articles in Greece (about ten over two years, making biotechnology roughly a quarterly news event); but at the other extreme, there were very many articles in the UK (1567 over two years, making biotechnology roughly a daily news event). The median coverage across Europe was about 200 articles over two years, or one to two articles per week in the quality press. We may say, therefore, that biotechnology in Europe as a whole had become a weekly news event by 1995–96.

We may characterise the media coverage of biotechnology across Europe on five dimensions: the intensity of coverage; the level of 'risk only' or 'benefit only' reporting (1995–96); and the framing of stories in terms of 'ethics' or 'progress (1992–96)'. Table 3 shows the countries ranked by media intensity (1995–96) and the grouping of each country on these 'other analytic' variables. Although our database is much richer, these comparative dimensions give a flavour of the patterns of biotechnology news across Europe in recent years.

The first and most striking finding is that press reporting of biotechnology is clearly and consistently dominated by benefit stories and the progress frame. Benefit-only stories have a median of 44% of all articles, and the progress frame has a median of 50%

Table 3. Media coverage: crossnational comparisons

Country	Intensity	Media evaluations		Media frames	
		Risk only	Benefit only	Ethics	Progress
UK	1	2	2	2	2
France	1	2	2	2	2
Austria	1	2	2	1	2
Netherlands	2	3	3	2	3
Switzerland	2	1	3	1	3
Germany	2	3	1	3	2
Sweden	2	2	2	2	2
Italy	2	2	2	2	2
Finland	3	2	1	3	2
Poland	3	3	2	3	1
Denmark	3	1	3	1	3
Greece	3	2	1	2	1

1 = high, 3 = low

of all articles. Risk-only coverage is relatively low across Europe, with a median of only 14% of all articles. The progress frame, although most important across Europe, varies from 26% of all articles in the Netherlands to 85% of all articles in Poland. Finally, we note that the ethics frame is only just emerging in media coverage in recent years, with a median value of just 14% of all articles.

The reported intensities of coverage correspond to a situation in which biotechnology is, roughly, a daily news event in the Austria, France and the UK; a weekly news events in Germany, Italy, the Netherlands, and Switzerland; and a monthly or even less frequent news events in Denmark, Finland, Greece and Poland. It should be remembered, of course, that this pattern reflects the situation in 1995–96. Since that time, the media in several countries (e.g. Greece and Poland) appear to have 'awakened' to biotechnology as an issue for the first time, while the media in other countries (e.g. Denmark) appear to have 'reawakened' to biotechnology after a period of relative lack of interest.

The relationship between the intensity and the nature of coverage is not consistent. High-intensity coverage is associated with middle-of-the-road content, while moderate- to low-intensity coverage is linked with a variety of types of content. We suggest that in Austria, France and the UK, high-intensity coverage and a middle-of-the-road form of coverage are indicative of a 'news value discourse' in which biotechnology is perceived to possess intrinsic news value. (As an aside, we note that in 1998 two major European car manufacturers chose to use genetic and evolutionary images in national advertising campaigns.) The combination of high intensity of coverage and high levels of ethics reporting makes Austria particularly interesting here. In Italy and Sweden, the absolute intensity of media coverage is somewhat lower, but the style of coverage is much the same as in Austria, France and the UK. Finally, it is of interest that – with the notable exception of Denmark – the lowest intensities of coverage of biotechnology are associated with an 'advocacy' style of reporting that emphasises benefits and progress to the virtual exclusion of practical and moral concerns.

Media coverage of biotechnology in Finland, Germany, Greece and Poland converges on a pattern which we define as a 'modernisation discourse'. Media coverage in these four countries is characterised by high benefit, low risk, low ethics, and moderate–high progress. Poland is the proto-typical case, with the most extreme biotechnology coverage. This makes a certain amount of sense for economically peripheral countries such as Finland, Greece and Poland. Here the media coverage of biotechnology is generally low intensity, and it supports the impetus to modernisation in the country as a whole. The pattern is more surprising in the German case, where coverage is generally moderate–high intensity, and strongly positive in overall tone. We suggest that this generally modern-ising discourse should be interpreted in the context of the refocusing of the German biotechnology debate in the mid 1990s, as Germany sought to cope with the economic aftermath of reunification and engaged in an internal debate about a supposedly *technikfeindliche Öffentlichkeit* (anti-technological public).

Austria, Denmark and Switzerland all have a pattern of media coverage in which considerable emphasis is placed upon ethical issues. In Italy, Denmark and Switzerland, we find the highest levels of risk-only stories. The media coverage in Denmark and Switzerland, therefore, is interesting by virtue of the fact that it gives considerable emphasis to both practical concerns: risk reporting and ethics report-ing. Hence, we may characterise these two countries as having a 'double concern discourse'. Such a discourse is markedly different from the modernising discourse of, say, Germany. In contrast to the German media, the Danish and Swiss media tend to downplay benefits relative to risks, to frame less in terms of progress, and to highlight ethical issues. In all probability, this represents the pattern of coverage under conditions of public controversy, such as those in Switzerland in the run-up to the national referendum in June 1998.

Austria and the Netherlands reveal rather indi-vidual patterns of media coverage. In Austria, we find high-intensity coverage together with middle-range risk and benefit stories and middle-range progress framing. However, what stands out for

Austria is a high level of ethics coverage. This defines a 'discourse of moral concern'. This, we suggest, may reflect the intense character of the media debate about biotechnology in Austria, particularly in the context of the popular petition that was circulated in 1996. By contrast, the Netherlands has moderate-intensity coverage and low scores on all other indicators except ethics. We define this pattern as 'a discourse of little concern'.

Public perceptions across Europe

In the chapter on the findings from the Eurobarometer survey, we developed a segmentation of the European public identifying the characteristics of those people who used the logics of support, risk-tolerant support or opposition. These profiles focus on individuals' background characteristics and their attitudes and values. This analysis uses data from individual respondents, and as such it inevitably ignores any national differences that may exist. A complementary analysis involves examining the data at the level of the country, and exploring national characteristics that are associated with different degrees of support and opposition in relation to the applications of biotechnology.

Table 4 contrasts the European countries on a number of key indicators. Since the Eurobarometer survey set out to chart how support for and opposition to biotechnology are associated with other variables such as knowledge and attitudes, we have taken the level of encouragement or support for biotechnology as the criterion of ranking. The column labelled 'support' is based on the mean score for each country across the six applications of biotechnology. The level of support is shown in descending country order from the most supportive (Finland) to the least supportive (Austria). Each country is categorised into one of three groups: high support, neutral and low support. However, within the supportive countries, Finland and Greece show substantially higher levels of support and as such are grouped separately from France and Italy. Equally, among the less supportive countries, Austria and Switzerland are considerably more extreme than Germany, Denmark and Sweden. This classification identifies two countries (the UK and the Netherlands) in the middle category. These two countries have very similar mean scores over the six applications of biotechnology and are on the supportive side of neutral.

The first striking observation is that, with the exception of Finland, the supportive countries are all from the south of Europe. In contrast, the less supportive countries are concentrated in Scandinavia (Sweden and Denmark) and the German speaking world (Austria, Switzerland and Germany). Of course, a north/south split is in no way an

Table 4. Public perceptions: crossnational comparisons

Country	Support	Technological optimism	Relaxed attitude to risk	Level of engagement	Knowledge
Finland	1	1	1	1	2
Greece	1	2	1	3	3
France	1	2	3	2	2
Italy	1	1	2	2	2
Netherlands	2	1	1	2	1
UK	2	1	1	2	1
Denmark	3	2	3	1	1
Germany	3	3	2	1	2
Sweden	3	1	3	1	1
Austria	3	3	3	1	3
Switzerland	3	2	3	1	1

1 = high, 3 = low

explanation: rather, it is merely a redescription. In order to interpret the pattern of differences, we need to examine other aspects of the table. First, consider technological optimism. This is an index based on the survey question which measures the extent to which people are optimistic about the contribution of five technologies (excluding biotechnology) to everyday life. While the supportive countries are moderate to high in technological optimism, the less supportive countries (with the exception of Sweden) are low to moderate in optimism. In other words, there appears to be a difference in the general climate for the reception of technology which carries over to biotechnology.

This difference is reflected in the contrast between the more supportive and less supportive countries in public attitudes to risk and regulations on biotechnology. The supportive countries tend to think that current regulations are sufficient to protect against risks and are generally prepared to accept some risk in return for progress. By contrast, in the less supportive countries we find much more concern about such issues; there is greater risk aversion, as well as greater concern about the adequacy of existing regulations.

Regardless of absolute levels of media coverage, our measure of engagement with biotechnology captures the salience of biotechnology amongst the public. The more highly engaged were more likely to have read about and talked about biotechnology in recent months (i.e. during the second half of 1996). Here again, we see a contrast. In the less supportive countries, there is always a high level of engagement.

For the supportive countries, the picture is not so clear. Finland has high engagement, while the other countries are moderate to low. Perhaps the Fins are interested in biotechnology because of its perceived potential for economic development, while in countries like Denmark, Sweden, Germany, Switzerland and Austria, with lower levels of technological optimism and greater concerns about risk and regulations, people are interested in biotechnology because it is perceived as a challenge (or even a threat) about which people feel they ought to know.

Finally, the relations between knowledge and support are complex. The more knowledgeable countries (UK, Netherlands, Denmark, Sweden and Switzerland) cover a wide range in terms of support. Switzerland is one of the most negative countries, while the UK and the Netherlands show modest levels of support. Low knowledge in Greece and Austria is associated with extreme positions on the scale of support. With the exception of Austria, as might be expected, there is a modest positive association between engagement and knowledge. The higher the level of engagement, the higher is the level of knowledge.

The overall pattern of results points to the following tentative conclusion. In the northerly group of countries, with the exception of Finland, there are doubts about the assumption that new technology equals progress, and biotechnology is a familiar yet problematic part of life. By contrast, in

the southerly group of countries (including Spain and Portugal, which featured in the survey but not in the media and policy analyses) there is a supportive environment for technology. Although biotechnology is little understood, people are confident in current regulations and not averse to risk. As such, they are supportive of this new but relatively unfamiliar development.

Policy, media and public perceptions across Europe

We turn, finally, to the integration of the results from the policy, media and public perception studies. At the highest level of comparative generalisation possible within this study, what can be said about biotechnology in the European public spheres?

Table 5 brings together the key indicators from the separate policy, media and public perception comparisons (see Tables 2–4). It is arranged by comparative length of public engagement with biotechnology, as evidenced by the date of entry of each country into the policy and media debates ('entry policy', and 'entry media'). These two variables show very high concordance; only in Italy and Switzerland do the dates of entry into the policy debate differ from the dates of entry into the media debate. Table 5 places early entrants (Sweden and UK) in the upper rows, mid-level entrants (Denmark, France, Germany, the Netherlands, Switzerland) in

Table 5. Policy, media coverage and public perceptions: crossnational comparisons

Country	Entry into the policy debate[a]	Entry into the media debate[a]	Size of biotechnology industry[b]	Number of releases of GMOs[b]	Media intensity[b]	Media tone[c,d]	Support[d]	Knowledge[b]
Sweden	1	1	1	2	2	2	3	1
UK	1	1	1	1	1	1	2	1
Denmark	2	2	2	2	3	3	3	1
France	2	2	1	1	1	1	2	1
Germany	2	2	1	2	2	1	3	2
Netherlands	2	2	1	2	2	1	2	1
Switzerland	2	3	1	3	2	2	3	1
Austria	3	3	3	3	1	3	3	3
Finland	3	3	2	3	3	1	1	2
Greece	3	3	3	3	3	1	1	3
Italy	3	2	2	1	2	2	1	2
Poland	3	3	3	3	3	1	(–)[e]	(–)[e]

a 1 = early, 3 = late
b 1 = high, 3 = low
c Media tone is a variable constructed on an overall assessment based on the country data and national profiles.
d 1 = more positive, 3 = more negative
e No survey was done in Poland.

the middle rows, and late entrants (Austria, Finland, Greece, Italy, Poland) in the lower rows.

In general, we would expect countries with a longer history of public engagement with biotechnology (i.e. earlier entrants into the policy and media debates) to be those with a greater involvement in modern biotechnology itself. In other words, we hypothesise that the degree of public engagement is dependent upon the degree of professional engagement with biotechnology in any particular country. This hypothesis is borne out by a comparison between the rankings on date of entry and the two extrinsic measures: the size of national biotechnology industries; and the number of authorised releases of genetically modified organisms into the environment in the different countries up to 1997. Table 5 shows that generally the earlier entrants into the policy and media debates tend to be those countries which have the largest biotechnology industries and the greatest number of authorised releases by 1997, with the exception of Italy.

These results are consistent with the view that countries that are earlier entrants into the policy and media debates are those with relatively more 'mature' public debates about biotechnology. By speaking of maturity in this context, we do not mean to imply either that there is a 'natural' trajectory of public debate about biotechnology or that earlier entrants somehow enjoy a superior standard (because 'more rational', better informed, or whatever) of public debate. We mean simply that earlier entrants are countries with a longer-lasting, more extensive and wide-ranging public discourse about biotechnology; they are countries in which the public domain has had a greater opportunity to experience, learn about, deliberate on and (possibly) come to hold clear views about modern biotechnology.

This being the case, it is interesting to return to Table 5 in order to discover what other characteristics earlier entrants into the policy and media debates may share, apart from the relative maturity of their public debates. Fairly clearly, earlier entrants tend also to possess a number of characteristics in common. They tend to have a relatively high level of media interest in biotechnology (as reflected in the intensity of media coverage), a rather mixed media tone towards biotechnology, a relatively high level of public knowledge of biotechnology, and a relatively low level of public support for biotechnology. This is the pattern that we might expect to observe in what may be termed 'late-industrial' or 'late-modern' societies; that is, societies with a relatively high level of technological and industrial development that have moved beyond the first flush of enthusiasm for science-based wealth creation and are actively engaged in debate about benefits in relation to risks (for example, in relation to the environment), quality of

life, and other so-called 'post-industrial' values.

The finding that a more mature public debate is characterised by media coverage that is generally more intense but often more critical, together with public perceptions that are on the whole better informed but at the same time less supportive, may appear paradoxical at first sight. We should bear in mind, however, that one reason for higher levels of media coverage of biotechnology is the existence of a greater number of (frequently controversial) public policy issues associated with biotechnology. We should also remember that, even in the earlier entry countries, the balance of media coverage of biotechnology still tends to be generally positive; it is simply not as positive as the media coverage in later entry countries. Finally, we should recall that our analysis of the Eurobarometer data suggested that in many countries better-informed respondents were more likely to possess definite opinions about biotechnology, both positive and negative; it is this polarising effect of greater knowledge that tends to create a lower average level of public support for biotechnology in the earlier entrant countries.

Clearly, there are individual exceptions to the pattern described above for earlier entrants. For example, the UK follows the pattern for early entrants fairly well, except for the fact that it has a higher level of public support for biotechnology. Similarly, Denmark also follows the pattern for early entrants pretty closely, except for the fact that it has a relatively negative media tone towards biotechnology. Again, Germany is a moderately early entrant with many of the features of a late-industrial society but, as we have observed, with a striking contrast between relatively positive media coverage and relatively negative public perceptions of biotechnology. In Germany, it is as if the media are operating with a discourse of modernisation while the public are adopting values more typical of late-industrial societies. These examples serve to remind us that we are dealing with tendencies and trends here, not with fixed laws about the relationships among variables at the country level. Given the extraordinary complexity of biotechnology in the public sphere, we are gratified at the existence of any coherent patterns at the national level, not grudging in our acknowledgment of occasional anomalies and exceptions.

Turning next to the later entrants into the policy and media debates – Austria, Finland, Greece, Italy and Poland – we find that this group of countries tends to share a much smaller number of other features apart the relative immaturity of their public debates about biotechnology. In fact, from the available data it appears that the only other feature they possess in common is a relatively low level of public knowledge about biotechnology. (Interestingly, this is what we would expect if – and only if –

public knowledge of biotechnology is influenced to a significant extent by the prominence of biotechnology in the public sphere.) In other matters – including the level of media interest, the tone of media coverage, and the level of public support for biotechnology – these countries tend to show extreme variability.

It is interesting to inquire why the later entrants into the biotechnology debate are otherwise so diverse. The reason, we suggest, is that – unlike many of the earlier entrants – the later entrants are individually configured quite differently in relation to modern biotechnology. Thus Austria, by far the least enthusiastic of the later entrants, is a relatively wealthy country that has been reluctant to become involved in modern agricultural biotechnology. Austria is well-known for its generally rather cautious attitude towards new technologies, and in the case of modern biotechnology it appears to have become involved largely through its accession to the European Union in 1995, which obliged it to implement a number of EU directives in this area.

Turning to the other extreme, we find that, so far as general attitudes are concerned, Finland is by far the most enthusiastic among the later entrants. Until relatively recently a largely agrarian and pre-industrial country, Finland industrialised rather rapidly in the 1970s and 1980s. Today, this country is generally known for its positive attitudes towards new technologies, and it has been quick to seize on the industrial and economic potential of modern biotechnology. The pattern of variables in Finland is consistent with a public debate that is dominated by the benefits that biotechnology is perceived to bring, at the same time that it is relatively untroubled by worries about risks and other concerns that have been so characteristic of many European countries.

As a former COMECON country, Poland is in a very different position from all the other countries in the study. It may be said to have had two more or less entirely distinct public spheres during the course of this study: the first under Soviet-style communism up until 1989, and the second under democratic rule thereafter. Although Poland has a number of relatively well-established traditional biotechnologies, it has been slow to build a base in modern biotechnology. There has been very little public debate about modern biotechnology in Poland until very recently. Such debate as has taken place since the collapse of communism appears to have been dominated by Poland's desire to obtain entry to the EU.

Italy is in a rather unusual position among this of group of late entrants. Despite its late entry into the debate, it has a rather close engagement with biotechnology as measured by the size of its industrial base and the number of releases of genetically modified organisms that had been authorised by 1997. This picture of relatively close engagement is confirmed by the existence of a moderate media tone and a moderate level of public knowledge about biotechnology. Italy appears to have much in common with the earlier entrants into the debate. However, what most clearly distinguishes Italy from these countries is its extremely positive public perception of biotechnology. In many ways, Italy in 1996 was that rare thing: a country active in biotechnology whose public had still to 'wake up' to the subject.

The final member of the group of late entrants into the debate is Greece. The Greek pattern of late entry, low intensity of media coverage, low intensity of policy activity and low level of public knowledge is suggestive of a country that in 1996 had scarcely begun to address modern biotechnology as a significant issue for public debate and public policy. All the available evidence from our study suggests that by 1996 there had been very little public debate of any kind in Greece about modern biotechnology. Since 1996, there have been signs of increased public attention to biotechnology in Greece and a number of other late or medium entrants to the public debate.

The national diversity in perceptions of biotechnology across Europe is clearly illustrated within the group of medium entrants to the biotechnology debate: Denmark, France, Germany, the Netherlands and Switzerland. Among these countries, Denmark, Germany and Switzerland are characterised by relatively negative public perceptions of modern biotechnology. But while the Danish media appear to resonate to negative Danish public perceptions with a generally critical tone, the Swiss media have a moderate tone and (as we have seen above) the German media a positive tone towards biotechnology. These contrasts raise fascinating questions about the role of the media in shaping public perceptions of new technologies in different cultural contexts.

In Denmark in 1996, there seems to be evidence that a certain phase in the public debate was coming to an end. For example, while media coverage was generally negative, we observe that its level was low and falling. However, the facts that food was a sensitive issue for the Danish public and that the level of policy activity was actually rising suggest that the potential existed for further active debate in the future. In France, by contrast, there are signs that if anything the level of public debate in 1996 was on the increase. For example, the level of media interest was high and rising. This, coupled with rising industrial activity in the area, suggests that France was entering a new, more intense phase of public debate about biotechnology at this time.

Finally, we note that in the wider comparative

context of Table 5 there appears to be a relationship between the form of policy response to biotechnology and the state of public opinion about biotechnology. On the whole, it is countries with more negative public attitudes towards biotechnology (such as Austria, Denmark, Germany and Sweden) that have tended to institute gene laws. This pattern suggests that gene laws have been passed in response to a perceived need to respond to public anxiety by instituting tough regulations. Once again, however, we should be cautious about over-generalising; for an obvious exception to this trend is Finland, the one Scandinavian country in the study which combines a relatively positive public attitude towards biotechnology with the existence of a national gene law.

Conclusion

The main purpose of this volume is to draw together a set of empirical sources arising from a comparative study of biotechnology and the European public. The data presented in the national profiles provides a rich context within which to try to understand better European public perceptions of biotechnology, as these are disclosed by the results of Eurobarometer 46.1. In this final chapter, we have started the process of contextual interpretation by exposing some of the more obvious relationships between policy debates, media coverage and public perceptions. If nothing else, this exercise has served to demonstrate the sheer complexity of public response to biotechnology in western Europe since 1970.

First and foremost, we have shown that there is no unified public discourse about biotechnology in Europe. In terms of public policy, we have found different European countries dealing with modern biotechnology over very different timescales and in very different ways. In terms of media coverage, we have found that a commonly held discourse of progress and benefit is paralleled by rather different patterns of media reportage in the European countries. Last but not least, we have seen from the results of Eurobarometer 46.1 that the different European countries tend to have widely differing levels of engagement with, knowledge about and attitudes towards biotechnology. In light of these results, it is a brave person indeed who would hazard a general conclusion about 'the European view of biotechnology'.

There are many more-or-less obvious reasons for the diversity of public discourses about biotechnology in Europe. One obvious reason is the complexity of biotechnology itself. We have noted, for example, that the European public as a whole tends to view different applications of biotechnology very differently. Given the way in which rDNA technology has found increasingly diverse applications in agriculture, medicine and industry over the past 25 years,

the existence of differentiated public perceptions may be considered both inevitable and appropriate. Only by means of the crudest caricature can the whole of modern biotechnology be captured in a single image.

It is not only the complexity of biotechnology itself, however, that underlies the diversity of public representations of this technology. For the national profiles reveal that the public in different European countries tend to view single applications very differently. The perceptual differences between countries may be explained in part by the very different place that particular biotechnologies occupy in different national economies, and in part by cultural differences which extend far wider and deeper than biotechnology itself. In this sense, we may say that a technologically deterministic view – that is, a view in which the nature of a particular technology itself determines the nature of the public representations of that technology – is falsified by our data.

Public representations of biotechnology in Europe are undoubtedly complex, but there is no reason to suppose that they are chaotic. Within our data, we have detected some striking patterns in the relationships between policy debates, media coverage and public perceptions at the national level. The very fact that such patterns exist is itself suggestive of the existence of a certain amount of orderliness and coherence in European public responses to biotechnology. This orderliness resists simplistic explanation in terms either of the character of the technology or of the character of individual nation states. If adequate explanations are to be found for the sorts of patterns that are described in this chapter, they are more likely to reside in a systemic understanding of the historical processes by which industrialised societies deal with the challenges posed by new technologies.

It is with the clarification of these historical processes that we and our colleagues in this research project are chiefly concerned. Our work has not ended with the production of this source-book: rather, it has just begun. In a second volume of analytical essays, we intend to draw on the data presented here in an attempt to understand better the dynamics of biotechnology in the public sphere. In these essays, we shall explore the factors that stimulate 'institutional learning' in both the scientific–industrial complex of biotechnology and in the public sphere. How do different forms of public discourse about biotechnology relate to wider processes of industrialisation in Europe? How do media and public perceptions of biotechnology contribute to agenda-setting for the policy domain? What are the interrelationships between types of regulation and public perception (including public confidence in the regulatory process)? To what extent does policy controversy influence media coverage and public perceptions, and especially the level of knowledge about biotechnology? And to what extent has ethics replaced risk as the prevailing frame of public debate about biotechnology in Europe? And, if so, why?

These are questions for our second volume. For now, we believe that we have demonstrated the value of documenting the evolution of a strategically significant and socially sensitive technology in the public sphere. We trust that this sourcebook will prove useful to other scholars and practitioners interested in the place of biotechnology in European culture. Western Europe is the region of the world that gave birth to modern science and technology in the sixteenth and seventeenth centuries. It is also the region of the world in which issues concerning the appropriate use of science and technology have come to be most keenly contested in the late twentieth century. There is much still to be learnt about the ways in which public interest and public concern are shaping the science and technology of the future.

Notes and references

1   Bud, R, *The Uses of Life: A History of Biotechnology* (Cambridge: Cambridge University Press, 1993).

2   Ernst & Young, *European Life Sciences '98: Continental Shift* (London: Ernst & Young, 1998).

3   Luhmann, N, *Die Wissenschaft der Gesellschaft* (Frankfurt am Main: Suhrkamp, 1990); Luhmann, N, *Social Systems* (Stanford: Stanford University Press, 1995).

4   Adapted from Ernst & Young, *European Life Sciences '98: Continental Shift* (London: Ernst & Young, 1998).

5   The index for the number of GMO releases has been constructed on the basis of *Genetic Engineering and Biotechnology Monitor 2* (1995), p4; Ernst & Young, *European Life Sciences '98: Continental Shift* (London: Ernst & Young, 1998); *GEO* (June 1989), p52.

Appendices

Appendix 1. Eurobarometer standard 46.1. Technical specifications

Between 18 October and 22 November 1996, INRA (Europe) a European network of market and public opinion research agencies, carried out series 46.1 of the Standard Eurobarometer at the request of the European Commission.

Eurobarometer 46.1 covers nationals of the EU Member States, aged 15 years and over, residing in the Member States of the European Union. The sampling principle applied in all Member States is a multi-stage, random (probability) one. In each EU country, a number of sampling points were drawn with probability proportional to population size (to cover the entire country) and population density.

To this end, the points were drawn systematically from all 'administrative regional units', after stratification by individual unit and type of region. They thus represent the entire territory of the Member States, according to the Eurostat-NUTS II and according to the distribution of the national resident population in terms of metropolitan, urban and rural areas. In each of these selected sampling points, a starting address was drawn at random. Further addresses were selected as every Nth address by 'random route' procedures from the initial address. In each household, the respondent was drawn at random. All interviews were carried out face-to-face in the respondent's home and in the appropriate national language (Table 1).

For each country, the sample was compared with the universe. The universe description was derived from EUROSTAT population data. For all EU Member States, national weighting was applied (using marginal and intercellular weighting), based on this universe description. In all countries, the minimum variables of sex, age, NUTS II regions and the size of locality were introduced in the iteration procedure. For international weighting (i.e. EU averages), INRA (Europe) uses the official population figures published by EUROSTAT in the Regional Statistics Yearbook of 1989. The total population figures introduced in this post-weighting procedure are given above.

The results of the Eurobarometer studies are analysed and reported in the form of tables, data files and analyses. For each question, a table of results is provided, with the full question text (English and French) at the head of the page. The results are expressed 1) as a percentage on total base and 2) as a percentage on the number of 'valid' responses (i.e. 'Don't know' and 'No answer' are excluded). All Eurobarometer data files are stored at the Zentral Archiv (Universität Köln, Bachemer Strasse 40, D-50869 Köln-Lindenthal). They are available to all member institutes of the European Consortium for Political Research (Essex), the Inter-University

Table 1. Details of fieldwork

Countries	Institutes	Number of interviews	Fieldwork dates	Population aged 15+(000)
Belgium	Marketing unit	1006	04/11–17/11	8,356
Denmark	GFK Denmark	1000	01/11–19/11	4,087
Germany (East)	INRA Germany (East)	1008	20/10–17/11	13,608
Germany (West)	INRA Germany (West)	1024	25/10–14/11	52,083
Greece	KEME	1012	01/11–14/11	7,474
Spain	CIMEI	1000	04/11–17/11	28,075
France	TMO	1003	26/10–17/11	43,590
Ireland	Lansdowne Market Research	1003	30/10–20/11	2,549
Italy	PRAGMA	1059	03/11–15/11	44,495
Luxembourg	ILRES	610	18/10–17/11	372
Netherlands	NIPO	1070	28/10–19/11	11,232
Portugal	METRIS	1003	30/10–14/11	7,338
Great Britain	NOP Corporate and Financial	1067	24/10–15/11	44,225
Northern Ireland	Ulster Marketing Surveys	324	03/11–17/11	1,159
Austria	SPECTRA	1009	04/11–17/11	6,044
Sweden	TEMO	1008	02/11–22/11	7,808
Finland	MARK. Development Centre	1040	01/11–19/11	4,017
Total number of interviews		*16246*		

Consortium for Political and Social Research (Michigan) and anyone interested in social science research. The results of the Eurobarometer surveys are analysed by the Public Opinion Surveys and Analyses Unit (Eurobarometer) of DG X/A of the European Commission, Rue de la Loi 200, B-1049 Brussels. They can be obtained from this address.

Readers are reminded that the survey results are estimations, the accuracy of which, all things being equal, depends on the sample size and the observed percentage. With samples of about 1000 interviews, real percentages vary within the following confidence intervals shown in Table 2.

Table 2. Confidence intervals for survey results

Observed percentages	10 or 90	20 or 80	30 or 70	40 or 60	50
Confidence intervals (%)	±1.9	±2.5	±2.7	±3.0	±3.1

Questionnaire

Q.6. Science and technology change the way we live. I am going to read out a list of areas in which new technologies are currently developing. For each of these areas, do you think it will improve our way of life in the next 20 years, it will have no effect, or it will make things worse? (**show card**)

Read out		Will improve	No effect	Will make things worse	Don't know
a) Solar energy	11	1	2	3	4
b) Computers & information technology	12	1	2	3	4
c) Split ballot A : biotechnology Split ballot B : genetic engineering	13	1	2	3	4
d) Telecommunications	14	1	2	3	4
e) New materials or substances	15	1	2	3	4
f) Space exploration	16	1	2	3	4

EB39.1 – Q.40. Trend

Q.7. You've just indicated to what degree you think various new technologies will change the way we live. Now, I would like to ask you what comes to mind when you think about modern biotechnology in a broad sense, that is including genetic engineering.

(**INT: write verbatims in full, prompt 'anything else?', after each word or phrase**)

1. . . . .

2. . . . .

3. . . . .

4. . . . .

5. . . . .

EB46.1. New

INT: Read out : 'For the rest of the interview we are using the term 'modern biotechnology' in a broad sense, that is including genetic engineering.'

Q. 8. Here are some statements. For each of them, please tell me whether you think it is true or false. If you don't know, say so and we will skip to the next statement.

Read out top/bottom and bottom/top alternately		True	False	Don't know
a) There are bacteria which live from waste water.	17	1	2	3
b) Ordinary tomatoes do not contain genes while genetically modified tomatoes do.	18	1	2	3
c) The cloning of living things produces exactly identical offspring.	19	1	2	3
d) By eating a genetically modified fruit, a person's genes could also become modified.	20	1	2	3
e) Viruses can be contaminated by bacteria.	21	1	2	3
f) Yeast for brewing beer consists of living organisms.	22	1	2	3
g) It is possible to find out in the first few months of pregnancy whether a child will have Down's Syndrome, trisomy, Mongolism. (**Use the one or two appropriate terms according to local language.**)	23	1	2	3
h) Genetically modified animals are always bigger than ordinary ones.	24	1	2	3
i) More than half of the human genes are identical to those of chimpanzees.	25	1	2	3
j) It is impossible to transfer animal genes into plants.	26	1	2	3

EB39.1 – Q.41. Trend modified

Q.9. There are differing views about whether people inherit particular characteristics, that is whether people are born with these characteristics or whether they acquire them mainly from their upbringing, or the conditions in which they lived. Please tell me whether you think each of the following characteristics is inherited or mainly the result of upbringing and living conditions.

Read out top/bottom and bottom/top alternately		Mainly inherited	Mainly living conditions	Don't know
a) Body size	27	1	2	3
b) Intelligence	28	1	2	3
c) Homosexual tendencies	29	1	2	3
d) Eye colour	30	1	2	3
e) Tendency to be happy	31	1	2	3
f) Criminal tendencies	32	1	2	3
g) Attitude to work	33	1	2	3
h) Athletic abilities	34	1	2	3
i) Susceptibility to mental illness	35	1	2	3
j) Musical abilities	36	1	2	3

EB46.1. New

Q.10. And now, some questions about various applications which are coming out of modern biotechnology. (**Show card with item 1, 4 dimensions and scale: ask a, then b, then c, then d.**)
(**INT. Code '1' if respondent says 'Definitely agree', code '2' if 'Tend to agreee', code '3' if 'Tend to disagree, code '4' if 'Definitely disagree' and code '5' if 'Don't know'**)

a) First of all, could you please tell me whether you definitely agree, tend to agree, tend to disagree or definitely disagree that it is useful for society to ...

b) And to what extent do you tend to agree or tend to disagree that this application is risky for society?

c) And to what extent do you tend to agree or tend to disagree that this application is morally acceptable?

d) And to what extent do you tend to agree or tend to disagree that this application should be encouraged?

(**Show card with item 2, four dimensions and scale: ask a, then b, then c, then d; then go to item 3, etc.**)

And what do you think of ...?
To what extent do you agree or disagree that ...?

a) This application is useful for society?

b) This application is risky for society?

c) This application is morally acceptable?

d) This application should be encouraged?

Read out	*a) Useful*	*b) Risky*	*c) Morally acceptable*	*d) Encouraged*
a) Use modern biotechnology in the production of foods, for example to make them higher in protein, keep longer or change the taste.	37	38	39	40
b) Taking genes from plant species and transferring them into crop plants, to make them more resistant to insect pests.	41	42	43	44
c) Introducing human genes into bacteria to produce medicines or vaccines, for example to produce insulin for diabetics.	45	46	47	48
d) Developing genetically modified animals for laboratory research studies, such as a mouse that has genes which cause it to develop cancer.	49	50	51	52
e) Introducing human genes into animals to produce organs for human transplants, such as into pigs for human heart transplants.	53	54	55	56
f) Using genetic testing to detect diseases we might have inherited from our parents such as cystic fibrosis, mucoviscidosis, thalassaemia. (**Use the best known example in each country.**)	57	58	59	60

EB46.1. New

Q.11. People have different views about the benefits and risks of modern biotechnology and about how it should be regulated and controlled. I am going to read you a number of statements. For each one, please tell me whether you tend to agree or tend to disagree. (Show card)

Read out top/bottom/top alternately		Tend to agree	Tend to disagree	Don't know
a) Current regulations are sufficient to protect people from any risks linked to modern biotechnology.	61	1	2	3
b) Irrespective of the regulations, biotechnologists will do whatever they like.	62	1	2	3
c) Only traditional breeding methods should be used, rather than changing the hereditary characteristics of plants and animals through modern biotechnology.	63	1	2	3
d) The regulation of modern biotechnology should be left mainly to industry.	64	1	2	3
e) Modern biotechnology is so complex that public consultation about it is a waste of time.	65	1	2	3
f) It is not worth putting special labels on genetically modified foods.	66	1	2	3
g) I would buy genetically modified fruits if they tasted better.	67	1	2	3
h) Religious organisations need to have their say in how modern biotechnology is regulated.	68	1	2	3
i) We have to accept some degree of risk from modern biotechnology if it enhances economic competitiveness in Europe.	69	1	2	3
j) Traditional breeding methods can be as effective as modern biotechnology, in changing the hereditary characteristics of plants and animals.	70	1	2	3

EB46.1 - New

Q.12. Which one of the following bodies do you think is best placed to regulate modern biotechnology? (Show card – read out – one answer only)

International organisations such as the United Nations (UN), the World Health Organisation (WHO)	71 1
Public bodies in (our country)	2
Ethics committees	3
Our national Parliament	4
The European Union, public bodies in the European Union	5
Scientific organisations	6
None of these (spontaneous)	7
Don't know	8

EB46.1 - New

Q.13. I am going to read you a list of ten things that might happen within the next 20 years as a result of developments in modern biotechnology. For each one, please tell me whether you think it is likely or unlikely to happen within the next 20 years. (**Show card with results from 'Modern biotechnology – likely – unlikely'**)

Read out – top/bottom and bottom/top alternately		Likely	Unlikely	Don't know
Substantially reducing environmental pollution	72	1	2	3
Allowing insurance companies to ask for a genetic test before they set a person's premium	73	1	2	3
Substantially reducing world hunger	74	1	2	3
Creating dangerous new diseases	75	1	2	3
Solving more crimes through genetic fingerprinting	76	1	2	3
Reducing the range of fruit and vegetables we can get	77	1	2	3
Curing most genetic diseases	78	1	2	3
Getting more out of natural resources in Third World countries	79	1	2	3
Producing designer babies	80	1	2	3
Replacing most existing food products with new varieties	81	1	2	3

EB46.1. NEW

Q.14. Split Ballot A. Now, I would like to know which of the following sources of information you have confidence in, to tell you the truth about modern biotechnology.
a) Please choose from the following list, the source of information you trust most.
(**show card – one answer only**)
b) Indicate also which other sources you would trust to tell you the truth about modern biotechnology.
(**show same card – multiple answer possible**)

Read out	a) Trust most		b) Other sources	
Consumer organisations	82	1	83	1
Environmental organisations		2		2
Animal welfare organisations		3		3
Political organisations		4		4
Trade unions		5		5
Religious organisations		6		6
Public authorities		7		7
Industry		8		8
School or university		9		9
Don't know		10		10

EB39.1 – Q.52. Trend slightly modified

Q.14. Split ballot B. a) Now, I would like to know which one of the following organisations you have confidence in, to tell you the truth about modern biotechnology. (Show card – one answer only)
b) And to tell you the truth about new genetically modified food crops grown in fields?
(Show same card – one answer only)
c) And to tell you the truth about introducing human genes into animals to produce organs for human transplants? (Show same card – one answer only)

Read out top/bottom and bottom/top alternately	a) Modern biotechnology		b) Genetically modified crops		c) Xeno transplants	
Consumer organisations	84	1	85	1	86	1
Environmental organisations		2		2		2
Animal welfare organisations		3		3		3
The medical profession		4		4		4
Farmers' organisations		5		5		5
Religious organisations		6		6		6
National public bodies		7		7		7
International public bodies		8		8		8
Industry		9		9		9
Universities		10		10		10
Political parties		11		11		11
Television and newspapers		12		12		12
None of these (spontaneous)		13		13		13
Don't know		14		14		14

EB39.1 – Q.52. Trend largely modified

Q.15. We've been discussing several issues to do with modern biotechnology. Some people think these issues are very important whilst others don't. How important are these issues to you person-ally? (Show card – read out). *If you think this is not at all important, you give a score of 1. if you think it is extremely important, you give a score of 10. The scores between 1 and 10 allow you to say how close to either side you are.*

Not at all important									Extremely important
87 1	2	3	4	·5	6	7	8	9	10

Refusal	88 1
Don't know	2

EB46.1. New

Q.16. Over the last three months, have you heard anything about issues involving modern biotechnology? (If yes) Was it in newspapers, in magazines, on television, or on the radio? (several answers possible)

No	89	1
Yes, in newspapers		2
Yes, in magazines		3
Yes, on television		4
Yes, on radio		5
Yes, does not remember (spontaneous)		6

EB46.1. New

Q.17. Before today, had you ever talked about modern biotechnology with someone? (If yes) Had you talked about it frequently, occasionally or only once or twice?

No, never	90	1
Yes, frequently		2
Yes, occasionally		3
Yes, only once or twice		4
Don't know		5

EB46.1 - New

Q.18. Which issue, do you think, will most influence your vote at the next general elections? (Show card – read out top/bottom and bottom/top alternately – four answers maximum)

Protection of the environment and nature, cutting down pollution	91	1
Fight for worker's rights		2
Protection of social benefits and health care		3
Education		4
Fight against racism		5
Protection of pensioners' rights		6
Fight against unemployment		7
Fight against homelessness and poverty		8
Fight against crime and delinquency		9
Taxation		10
Don't know		11

EB41 – Q.39. Trend strongly modified

Q.19. Which newspapers or magazines, if any, do you read at least once a week? (INT: record name of newspapers or magazines – verbatim)

1 None . . . .

2 . . . .

3 . . . .

4 . . . .

5 . . . .

EB46.1. New

Appendix 2. Eurobarometer 1996 survey results

Table 1. Responses to the question: 'I am going to read out a list of areas in which new technologies are currently developing. For each of these areas do you think it will improve our way of life in the next 20 years, it will have no effect, or it will make things worse?' (Eurobarometer code Q6)

Table 1a. Solar energy

Country	Will improve (%)	No effect (%)	Will make things worse (%)	Don't know (%)
Belgium	69.7	18.5	2.6	9.2
Denmark	81.6	12.3	1.8	4.3
Germany	71.3	20.6	1.3	6.7
Greece	69.0	5.9	14.7	10.4
Italy	73.5	9.6	9.3	7.5
Spain	81.1	5.2	5.1	8.6
France	64.4	24.6	3.9	7.1
Ireland	64.4	12.1	4.8	18.7
Luxembourg	73.5	13.3	3.8	9.3
Netherlands	87.7	7.9	2.3	2.1
Portugal	72.7	7.9	7.1	12.3
UK	76.9	15.9	1.7	5.5
Norway	66.8	14.7	4.0	14.4
Finland	77.8	16.5	1.3	4.4
Sweden	77.2	14.8	2.1	5.8
Austria	71.9	16.6	1.1	10.4
Switzerland	80.8	14.7	1.5	3.0
Mean	**74.2**	**13.6**	**4.0**	**8.2**

Table 1b. Computers and information technology

Country	Will improve (%)	No effect (%)	Will make things worse (%)	Don't know (%)
Belgium	76.0	8.3	10.5	5.1
Denmark	71.2	10.9	10.7	7.2
Germany	67.0	18.4	6.6	7.9
Greece	74.8	2.9	13.0	9.3
Italy	83.2	4.6	6.0	6.2
Spain	85.7	3.9	3.3	7.1
France	70.6	10.5	13.6	5.3
Ireland	83.6	4.1	5.1	7.2
Luxembourg	71.6	12.5	6.0	9.8
Netherlands	78.4	6.6	9.5	5.5
Portugal	83.8	3.5	5.6	7.1
UK	81.8	4.5	10.4	3.3
Norway	66.9	6.6	10.4	16.1
Finland	82.2	7.1	6.2	4.5
Sweden	79.6	7.3	8.2	4.9
Austria	67.1	13.1	8.5	11.3
Switzerland	69.2	13.6	11.8	5.4
Mean	**76.1**	**8.1**	**8.6**	**7.2**

Table 1c. Biotechnology and genetic engineering (mean over split ballot)

Country	Will improve (%)	No effect (%)	Will make things worse (%)	Don't know (%)
Belgium	51.0	9.8	16.8	22.4
Denmark	44.5	9.4	30.2	15.9
Germany	36.2	18.3	23.2	22.3
Greece	30.1	4.0	18.7	47.2
Italy	57.1	4.7	15.4	22.8
Spain	56.3	3.5	9.9	30.3
France	48.7	10.6	14.8	25.9
Ireland	43.6	7.0	16.8	32.5
Luxembourg	43.6	13.9	20.0	22.5
Netherlands	49.5	9.8	25.1	15.6
Portugal	54.7	5.6	8.4	31.3
UK	45.9	8.5	25.1	20.4
Norway	32.6	6.0	30.9	30.4
Finland	44.3	13.4	24.5	17.8
Sweden	49.5	9.3	17.6	23.5
Austria	27.9	12.9	36.5	22.6
Switzerland	37.7	11.2	32.1	19.0
Mean	**44.3**	**9.2**	**21.6**	**24.9**

Table 1d. Telecommunications

Country	Will improve (%)	No effect (%)	Will make things worse (%)	Don't know (%)
Belgium	76.2	11.0	5.9	6.8
Denmark	74.5	15.2	5.3	5.0
Germany	68.8	19.9	3.9	7.4
Greece	86.3	3.2	4.8	5.7
Italy	86.0	6.7	2.7	4.6
Spain	88.7	2.5	1.8	7.0
France	81.7	11.1	4.2	3.0
Ireland	87.3	3.3	1.2	8.2
Luxembourg	76.7	12.0	3.7	7.7
Netherlands	86.9	7.2	3.7	2.2
Portugal	91.2	2.4	1.7	4.7
UK	87.1	6.7	2.6	3.6
Norway	75.3	7.2	4.2	13.3
Finland	81.5	11.5	2.7	4.3
Sweden	83.8	8.3	3.3	4.5
Austria	62.0	18.8	6.3	12.9
Switzerland	74.0	17.2	5.6	3.2
Mean	**80.6**	**9.6**	**3.7**	**6.1**

Table 1e. New materials or substances

Country	Will improve (%)	No effect (%)	Will make things worse (%)	Don't know (%)
Belgium	69.8	12.8	3.8	13.6
Denmark	67.7	12.8	5.5	14.0
Germany	56.2	20.5	5.0	18.2
Greece	51.8	7.3	16.2	24.7
Italy	60.4	9.3	12.8	17.4
Spain	59.4	5.1	6.8	28.7
France	73.7	12.6	3.6	10.1
Ireland	65.1	7.1	6.6	21.2
Luxembourg	60.7	15.8	5.3	18.2
Netherlands	78.0	11.5	3.0	7.5
Portugal	62.9	9.5	5.7	21.9
UK	72.1	11.1	3.6	13.2
Norway	62.3	7.7	4.0	26.0
Finland	71.7	12.8	4.7	10.8
Sweden	71.8	12.8	4.5	10.9
Austria	45.6	20.7	11.1	22.5
Switzerland	65.5	15.9	4.7	13.9
Mean	**64.5**	**12.0**	**6.3**	**17.2**

Table 1f. Space exploration

Country	Will improve (%)	No effect (%)	Will make things worse (%)	Don't know (%)
Belgium	49.3	25.8	8.0	16.9
Denmark	46.6	35.4	7.5	10.5
Germany	42.3	35.8	6.4	15.5
Greece	56.4	13.1	11.7	18.8
Italy	57.6	19.1	8.0	15.3
Spain	64.8	12.6	4.7	17.9
France	47.0	34.8	8.0	10.1
Ireland	41.0	27.3	10.2	21.5
Luxembourg	46.9	29.6	8.7	14.8
Netherlands	52.0	31.1	8.1	8.8
Portugal	51.2	17.0	12.7	19.1
UK	42.0	41.5	7.4	9.1
Norway	46.0	21.1	6.7	26.1
Finland	44.1	38.5	6.1	11.3
Sweden	51.3	29.7	6.9	12.1
Austria	44.2	29.8	9.2	16.7
Switzerland	42.9	36.8	10.8	9.5
Mean	**48.6**	**28.2**	**8.3**	**14.9**

Table 1g. Split ballot   A: biotechnology

Country	Will improve (%)	No effect (%)	Will make things worse (%)	Don't know (%)
Belgium	56.6	9.8	12.2	21.3
Denmark	52.1	9.8	19.1	19.1
Germany	40.5	18.4	18.8	22.3
Greece	33.1	4.2	16.3	46.4
Italy	58.5	4.6	10.4	26.5
Spain	53.6	3.6	7.6	35.3
France	49.8	9.4	13.3	27.5
Ireland	47.4	6.0	12.6	34.0
Luxembourg	48.7	14.0	16.0	21.3
Netherlands	56.2	8.1	20.3	15.4
Portugal	54.1	4.8	8.3	32.8
UK	55.1	7.4	13.2	24.3
Norway	39.3	5.5	23.5	31.8
Finland	62.7	11.2	10.0	16.1
Sweden	57.1	8.5	11.5	22.8
Austria	36.8	11.6	27.7	23.9
Mean	**50.1**	**8.4**	**15.1**	**26.4**

B: genetic engineering

Country	Will improve (%)	No effect (%)	Will make things worse (%)	Don't know (%)
Belgium	45.5	9.7	21.3	23.5
Denmark	36.7	9.0	41.8	12.6
Germany	31.9	18.3	27.5	22.3
Greece	27.1	3.8	20.9	48.2
Italy	55.4	5.0	20.5	19.1
Spain	59.0	3.4	12.3	25.4
France	47.7	11.8	16.2	24.2
Ireland	40.2	8.1	20.7	31.1
Luxembourg	38.3	13.7	24.0	24.0
Netherlands	43.2	11.4	29.5	15.8
Portugal	55.3	6.4	8.5	29.8
UK	36.3	9.7	37.7	16.3
Norway	25.9	6.5	38.7	28.9
Finland	26.1	15.4	38.9	19.6
Sweden	41.6	10.4	23.8	24.2
Austria	17.7	14.4	46.8	21.1
Mean	**39.4**	**9.7**	**26.8**	**24.2**

Table 2. Responses to the question: 'Here are some statements. For each of them, please tell me whether you think it is true or false. If you don't know, say so and we will skip to the next statement.' (Eurobarometer code Q8)

Table 2a. There are bacteria which live from waste water

Country	True (%)	False (%)	Don't know (%)
Belgium	81.6	4.3	14.1
Denmark	91.3	2.1	6.6
Germany	81.2	5.1	13.6
Greece	72.0	6.2	21.8
Italy	82.6	3.1	14.3
Spain	79.2	1.8	19.0
France	81.0	5.8	13.2
Ireland	81.8	1.9	16.3
Luxembourg	85.8	2.0	12.2
Netherlands	96.3	1.2	2.5
Portugal	71.7	6.1	22.2
UK	92.6	1.6	5.8
Norway	81.3	3.8	14.9
Finland	94.6	1.3	4.1
Sweden	91.2	2.2	6.6
Austria	73.4	9.1	17.5
Switzerland	88.8	2.3	8.9
Mean	**83.9**	**3.6**	**12.6**

Table 2b. Ordinary tomatoes do not contain genes while genetically modified tomatoes do

Country	True (%)	False (%)	Don't know (%)
Belgium	27.1	30.8	42.1
Denmark	26.3	44.1	29.6
Germany	44.4	36.0	19.6
Greece	41.3	20.2	38.6
Italy	20.9	35.1	43.9
Spain	26.1	27.8	46.1
France	28.6	32.0	39.4
Ireland	28.5	20.2	51.3
Luxembourg	36.5	39.8	23.7
Netherlands	21.9	50.7	27.4
Portugal	31.9	27.0	41.1
UK	22.0	40.1	37.8
Norway	34.0	34.0	32.0
Finland	28.7	44.4	27.0
Sweden	30.1	46.0	23.9
Austria	43.6	33.8	22.7
Switzerland	30.6	48.5	20.9
Mean	**30.6**	**35.8**	**33.6**

Table 2c. The cloning of living things produces exactly identical offspring

Country	True (%)	False (%)	Don't know (%)
Belgium	40.7	22.7	36.6
Denmark	49.4	16.9	33.7
Germany	44.4	18.3	37.3
Greece	39.3	15.3	45.4
Italy	43.0	19.9	37.1
Spain	34.6	13.3	52.1
France	49.8	16.7	33.5
Ireland	38.7	20.9	40.5
Luxembourg	47.0	20.0	33.0
Netherlands	54.1	26.2	19.7
Portugal	33.7	18.7	47.6
UK	60.6	21.8	17.6
Norway	47.7	23.0	29.3
Finland	46.0	31.1	22.9
Sweden	39.5	24.2	36.2
Austria	38.4	20.2	41.4
Switzerland	70.2	16.7	13.1
Mean	**45.7**	**20.3**	**33.9**

Table 2d. By eating genetically modified fruit, a person's genes could also become modified

Country	True (%)	False (%)	Don't know (%)
Belgium	20.2	50.9	29.0
Denmark	21.4	56.8	21.8
Germany	29.5	38.0	32.5
Greece	29.7	35.8	34.4
Italy	18.1	57.8	24.0
Spain	27.2	39.7	33.2
France	22.3	52.3	25.4
Ireland	22.5	33.8	43.7
Luxembourg	24.2	48.0	27.8
Netherlands	9.7	74.5	15.8
Portugal	37.0	31.8	31.2
UK	14.9	55.0	30.1
Norway	19.7	46.5	33.8
Finland	26.6	53.9	19.5
Sweden	17.6	62.3	20.2
Austria	39.1	28.6	32.3
Switzerland	14.9	59.7	25.4
Mean	**23.2**	**48.6**	**28.3**

Table 2e. Viruses can be contaminated by bacteria

Country	True (%)	False (%)	Don't know (%)
Belgium	47.8	20.7	31.5
Denmark	41.9	29.2	28.9
Germany	51.3	16.7	32.0
Greece	49.1	12.9	38.0
Italy	34.7	23.0	42.3
Spain	48.3	14.8	36.9
France	48.5	17.3	34.2
Ireland	64.4	10.0	25.6
Luxembourg	40.9	23.0	36.1
Netherlands	44.7	37.4	17.9
Portugal	48.4	14.4	37.1
UK	60.6	16.3	23.1
Norway	33.4	23.8	42.8
Finland	50.5	27.4	22.1
Sweden	32.5	37.9	29.6
Austria	45.5	16.2	38.3
Switzerland	42.4	27.0	30.5
Mean	**46.3**	**21.6**	**32.1**

Table 2f. Yeast for brewing beer consists of living organisms

Country	True (%)	False (%)	Don't know (%)
Belgium	71.2	11.5	17.3
Denmark	89.9	4.1	6.0
Germany	74.9	10.0	15.0
Greece	49.7	18.2	32.1
Italy	65.2	15.6	19.1
Spain	45.9	16.2	37.9
France	65.9	14.0	20.1
Ireland	69.6	7.9	22.5
Luxembourg	66.7	13.8	19.5
Netherlands	63.8	21.8	14.4
Portugal	41.7	21.8	36.5
UK	85.3	4.4	10.3
Norway	75.5	11.1	13.4
Finland	70.3	14.0	15.7
Sweden	86.2	5.9	7.9
Austria	58.4	14.8	26.8
Switzerland	77.8	8.4	13.8
Mean	**68.2**	**12.5**	**19.3**

Table 2g. If it is possible to find out in the first few months of pregnancy whether a child will have Down's Syndrome, trisomy or Mongolism (term used according to local language)

Country	True (%)	False (%)	Don't know (%)
Belgium	75.4	8.8	15.8
Denmark	89.6	4.0	6.4
Germany	77.2	9.0	13.8
Greece	68.0	7.0	25.0
Italy	80.9	8.7	10.3
Spain	76.2	4.7	19.1
France	86.7	6.8	6.5
Ireland	74.5	5.4	20.1
Luxembourg	74.5	11.2	14.3
Netherlands	90.5	5.3	4.2
Portugal	65.3	18.2	16.5
UK	86.7	2.8	10.5
Norway	81.9	5.3	12.8
Finland	80.6	5.8	13.6
Sweden	87.3	4.5	8.2
Austria	68.7	10.6	20.7
Switzerland	86.7	5.8	7.5
Mean	**79.6**	**7.2**	**13.2**

Table 2h. Genetically modified animals are always bigger than ordinary ones

Country	True (%)	False (%)	Don't know (%)
Belgium	33.3	31.3	35.4
Denmark	16.4	53.9	29.7
Germany	35.7	37.4	26.9
Greece	54.1	15.8	30.1
Italy	35.3	34.6	30.2
Spain	35.4	28.2	36.4
France	38.2	34.3	27.5
Ireland	43.9	17.3	38.8
Luxembourg	43.8	30.2	26.0
Netherlands	21.0	55.6	23.4
Portugal	40.8	23.7	35.5
UK	26.6	40.0	33.3
Norway	31.1	40.0	28.9
Finland	38.8	42.5	18.7
Sweden	37.1	48.7	14.1
Austria	41.9	24.8	33.3
Switzerland	19.5	51.9	28.7
Mean	**34.6**	**36.1**	**29.3**

Table 2i. More than half of the human genes are identical to those of chimpanzees

Country	True (%)	False (%)	Don't know (%)
Belgium	39.6	18.2	42.3
Denmark	63.9	10.4	25.7
Germany	46.1	14.0	39.9
Greece	44.4	14.6	41.1
Italy	54.4	13.6	32.0
Spain	48.0	12.0	40.0
France	50.7	16.5	32.8
Ireland	47.2	12.5	40.3
Luxembourg	47.3	16.2	36.5
Netherlands	63.5	17.0	19.5
Portugal	49.0	20.3	30.7
UK	57.8	8.4	33.8
Norway	45.8	13.7	40.5
Finland	50.5	22.7	26.8
Sweden	57.5	17.5	25.1
Austria	35.2	19.3	45.5
Switzerland	47.6	18.5	33.9
Mean	**50.0**	**15.6**	**34.4**

Table 2j. It is impossible to transfer animal genes into plants

Country	True (%)	False (%)	Don't know (%)
Belgium	31.9	23.6	44.6
Denmark	38.8	28.0	33.2
Germany	29.1	23.7	47.2
Greece	32.1	18.0	49.9
Italy	31.0	29.9	39.1
Spain	24.3	22.1	53.6
France	36.6	22.4	41.0
Ireland	26.7	26.1	47.2
Luxembourg	30.3	25.0	44.7
Netherlands	39.0	40.8	20.2
Portugal	29.8	25.1	45.1
UK	18.1	33.8	48.1
Norway	30.2	25.0	44.8
Finland	38.7	37.5	23.8
Sweden	34.7	37.2	28.0
Austria	27.0	22.9	50.1
Switzerland	31.1	29.1	39.8
Mean	**31.2**	**27.7**	**41.1**

Table 2k. Mean values of 'image scale' (sum of items 2b, 2d and 2h. Scale 0–3, where 3 = high menacing images) and 'trend knowledge scale' (sum of items 2a, 2c, 2e, 2f and 2g. Scale 0–5, where 5 = high knowledge)

Country	Image scale	Trend knowledge scale
Belgium	0.80	2.98
Denmark	0.64	3.51
Germany	1.09	3.01
Greece	1.25	2.59
Italy	0.74	3.03
Spain	0.89	2.71
France	0.89	3.04
Ireland	0.95	2.94
Luxembourg	1.05	3.11
Netherlands	0.53	3.42
Portugal	1.10	2.46
UK	0.64	3.45
Norway	0.84	3.17
Finland	0.94	3.20
Sweden	0.85	3.42
Austria	1.25	2.71
Switzerland	0.65	3.52
Mean	**0.88**	**3.08**

Table 3. Responses to the question: 'There are differing views about whether people inherit particular characteristics or whether they acquire them mainly from their upbringing or the conditions in which they lived. Please tell me whether you think each of the following characteristics is inherited or mainly the result of upbringing and living conditions.' (Eurobarometer code Q9)

Table 3a. Body size

Country	Mainly inherited (%)	Mainly living conditions (%)	Don't know (%)
Belgium	80.5	14.7	4.8
Denmark	76.8	18.3	4.9
Germany	93.2	4.5	2.3
Greece	73.0	24.5	2.5
Italy	83.3	15.1	1.6
Spain	81.6	16.2	2.2
France	84.2	13.5	2.3
Ireland	85.3	11.5	3.2
Luxembourg	83.0	11.2	5.8
Netherlands	92.4	6.2	1.4
Portugal	67.6	28.8	3.6
UK	80.5	15.9	3.6
Norway	75.8	19.2	5.0
Finland	94.1	4.2	1.7
Sweden	90.0	7.0	3.0
Austria	82.1	13.0	4.9
Switzerland	85.9	11.8	2.3
Mean	**82.9**	**13.9**	**3.2**

Table 3b. Intelligence

Country	Mainly inherited (%)	Mainly living conditions (%)	Don't know (%)
Belgium	52.9	40.9	6.2
Denmark	69.6	25.4	5.0
Germany	70.3	26.2	3.5
Greece	51.2	45.6	3.2
Italy	60.2	36.9	2.9
Spain	53.3	42.2	4.5
France	38.9	55.2	5.9
Ireland	78.2	17.7	4.1
Luxembourg	59.5	34.7	5.8
Netherlands	77.2	19.1	3.7
Portugal	47.9	49.7	2.4
UK	60.4	32.0	7.5
Norway	74.7	17.5	7.9
Finland	75.9	19.3	4.8
Sweden	68.8	25.1	6.1
Austria	68.3	26.7	5.0
Switzerland	50.0	46.7	3.2
Mean	**62.2**	**33.0**	**4.8**

Table 3c. Homosexual tendencies

Country	Mainly inherited (%)	Mainly living conditions (%)	Don't know (%)
Belgium	18.3	56.5	25.3
Denmark	26.0	52.3	21.7
Germany	33.6	45.4	21.0
Greece	25.2	63.9	10.9
Italy	19.5	61.0	19.5
Spain	20.7	61.7	17.6
France	12.9	66.0	21.1
Ireland	29.1	40.4	30.4
Luxembourg	22.6	59.6	17.8
Netherlands	39.0	41.5	19.5
Portugal	17.1	67.1	15.8
UK	26.8	40.3	32.9
Norway	30.7	39.7	29.6
Finland	26.1	50.7	23.3
Sweden	27.1	50.5	22.9
Austria	23.3	54.5	22.2
Switzerland	24.3	57.9	17.8
Mean	**24.9**	**53.3**	**21.8**

Table 3d. Eye colour

Country	Mainly inherited (%)	Mainly living conditions (%)	Don't know (%)
Belgium	90.6	5.2	4.2
Denmark	95.5	2.5	2.0
Germany	93.6	3.2	3.2
Greece	96.4	3.3	0.3
Italy	95.0	3.6	1.4
Spain	96.2	2.6	1.2
France	93.6	5.3	1.1
Ireland	96.4	1.5	2.1
Luxembourg	90.5	4.7	4.8
Netherlands	96.6	2.7	0.7
Portugal	90.0	8.2	1.8
UK	94.8	2.1	3.1
Norway	94.7	1.9	3.4
Finland	97.5	0.9	1.6
Sweden	97.4	1.6	1.0
Austria	91.1	5.0	3.9
Switzerland			
Mean	**94.5**	**3.4**	**2.2**

Table 3e. Tendency to be happy

Country	Mainly inherited (%)	Mainly living conditions (%)	Don't know (%)
Belgium	16.8	74.8	8.4
Denmark	19.6	72.9	7.5
Germany	18.0	76.3	5.7
Greece	16.7	81.4	1.9
Italy	26.6	63.4	9.9
Spain	15.3	77.4	7.3
France	18.7	72.3	9.0
Ireland	33.4	60.3	6.3
Luxembourg	22.0	68.2	9.8
Netherlands	13.4	82.1	4.5
Portugal	13.5	82.6	3.9
UK	26.9	68.2	4.9
Norway	28.3	60.6	11.1
Finland	28.0	66.3	5.7
Sweden	8.1	85.2	6.7
Austria	42.6	51.7	5.7
Switzerland	12.4	84.7	2.9
Mean	**21.2**	**72.4**	**6.5**

Table 3f. Criminal tendencies

Country	Mainly inherited (%)	Mainly living conditions (%)	Don't know (%)
Belgium	14.0	70.4	15.6
Denmark	10.4	85.1	4.5
Germany	16.4	74.6	8.9
Greece	19.3	76.1	4.6
Italy	24.1	66.1	9.8
Spain	15.8	75.1	9.1
France	15.3	71.7	13.0
Ireland	18.5	74.6	6.9
Luxembourg	15.8	74.7	9.5
Netherlands	14.5	81.0	4.5
Portugal	11.4	85.3	3.3
UK	12.8	78.7	8.5
Norway	11.0	78.9	10.2
Finland	14.7	80.4	4.9
Sweden	9.0	85.5	5.5
Austria	23.2	68.6	8.2
Switzerland	11.7	82.7	5.7
Mean	**15.2**	**77.1**	**7.8**

Table 3g. Attitude to work

Country	Mainly inherited (%)	Mainly living conditions (%)	Don't know (%)
Belgium	14.9	76.1	8.9
Denmark	13.8	82.5	3.7
Germany	13.2	82.7	4.1
Greece	17.8	81.0	1.2
Italy	15.4	79.4	5.1
Spain	17.0	76.7	6.3
France	18.1	76.2	5.7
Ireland	32.0	62.0	6.0
Luxembourg	17.8	74.7	7.5
Netherlands	11.0	85.6	3.4
Portugal	17.4	79.4	3.2
UK	17.1	74.7	8.2
Norway	17.4	71.3	11.3
Finland	22.4	74.4	3.2
Sweden	12.7	83.7	3.6
Austria	21.3	71.6	7.1
Switzerland	8.8	89.2	2.0
Mean	**16.9**	**77.8**	**5.3**

Table 3h. Athletic abilities

Country	Mainly inherited (%)	Mainly living conditions (%)	Don't know (%)
Belgium	44.7	44.9	10.3
Denmark	57.2	35.7	7.1
Germany	53.5	40.5	6.0
Greece	35.5	59.8	4.7
Italy	27.8	68.3	3.9
Spain	31.2	61.6	7.2
France	32.8	62.3	4.9
Ireland	70.4	24.6	5.0
Luxembourg	46.5	45.3	8.2
Netherlands	52.8	43.8	3.4
Portugal	30.1	67.2	2.7
UK	65.4	26.2	8.4
Norway	35.6	51.7	12.7
Finland	59.8	35.8	4.4
Sweden	55.2	39.6	5.2
Austria	43.4	50.4	6.2
Switzerland	43.5	52.4	4.1
Mean	**46.2**	**47.7**	**6.1**

Table 3i. Susceptibility to mental illness

Country	Mainly inherited (%)	Mainly living conditions (%)	Don't know (%)
Belgium	60.6	27.3	12.0
Denmark	65.7	26.3	8.0
Germany	70.4	21.1	8.4
Greece	53.0	42.8	4.2
Italy	59.1	31.9	9.0
Spain	38.7	48.9	12.4
France	51.7	36.4	11.9
Ireland	77.5	14.9	7.6
Luxembourg	50.0	39.2	10.8
Netherlands	76.0	17.7	6.3
Portugal	47.5	45.2	7.3
UK	67.5	19.3	13.2
Norway	63.3	21.1	15.6
Finland	73.6	19.4	7.0
Sweden	68.4	21.9	9.6
Austria	76.7	14.4	8.9
Switzerland	68.9	25.4	5.8
Mean	**63.2**	**27.6**	**9.3**

Table 3j. Musical abilities

Country	Mainly inherited (%)	Mainly living conditions (%)	Don't know (%)
Belgium	57.0	34.4	8.6
Denmark	71.9	24.6	3.5
Germany	69.4	23.7	6.9
Greece	52.1	43.0	4.9
Italy	49.3	45.1	5.5
Spain	32.6	58.7	8.7
France	39.6	54.2	6.2
Ireland	78.2	15.8	6.0
Luxembourg	62.0	28.7	9.3
Netherlands	70.1	27.4	2.5
Portugal	36.1	58.5	5.4
UK	66.5	26.2	7.3
Norway	75.1	16.5	8.3
Finland	89.8	8.1	2.1
Sweden	83.0	13.6	3.4
Austria	66.0	27.8	6.2
Switzerland	62.1	33.1	4.8
Mean	**62.4**	**31.8**	**5.8**

Table 4. Responses to the question on usefulness of biotechnology: 'Do you definitely agree, tend to agree, tend to disagree or definitely disagree that it is . . .' (Eurobarometer code Q10)

Table 4a. 'Useful for society to use modern biotechnology in the production of foods, for example to make them higher in protein, keep longer or change the taste?'

Country	Definitely agree (%)	Tend to agree (%)	Tend to disagree (%)	Definitely disagree (%)	Don't know (%)
Belgium	19.4	34.0	16.8	13.9	15.9
Denmark	17.5	31.4	18.9	28.3	4.0
Germany	17.8	33.6	22.3	15.4	10.8
Greece	22.8	25.2	14.1	27.3	10.6
Italy	19.1	33.6	20.5	22.6	4.3
Spain	35.1	27.3	12.0	12.3	13.3
France	16.0	32.2	23.8	20.5	7.5
Ireland	27.1	30.7	12.4	13.4	16.4
Luxembourg	26.5	19.7	22.5	19.4	11.9
Netherlands	15.7	54.0	11.8	14.9	3.6
Portugal	18.0	43.9	18.5	11.0	8.5
UK	20.4	41.3	16.6	11.6	10.1
Norway	11.1	31.9	19.3	26.1	11.5
Finland	28.1	41.1	17.1	6.9	6.8
Sweden	9.2	31.2	24.5	31.1	4.0
Austria	15.3	15.9	17.8	35.3	15.7
Switzerland	13.7	30.3	30.4	21.7	3.9
Mean	**19.4**	**33.1**	**18.7**	**19.5**	**9.3**

Table 4b. 'Useful for society to take genes from plant species and transfer them into crop plants, to make them more resistant to insect pests?'

Country	Definitely agree (%)	Tend to agree (%)	Tend to disagree (%)	Definitely disagree (%)	Don't know (%)
Belgium	37.5	33.6	9.5	6.4	13.0
Denmark	43.3	26.7	9.9	16.5	3.6
Germany	24.3	38.1	16.5	10.5	10.5
Greece	37.7	30.1	9.1	12.0	11.0
Italy	40.2	35.9	10.0	8.7	5.1
Spain	38.3	27.3	8.4	11.0	14.9
France	36.3	34.8	16.2	8.0	4.8
Ireland	32.6	33.6	9.3	7.8	16.7
Luxembourg	34.4	22.9	15.7	15.2	11.9
Netherlands	39.9	39.6	7.6	10.3	2.5
Portugal	31.0	44.9	9.5	4.4	10.2
UK	34.8	39.6	8.8	6.1	10.7
Norway	16.6	34.5	17.2	20.4	11.2
Finland	52.0	28.0	8.8	4.5	6.7
Sweden	27.2	35.1	15.7	15.6	6.3
Austria	19.4	16.8	16.8	31.0	16.0
Switzerland	23.0	37.9	19.6	15.4	4.0
Mean	**33.4**	**33.2**	**12.2**	**11.9**	**9.3**

Table 4c. 'Useful for society to introduce human genes into bacteria to produce medicines or vaccines, for example to produce insulin for diabetics?'

Country	Definitely agree (%)	Tend to agree (%)	Tend to disagree (%)	Definitely disagree (%)	Don't know (%)
Belgium	52.7	28.4	6.4	4.0	8.5
Denmark	69.4	17.4	3.8	5.3	4.1
Germany	37.4	37.8	8.4	6.4	10.0
Greece	44.2	32.0	5.0	6.2	12.6
Italy	50.3	29.1	7.1	6.9	6.6
Spain	51.5	24.7	4.4	5.3	14.1
France	57.3	26.6	5.9	4.8	5.4
Ireland	44.5	29.4	4.4	6.8	14.8
Luxembourg	56.3	20.9	7.0	4.5	11.2
Netherlands	56.5	32.1	3.5	4.9	3.0
Portugal	39.6	45.3	5.5	2.5	7.2
UK	53.6	30.9	3.7	3.1	8.7
Norway	34.4	34.3	5.6	8.0	17.6
Finland	57.1	23.7	5.9	5.7	7.5
Sweden	57.6	25.3	5.4	6.0	5.7
Austria	37.3	22.7	11.8	12.2	16.0
Switzerland	41.2	38.7	9.2	6.5	4.5
Mean	**49.3**	**29.6**	**6.0**	**5.9**	**9.2**

Table 4d. 'Useful for society to develop genetically modified animals for laboratory research studies, such as a mouse that has genes which cause it to develop cancer?'

Country	Definitely agree (%)	Tend to agree (%)	Tend to disagree (%)	Definitely disagree (%)	Don't know (%)
Belgium	27.1	30.6	11.7	19.9	10.7
Denmark	50.3	24.3	6.3	15.3	3.8
Germany	16.3	29.9	20.1	23.6	10.0
Greece	33.0	27.4	8.3	15.3	16.0
Italy	30.8	33.2	13.4	16.4	6.2
Spain	37.8	25.2	8.2	12.5	16.3
France	34.9	33.1	11.8	13.8	6.4
Ireland	25.7	27.8	10.5	19.0	17.1
Luxembourg	32.9	20.4	16.9	16.4	13.5
Netherlands	19.3	37.2	13.1	28.8	1.6
Portugal	32.3	43.6	8.6	7.4	8.1
UK	21.5	34.1	12.8	21.7	9.9
Norway	22.0	31.8	12.3	21.5	12.3
Finland	32.4	30.4	13.7	13.5	9.9
Sweden	19.0	28.1	16.2	31.7	5.0
Austria	21.1	17.2	15.4	30.3	16.1
Switzerland	20.0	32.3	18.4	25.9	3.4
Mean	**27.9**	**30.0**	**12.7**	**19.7**	**9.7**

Table 4e. 'Useful for society to introduce human genes into animals to produce organs for human transplants, such as into pigs for human heart transplants?'

Country	Definitely agree (%)	Tend to agree (%)	Tend to disagree (%)	Definitely disagree (%)	Don't know (%)
Belgium	31.7	25.0	13.5	20.3	9.5
Denmark	35.4	21.6	9.2	28.0	5.8
Germany	16.3	31.0	17.3	24.3	11.0
Greece	26.2	29.5	9.7	18.5	16.1
Italy	24.3	27.9	15.9	23.7	8.1
Spain	39.4	29.1	6.2	10.3	15.0
France	27.7	31.3	13.9	18.5	8.5
Ireland	19.3	22.1	13.4	25.8	19.4
Luxembourg	32.7	18.4	15.4	20.9	12.7
Netherlands	18.2	32.7	12.5	33.5	3.1
Portugal	24.8	37.3	14.0	12.7	11.2
UK	21.2	32.2	16.2	19.9	10.4
Norway	13.5	24.9	14.2	34.5	12.8
Finland	22.3	25.5	19.0	22.5	10.8
Sweden	18.4	30.5	16.5	27.2	7.5
Austria	16.3	15.9	18.4	33.0	16.4
Switzerland	16.4	26.7	22.8	29.4	4.8
Mean	**23.6**	**27.4**	**14.6**	**23.8**	**10.7**

Table 4f. 'Useful for society to use genetic testing to detect diseases we might have inherited from our parents such as cystic fibrosis, mucoviscidosis or thalassaemia?'

Country	Definitely agree (%)	Tend to agree (%)	Tend to disagree (%)	Definitely disagree (%)	Don't know (%)
Belgium	51.7	29.9	5.4	3.5	9.5
Denmark	63.2	20.1	6.2	5.3	5.2
Germany	34.9	39.9	10.7	5.7	8.7
Greece	70.5	19.4	2.5	1.6	6.0
Italy	63.5	24.7	2.8	3.8	5.2
Spain	53.2	24.8	3.0	4.5	14.5
France	62.6	25.2	4.3	4.2	3.7
Ireland	51.0	28.3	2.6	4.8	13.3
Luxembourg	53.9	20.5	8.8	5.2	11.5
Netherlands	55.2	32.8	4.1	4.8	3.1
Portugal	45.4	45.0	3.2	1.3	5.0
UK	55.6	32.8	2.5	2.2	6.9
Norway	34.1	32.0	6.0	13.0	14.9
Finland	57.7	26.6	4.1	2.4	9.1
Sweden	52.3	29.8	7.3	4.0	6.7
Austria	31.2	20.2	13.8	15.3	19.5
Switzerland	40.2	37.4	10.8	8.0	3.7
Mean	**51.5**	**29.0**	**5.7**	**5.3**	**8.5**

Table 5. Responses to the question on risk of biotechnology: 'Do you definitely agree, tend to agree, tend to disagree or definitely disagree that it is . . .' (Eurobarometer code Q10)

Table 5a. 'Risky to use modern biotechnology in the production of foods, for example to make them higher in protein, keep longer or change the taste?'

Country	Definitely agree (%)	Tend to agree (%)	Tend to disagree (%)	Definitely disagree (%)	Don't know (%)
Belgium	21.2	27.8	22.6	12.4	16.0
Denmark	34.5	32.5	16.9	10.4	5.7
Germany	25.2	35.1	22.7	6.1	10.9
Greece	34.1	25.8	16.4	9.4	14.3
Italy	26.7	34.3	18.0	9.7	11.2
Spain	17.8	31.4	18.4	13.1	19.3
France	30.8	34.9	16.2	9.8	8.4
Ireland	22.3	32.6	16.5	11.6	17.0
Luxembourg	33.4	19.7	21.2	13.0	12.5
Netherlands	27.4	46.8	11.4	10.2	4.1
Portugal	17.0	44.2	16.7	4.8	17.3
UK	21.6	41.6	20.2	5.3	11.2
Norway	27.0	34.0	15.5	9.5	13.9
Finland	10.0	30.2	32.0	17.1	10.6
Sweden	31.1	39.1	15.4	9.8	4.7
Austria	30.2	21.2	16.1	13.0	19.5
Switzerland	27.4	39.4	20.0	6.7	6.4
Mean	**25.6**	**33.9**	**18.5**	**10.0**	**11.9**

Table 5b. 'Risky to take genes from plant species and transfer them into crop plants, to make them more resistant to insect pests?'

Country	Definitely agree (%)	Tend to agree (%)	Tend to disagree (%)	Definitely disagree (%)	Don't know (%)
Belgium	16.8	22.2	24.2	18.3	18.4
Denmark	27.8	34.1	15.3	15.4	7.4
Germany	15.8	28.0	33.6	10.3	12.2
Greece	16.7	21.6	23.0	19.9	18.8
Italy	16.9	27.2	26.0	16.7	13.3
Spain	15.5	27.7	20.2	15.1	21.5
France	18.7	33.2	22.0	13.6	12.5
Ireland	17.7	27.2	20.8	13.9	20.4
Luxembourg	28.6	20.1	19.6	15.7	16.1
Netherlands	19.0	45.2	14.8	16.1	5.0
Portugal	10.9	36.2	24.9	8.5	19.5
UK	14.0	39.5	24.8	8.5	13.1
Norway	22.0	33.7	19.3	8.9	16.1
Finland	7.5	23.5	33.4	25.9	9.7
Sweden	19.9	35.5	21.7	14.8	8.1
Austria	27.4	21.4	18.7	13.7	18.8
Switzerland	19.7	37.5	25.5	10.0	7.4
Mean	**18.3**	**30.5**	**22.9**	**14.4**	**14.0**

Table 5c. *'Risky to introduce human genes into bacteria to produce medicines or vaccines, for example to produce insulin for diabetics?'*

Country	Definitely agree (%)	Tend to agree (%)	Tend to disagree (%)	Definitely disagree (%)	Don't know (%)
Belgium	14.1	27.6	22.7	17.3	18.2
Denmark	13.7	27.5	17.5	31.1	10.2
Germany	12.8	29.6	34.5	10.6	12.5
Greece	11.3	19.8	24.1	21.7	23.2
Italy	14.1	29.7	24.8	13.9	17.6
Spain	17.1	28.0	20.2	15.2	19.5
France	19.0	34.0	19.0	13.6	14.4
Ireland	16.8	28.2	20.2	16.4	18.4
Luxembourg	17.2	16.7	25.5	24.5	16.0
Netherlands	20.7	50.9	10.9	11.9	5.6
Portugal	12.8	40.8	19.8	6.9	19.8
UK	12.8	38.8	22.9	11.0	14.5
Norway	14.0	31.5	16.3	12.6	25.5
Finland	9.3	19.5	27.8	29.8	13.6
Sweden	16.2	29.8	22.4	20.9	10.7
Austria	14.8	24.5	22.6	17.9	20.2
Switzerland	17.9	35.3	27.9	11.7	7.2
Mean	**14.9**	**30.5**	**22.2**	**16.7**	**15.7**

Table 5d. *'Risky to develop genetically modified animals for laboratory research studies, such as a mouse that has genes which cause it to develop cancer?'*

Country	Definitely agree (%)	Tend to agree (%)	Tend to disagree (%)	Definitely disagree (%)	Don't know (%)
Belgium	22.8	29.3	18.5	15.2	14.2
Denmark	22.4	24.6	19.2	27.5	6.3
Germany	19.3	31.5	25.5	11.5	12.1
Greece	22.3	19.7	19.8	18.2	20.0
Italy	24.0	34.1	17.2	12.4	12.3
Spain	19.0	30.0	17.9	13.0	20.0
France	25.1	31.3	19.3	13.5	10.8
Ireland	21.2	26.2	19.1	13.8	19.8
Luxembourg	23.2	15.7	22.7	22.2	16.2
Netherlands	29.3	37.8	13.9	15.7	3.4
Portugal	17.7	41.9	18.0	5.8	16.7
UK	29.4	30.9	17.8	12.1	9.7
Norway	21.9	26.1	17.6	16.5	17.8
Finland	15.9	24.5	30.5	18.2	10.8
Sweden	26.8	29.6	19.5	17.2	6.9
Austria	22.4	19.7	22.1	18.2	17.6
Switzerland	26.0	30.8	25.7	11.3	6.2
Mean	**22.9**	**28.8**	**20.2**	**15.3**	**12.9**

Table 5e. 'Risky to introduce human genes into animals to produce organs for human transplants, such as into pigs for human heart transplants?'

Country	Definitely agree (%)	Tend to agree (%)	Tend to disagree (%)	Definitely disagree (%)	Don't know (%)
Belgium	31.8	26.4	14.7	12.0	15.1
Denmark	32.7	24.8	15.6	18.0	8.9
Germany	24.7	30.3	21.1	11.4	12.4
Greece	29.9	20.3	15.8	13.5	20.5
Italy	33.2	31.4	13.9	8.0	13.4
Spain	22.0	30.8	16.5	11.7	19.0
France	32.4	30.5	14.7	10.5	12.0
Ireland	32.2	24.5	10.5	13.0	19.7
Luxembourg	27.7	16.9	18.2	20.2	17.0
Netherlands	35.5	38.4	8.5	14.1	3.5
Portugal	26.9	39.1	11.7	4.4	17.9
UK	35.6	34.7	13.4	8.0	8.3
Norway	33.7	24.7	12.0	11.1	18.5
Finland	23.7	28.1	22.4	14.1	11.7
Sweden	27.4	31.2	16.4	15.7	9.3
Austria	25.5	20.2	18.0	17.8	18.5
Switzerland	36.0	31.3	16.7	9.6	6.3
Mean	**30.1**	**28.7**	**15.2**	**12.4**	**13.6**

Table 5f. 'Risky to use genetic testing to detect diseases we might have inherited from our parents such as cystic fibrosis, mucoviscidosis or thalassaemia?'

Country	Definitely agree (%)	Tend to agree (%)	Tend to disagree (%)	Definitely disagree (%)	Don't know (%)
Belgium	14.0	23.5	21.4	20.5	20.7
Denmark	14.7	23.8	18.2	34.1	9.1
Germany	11.3	24.5	38.1	14.4	11.6
Greece	6.7	8.4	24.2	47.4	13.3
Italy	14.1	25.3	24.5	22.0	14.2
Spain	15.5	22.8	21.6	19.0	21.2
France	12.9	29.6	24.8	21.4	11.3
Ireland	18.3	23.7	20.7	18.4	18.9
Luxembourg	19.3	12.5	26.3	26.0	16.0
Netherlands	18.0	44.4	13.3	18.8	5.4
Portugal	13.4	37.4	20.6	10.4	18.2
UK	11.8	30.9	29.4	14.5	13.4
Norway	16.2	25.2	19.0	17.8	21.8
Finland	5.1	15.4	33.7	31.0	14.8
Sweden	12.1	31.8	21.4	22.2	12.4
Austria	15.1	23.1	22.2	19.2	20.3
Switzerland	14.2	25.9	32.2	19.1	8.6
Mean	**13.6**	**25.5**	**24.2**	**22.0**	**14.7**

Table 6. Responses to the question on moral acceptability: 'Do you definitely agree, tend to agree, tend to disagree or definitely disagree that it is . . .' (Eurobarometer code Q10)

Table 6a. 'Morally acceptable to use modern biotechnology in the production of foods, for example to make them higher in protein, keep longer or change the taste?'

Country	Definitely agree (%)	Tend to agree (%)	Tend to disagree (%)	Definitely disagree (%)	Don't know (%)
Belgium	19.9	33.9	16.5	13.4	16.4
Denmark	14.2	26.2	21.9	30.3	7.4
Germany	11.3	31.8	26.0	17.7	13.1
Greece	13.2	26.4	20.8	24.4	15.2
Italy	20.4	34.5	21.0	17.1	7.0
Spain	23.3	30.7	14.6	14.4	17.1
France	15.3	35.5	23.9	16.2	9.1
Ireland	19.6	30.8	17.1	13.1	19.3
Luxembourg	17.5	27.4	20.5	18.4	16.2
Netherlands	15.4	51.3	16.2	12.8	4.3
Portugal	13.8	44.5	20.4	8.9	12.5
UK	15.6	37.4	23.1	11.5	12.4
Norway	9.4	29.1	21.6	27.3	12.6
Finland	20.1	37.5	21.3	10.7	10.3
Sweden	16.2	26.5	23.0	29.8	4.5
Austria	9.4	16.2	24.3	32.3	17.7
Switzerland	10.9	33.5	29.6	19.3	6.8
Mean	**15.6**	**32.7**	**21.3**	**18.7**	**11.8**

Table 6b. 'Morally acceptable to take genes from plant species and transfer them into crop plants, to make them more resistant to insect pests?'

Country	Definitely agree (%)	Tend to agree (%)	Tend to disagree (%)	Definitely disagree (%)	Don't know (%)
Belgium	29.8	38.1	10.6	6.5	15.0
Denmark	24.5	29.8	17.0	22.2	6.4
Germany	15.6	39.3	20.2	12.6	12.2
Greece	25.8	33.2	12.0	12.5	16.5
Italy	31.8	38.8	13.4	8.7	7.3
Spain	29.9	29.2	11.8	11.2	17.9
France	24.8	40.2	19.0	8.2	7.7
Ireland	23.0	34.4	13.2	9.7	19.7
Luxembourg	24.2	29.0	15.4	16.9	14.5
Netherlands	21.4	51.6	12.7	10.3	4.0
Portugal	20.8	51.5	12.1	3.0	12.6
UK	19.5	44.5	14.8	7.0	14.3
Norway	10.4	33.2	19.5	22.2	14.7
Finland	34.6	35.7	13.4	7.0	9.2
Sweden	27.9	32.0	17.7	17.1	5.3
Austria	11.5	16.7	25.5	28.5	17.8
Switzerland	15.5	40.0	23.1	15.5	5.9
Mean	**23.0**	**36.5**	**16.0**	**12.8**	**11.8**

Table 6c. 'Morally acceptable to introduce human genes into bacteria to produce medicines or vaccines, for example to produce insulin for diabetics?'

Country	Definitely agree (%)	Tend to agree (%)	Tend to disagree (%)	Definitely disagree (%)	Don't know (%)
Belgium	37.6	36.6	9.7	3.7	12.4
Denmark	46.9	31.3	7.6	9.1	5.0
Germany	21.8	44.1	13.3	9.1	11.6
Greece	31.5	33.1	9.3	7.3	18.8
Italy	34.2	37.2	11.1	8.9	8.6
Spain	34.9	29.1	9.9	9.2	16.9
France	32.0	42.1	12.8	4.8	8.4
Ireland	30.3	33.9	8.5	9.2	18.1
Luxembourg	39.2	27.8	12.2	7.0	13.8
Netherlands	27.3	53.8	7.8	7.2	3.9
Portugal	23.0	51.9	11.8	3.0	10.3
UK	32.4	41.1	8.5	4.5	13.5
Norway	21.0	33.7	13.2	10.4	21.7
Finland	36.7	33.8	11.3	8.4	9.7
Sweden	46.5	31.9	7.9	7.6	6.2
Austria	22.3	24.5	20.8	14.2	18.2
Switzerland	26.3	43.7	15.6	8.2	6.2
Mean	**31.8**	**37.3**	**11.2**	**7.8**	**11.9**

Table 6d. 'Morally acceptable to develop genetically modified animals for laboratory research studies, such as a mouse that has genes which cause it to develop cancer?'

Country	Definitely agree (%)	Tend to agree (%)	Tend to disagree (%)	Definitely disagree (%)	Don't know (%)
Belgium	14.7	23.6	19.8	29.1	12.7
Denmark	19.6	24.6	18.7	33.3	3.9
Germany	6.8	21.2	26.3	34.9	10.8
Greece	17.4	27.0	16.2	20.4	19.0
Italy	15.7	27.7	23.2	25.7	7.7
Spain	19.9	25.1	17.8	18.2	19.0
France	16.2	30.8	26.0	20.2	6.7
Ireland	14.0	22.1	16.1	26.6	21.3
Luxembourg	18.7	21.0	18.8	24.8	16.7
Netherlands	10.5	40.5	18.6	27.8	2.6
Portugal	14.2	43.0	20.6	10.8	11.4
UK	10.8	24.8	23.8	30.4	10.3
Norway	14.5	26.4	18.7	24.8	15.5
Finland	14.9	29.4	23.7	23.6	8.3
Sweden	14.1	26.2	18.8	36.9	4.0
Austria	13.3	16.8	22.0	32.1	15.8
Switzerland	9.3	24.6	23.2	38.0	4.9
Mean	**14.3**	**26.9**	**20.8**	**27.0**	**11.1**

Table 6e. *'Morally acceptable to introduce human genes into animals to produce organs for human transplants, such as into pigs for human heart transplants?'*

Country	Definitely agree (%)	Tend to agree (%)	Tend to disagree (%)	Definitely disagree (%)	Don't know (%)
Belgium	15.6	23.7	20.9	27.4	12.4
Denmark	15.5	21.9	15.2	41.5	5.9
Germany	7.8	20.6	23.8	35.9	11.8
Greece	12.5	21.8	16.0	27.9	21.7
Italy	12.8	22.7	22.7	33.0	8.8
Spain	20.3	26.2	15.7	18.4	19.4
France	12.6	26.8	23.3	27.9	9.5
Ireland	9.1	14.5	18.0	34.5	23.9
Luxembourg	20.2	18.5	17.2	29.2	15.0
Netherlands	11.1	33.9	16.1	35.1	3.8
Portugal	8.2	35.2	25.1	18.2	13.3
UK	9.6	23.6	21.6	33.5	11.6
Norway	10.4	18.5	17.7	37.4	15.9
Finland	10.6	21.7	22.5	35.2	9.9
Sweden	12.4	29.4	19.8	33.3	5.1
Austria	10.0	17.0	22.6	33.9	16.6
Switzerland	6.9	19.2	26.5	41.3	6.1
Mean	**11.9**	**23.4**	**20.4**	**32.1**	**12.3**

Table 6f. *'Morally acceptable to use genetic testing to detect diseases we might have inherited from our parents such as cystic fibrosis, mucoviscidosis or thalassaemia?'*

Country	Definitely agree (%)	Tend to agree (%)	Tend to disagree (%)	Definitely disagree (%)	Don't know (%)
Belgium	38.0	34.7	9.2	5.3	12.8
Denmark	42.1	31.1	11.3	10.0	5.5
Germany	21.2	43.4	15.3	8.9	11.1
Greece	61.3	23.5	3.4	2.0	9.8
Italy	43.8	37.2	7.0	6.2	5.7
Spain	34.0	32.1	8.6	7.9	17.4
France	40.6	39.3	10.3	3.9	5.9
Ireland	32.1	35.2	7.1	7.1	18.5
Luxembourg	36.4	27.9	13.0	8.0	14.7
Netherlands	31.4	48.9	8.1	7.0	4.6
Portugal	26.8	54.3	8.9	1.6	8.4
UK	34.7	45.0	5.9	3.5	10.8
Norway	23.7	28.8	14.0	15.2	18.2
Finland	36.6	38.3	8.5	5.0	11.6
Sweden	39.3	36.1	12.6	5.7	6.3
Austria	21.5	23.6	19.8	15.7	19.4
Switzerland	26.9	42.1	12.6	12.7	5.8
Mean	**34.7**	**36.8**	**10.3**	**7.4**	**10.9**

Table 7. Responses to the question on encouraging biotechnology: 'Do you definitely agree, tend to agree, tend to disagree or definitely disagree that society . . .' (Eurobarometer code Q10)

Table 7a. 'Should be encouraged to use modern biotechnology in the production of foods, for example to make them higher in protein, keep longer or change the taste?'

Country	Definitely agree (%)	Tend to agree (%)	Tend to disagree (%)	Definitely disagree (%)	Don't know (%)
Belgium	16.9	29.9	15.8	19.2	18.2
Denmark	11.6	18.6	16.2	44.8	8.9
Germany	11.5	29.2	22.7	23.4	13.2
Greece	16.5	26.4	13.9	30.0	13.3
Italy	17.9	29.2	20.2	24.5	8.1
Spain	24.9	27.7	13.4	13.9	20.0
France	12.0	26.5	25.4	25.7	10.4
Ireland	18.5	26.5	14.7	19.3	21.1
Luxembourg	13.4	20.4	20.6	22.7	22.9
Netherlands	12.5	43.8	18.2	21.4	4.1
Portugal	14.5	41.0	20.1	13.1	11.4
UK	13.8	31.0	21.7	19.3	14.3
Norway	6.8	24.3	21.2	33.0	14.7
Finland	26.3	32.5	17.8	13.6	9.8
Sweden	11.1	22.6	21.6	40.2	4.5
Austria	7.3	10.6	19.2	44.7	18.2
Switzerland	9.6	23.4	31.0	29.9	6.1
Mean	**14.4**	**27.4**	**19.6**	**25.9**	**12.7**

Table 7b. 'Should be encouraged to take genes from plant species and transfer them into crop plants, to make them more resistant to insect pests?'

Country	Definitely agree (%)	Tend to agree (%)	Tend to disagree (%)	Definitely disagree (%)	Don't know (%)
Belgium	27.8	34.4	12.5	8.2	17.0
Denmark	22.4	25.6	14.7	29.1	8.2
Germany	16.4	34.0	20.1	15.8	13.7
Greece	29.5	30.6	10.1	15.9	13.9
Italy	29.5	40.2	11.5	10.7	8.0
Spain	27.9	28.4	11.2	12.9	19.7
France	23.1	33.9	19.9	13.4	9.7
Ireland	22.1	31.0	14.1	12.6	20.3
Luxembourg	19.4	24.7	15.4	19.7	20.9
Netherlands	22.7	43.7	14.8	15.5	3.3
Portugal	24.7	47.5	9.7	5.4	12.6
UK	21.2	38.0	14.9	11.6	14.3
Norway	9.7	29.3	20.6	25.3	15.1
Finland	37.1	34.7	11.2	7.8	9.2
Sweden	21.9	32.1	16.7	23.0	6.4
Austria	9.0	13.9	22.9	36.5	17.7
Switzerland	13.4	30.7	27.1	22.8	6.0
Mean	**22.3**	**32.7**	**15.8**	**16.8**	**12.5**

Table 7c. 'Should be encouraged to introduce human genes into bacteria to produce medicines or vaccines, for example to produce insulin for diabetics?'

Country	Definitely agree (%)	Tend to agree (%)	Tend to disagree (%)	Definitely disagree (%)	Don't know (%)
Belgium	40.1	33.6	9.3	4.2	12.9
Denmark	46.5	29.8	6.2	9.5	8.0
Germany	24.9	42.0	11.4	9.5	12.2
Greece	35.1	33.7	7.6	7.6	16.0
Italy	37.2	36.0	9.2	8.6	8.8
Spain	36.6	30.7	7.9	7.2	17.5
France	39.2	37.0	8.6	7.0	8.1
Ireland	32.4	32.8	7.1	9.8	17.9
Luxembourg	42.0	23.8	9.5	5.7	18.9
Netherlands	32.9	44.6	9.2	9.6	3.8
Portugal	26.0	52.1	8.0	3.8	10.0
UK	34.9	40.3	6.9	5.5	12.3
Norway	22.1	35.5	10.5	10.1	21.7
Finland	42.8	30.4	8.8	7.8	10.1
Sweden	42.8	31.5	9.5	9.6	6.6
Austria	23.6	21.6	17.1	18.8	18.9
Switzerland	25.5	38.5	17.3	10.5	8.2
Mean	**34.1**	**35.1**	**9.6**	**8.6**	**12.3**

Table 7d. 'Should be encouraged to develop genetically modified animals for laboratory research studies, such as a mouse that has genes which cause it to develop cancer?'

Country	Definitely agree (%)	Tend to agree (%)	Tend to disagree (%)	Definitely disagree (%)	Don't know (%)
Belgium	15.3	24.4	20.9	25.5	13.9
Denmark	20.3	24.4	14.7	30.5	10.0
Germany	8.1	22.8	22.6	34.1	12.3
Greece	23.5	27.2	11.3	18.9	19.1
Italy	19.6	31.4	18.9	20.7	9.4
Spain	21.5	27.9	13.8	16.0	20.8
France	21.1	30.6	19.4	18.3	10.7
Ireland	16.3	23.1	13.4	26.1	21.1
Luxembourg	21.2	17.1	17.1	21.4	23.2
Netherlands	11.9	35.0	17.7	32.6	2.7
Portugal	19.0	46.7	13.9	9.4	10.9
UK	11.3	26.8	20.2	29.8	12.0
Norway	14.9	24.1	15.7	25.2	20.1
Finland	20.3	27.4	20.5	19.8	11.9
Sweden	12.8	24.3	18.6	39.0	5.3
Austria	13.0	16.1	17.8	35.3	17.8
Switzerland	10.7	21.6	24.5	38.3	4.8
Mean	**16.4**	**26.8**	**17.7**	**26.1**	**13.1**

Table 7e. 'Should be encouraged to introduce human genes into animals to produce organs for human transplants, such as into pigs for human heart transplants?'

Country	Definitely agree (%)	Tend to agree (%)	Tend to disagree (%)	Definitely disagree (%)	Don't know (%)
Belgium	18.3	23.2	17.6	26.6	14.2
Denmark	18.5	18.8	11.1	41.0	10.6
Germany	9.3	22.7	20.3	34.5	13.2
Greece	17.9	23.2	12.1	24.9	21.8
Italy	14.9	26.0	19.4	29.5	10.2
Spain	24.7	28.2	11.2	14.5	21.5
France	15.2	28.3	20.8	23.3	12.4
Ireland	12.1	15.8	14.6	33.8	23.7
Luxembourg	20.1	16.8	15.3	27.3	20.5
Netherlands	12.7	31.4	15.0	38.1	2.8
Portugal	14.8	37.6	18.4	15.8	13.5
UK	12.6	22.3	17.8	32.9	14.4
Norway	11.6	18.3	18.2	37.1	14.8
Finland	15.0	21.8	21.8	30.0	11.3
Sweden	13.0	26.7	18.7	34.1	7.6
Austria	10.3	15.4	19.8	36.3	18.2
Switzerland	7.5	19.6	23.0	43.6	6.4
Mean	**14.5**	**23.4**	**17.4**	**30.9**	**13.8**

Table 7f. 'Should be encouraged to use genetic testing to detect diseases we might have inherited from our parents such as cystic fibrosis, mucoviscidosis or thalassaemia?'

Country	Definitely agree (%)	Tend to agree (%)	Tend to disagree (%)	Definitely disagree (%)	Don't know (%)
Belgium	40.1	31.7	9.8	5.3	13.1
Denmark	40.3	28.4	10.8	10.7	9.8
Germany	24.1	40.1	14.1	9.0	12.6
Greece	65.8	21.6	2.3	2.0	8.3
Italy	50.7	32.7	5.4	5.1	6.1
Spain	37.4	30.4	7.2	6.7	18.3
France	47.0	36.3	6.3	4.3	6.1
Ireland	37.0	31.7	4.6	8.5	18.2
Luxembourg	42.6	21.7	11.5	6.3	17.9
Netherlands	34.5	43.5	8.7	9.8	3.5
Portugal	34.1	50.6	5.2	1.9	8.2
UK	39.5	40.8	5.1	4.1	10.5
Norway	22.1	31.0	11.5	16.2	19.1
Finland	41.4	35.6	7.3	4.6	11.1
Sweden	39.2	33.4	11.8	7.8	7.8
Austria	23.5	22.3	17.1	17.9	19.2
Switzerland	27.1	36.6	15.5	13.5	7.3
Mean	**37.9**	**33.7**	**9.0**	**7.9**	**11.4**

Table 8. Means for six applications of modern biotechnology according to perceived usefulness, risk, moral acceptability, and whether they should be encouraged. (Eurobarometer code Q10)

Table 8a. Usefulness (scale: −2 = low, to +2 = high)

Country	Food production	Crop plants	Medicines and vaccines	Research animals	Xeno transplants	Genetic testing
Belgium	0.28	0.86	1.20	0.33	0.34	1.21
Denmark	−0.09	0.70	1.42	0.88	0.27	1.29
Germany	0.16	0.49	0.91	−0.05	−0.02	0.87
Greece	0.02	0.72	1.03	0.54	0.35	1.55
Italy	0.06	0.89	1.09	0.49	0.13	1.41
Spain	0.61	0.73	1.13	0.68	0.81	1.20
France	0.01	0.75	1.26	0.64	0.36	1.38
Ireland	0.46	0.74	1.01	0.31	0.04	1.18
Luxembourg	0.11	0 46	1.17	0.37	0.27	1.09
Netherlands	0.44	0.91	1.32	0.05	−0.10	1.30
Portugal	0.39	0.89	1.14	0.85	0.48	1.30
UK	0.42	0.88	1.28	0.21	0.19	1.37
Norway	−0.17	0.10	0.81	0.20	−0.31	0.68
Finland	0.66	1.14	1.21	0.54	0.06	1.33
Sweden	−0.37	0.43	1.23	−0.13	−0.03	1.19
Austria	−0.42	−0.23	0.61	−0.17	−0.36	0.38
Switzerland	−0.16	0.34	0.99	0.02	−0.22	0.91
Mean	**0.14**	**0.64**	**1.10**	**0.34**	**0.12**	**1.16**

Table 8b. Risk[a] (scale: −2 = low, to +2 = high)

Country	Food production	Crop plants	Medicines and vaccines	Research animals	Xeno transplants	Genetic testing
Belgium	0.23	−0.05	0.02	0.26	0.51	−0.11
Denmark	0.64	0.44	−0.25	−0.04	0.39	−0.33
Germany	0.51	0.05	0.00	0.22	0.36	−0.20
Greece	0.59	−0.08	−0.25	0.08	0.37	−0.97
Italy	0.50	0.02	0.05	0.40	0.68	−0.15
Spain	0.22	0.08	0.12	0.24	0.35	−0.06
France	0.61	0.21	0.26	0.35	0.60	−0.12
Ireland	0.38	0.14	0.08	0.22	0.52	0.03
Luxembourg	0.39	0.26	−0.24	−0.05	0.14	−0.27
Netherlands	0.70	0.36	0.58	0.51	0.73	0.30
Portugal	0.52	0.16	0.33	0.47	0.72	0.23
UK	0.54	0.26	0.20	0.48	0.76	−0.04
Norway	0.54	0.41	0.18	0.19	0.58	0.03
Finland	−0.16	−0.46	−0.49	−0.11	0.25	−0.70
Sweden	0.66	0.24	−0.02	0.29	0.38	−0.10
Austria	0.40	0.30	−0.04	0.06	0.17	−0.07
Switzerland	0.61	0.31	0.20	0.34	0.67	−0.16
Mean	**0.46**	**0.15**	**0.05**	**0.24**	**0.49**	**−0.16**

a In the figures depicting national and European attitudes (Figure 3 in Part III and Figure 11 in Part III respectively), the risk scale has been reversed for ease of interpretation.

Table 8c. Moral acceptability (scale: −2 = low, to +2 = high)

Country	Food production	Crop plants	Medicines and vaccines	Research animals	Xeno transplants	Genetic testing
Belgium	0.30	0.74	0.95	−0.25	−0.21	0.91
Denmark	−0.28	0.17	0.99	−0.22	−0.45	0.84
Germany	−0.08	0.25	0.56	−0.61	−0.59	0.52
Greece	−0.17	0.48	0.72	0.05	−0.25	1.39
Italy	0.20	0.71	0.77	−0.15	−0.40	1.05
Spain	0.34	0.55	0.71	0.11	0.14	0.76
France	0.10	0.54	0.84	−0.03	−0.27	1.02
Ireland	0.27	0.48	0.67	−0.19	−0.54	0.78
Luxembourg	0.05	0.28	0.80	−0.10	−0.17	0.72
Netherlands	0.40	0.61	0.86	−0.13	−0.30	0.90
Portugal	0.34	0.75	0.80	0.29	−0.10	0.96
UK	0.22	0.55	0.88	−0.38	−0.46	1.01
Norway	−0.28	−0.10	0.42	−0.13	−0.53	0.32
Finland	0.35	0.78	0.79	−0.12	−0.50	0.93
Sweden	−0.24	0.36	1.02	−0.38	−0.32	0.91
Austria	−0.54	−0.43	0.20	−0.43	−0.53	0.15
Switzerland	−0.13	0.17	0.64	−0.56	−0.76	0.58
Mean	**0.05**	**0.41**	**0.74**	**−0.19**	**−0.37**	**0.81**

Table 8d. Whether they should be encouraged (scale: −2 = low, to +2 = high)

Country	Food production	Crop plants	Medicines and vaccines	Research animals	Xeno transplants	Genetic testing
Belgium	0.09	0.61	0.96	−0.17	−0.11	0.91
Denmark	−0.64	0.03	0.98	−0.11	−0.37	0.77
Germany	−0.17	0.15	0.61	−0.52	−0.48	0.56
Greece	−0.15	0.48	0.81	0.25	−0.03	1.47
Italy	−0.04	0.66	0.84	0.10	−0.23	1.19
Spain	0.36	0.47	0.82	0.25	0.37	0.85
France	−0.26	0.33	0.93	0.17	−0.09	1.16
Ireland	0.10	0.36	0.71	−0.10	−0.42	0.84
Luxembourg	−0.19	0.09	0.87	0.00	−0.13	0.83
Netherlands	0.07	0.43	0.82	−0.24	−0.35	0.84
Portugal	0.24	0.76	0.89	0.52	0.17	1.10
UK	−0.02	0.42	0.92	−0.30	−0.36	1.07
Norway	−0.49	−0.22	0.49	−0.12	−0.51	0.31
Finland	0.40	0.82	0.92	0.07	−0.30	1.02
Sweden	−0.57	0.13	0.88	−0.47	−0.34	0.84
Austria	−0.83	−0.64	0.14	−0.46	−0.56	0.16
Switzerland	−0.48	−0.15	0.51	−0.58	−0.76	0.48
Mean	**−0.15**	**0.28**	**0.77**	**−0.10**	**−0.27**	**0.85**

Table 8e. Mean values summed over the six applications (scale: −12 = low, to +12 = high)

Country	Useful (%)	Risky (%)	Acceptable (%)	Encouraged (%)
Belgium	4.23	0.82	2.44	2.30
Denmark	4.48	0.83	1.06	0.60
Germany	2.36	0.94	0.10	0.16
Greece	4.22	−0.26	2.22	2.83
Italy	4.07	1.50	2.18	2.51
Spain	5.15	0.95	2.60	3.12
France	4.38	1.90	2.20	2.24
Ireland	3.64	1.37	1.47	1.49
Luxembourg	3.46	0.22	1.59	1.46
Netherlands	3.91	3.17	2.34	1.59
Portugal	5.04	2.44	3.04	3.68
UK	4.35	2.19	1.83	1.73
Norway	1.31	1.91	−0.31	−0.55
Finland	4.93	−1.67	2.23	2.93
Sweden	2.27	1.44	1.34	0.47
Austria	−0.18	0.82	−1.57	−2.20
Switzerland	1.86	1.96	−0.10	1.37
Mean	**3.50**	**1.23**	**1.44**	**1.37**

Table 9. Responses to the question: 'People have different views about the benefits and risks of modern biotechnology and about how it should be regulated and controlled.' (Eurobarometer code Q11)

Table 9a. Current regulations are sufficient to protect people from any risks linked to modern biotechnology

Country	Tend to agree (%)	Tend to disagree (%)	Don't know (%)
Belgium	23.3	54.7	21.9
Denmark	23.5	61.9	14.6
Germany	26.7	58.1	15.2
Greece	17.8	51.9	30.3
Italy	18.0	46.7	35.3
Spain	21.8	44.2	34.0
France	21.5	60.3	18.2
Ireland	22.8	39.2	37.9
Luxembourg	26.3	54.6	19.1
Netherlands	35.6	48.3	16.2
Portugal	23.1	38.9	38.0
UK	24.2	53.5	22.3
Norway	29.3	42.5	28.2
Finland	32.6	49.6	17.8
Sweden	22.2	49.2	28.6
Austria	20.7	67.3	12.0
Switzerland	18.1	66.7	15.1
Mean	**23.9**	**52.2**	**23.9**

Table 9b. Irrespective of the regulations, biotechnologists will do whatever they like

Country	Tend to agree (%)	Tend to disagree (%)	Don't know (%)
Belgium	57.8	31.3	10.9
Denmark	71.1	25.8	3.1
Germany	59.7	26.8	13.5
Greece	40.6	36.8	22.6
Italy	53.1	30.5	16.4
Spain	46.1	30.6	23.2
France	59.8	30.8	9.4
Ireland	55.6	19.5	24.9
Luxembourg	59.7	30.2	10.2
Netherlands	55.5	39.9	4.6
Portugal	37.0	38.3	24.7
UK	59.6	30.9	9.4
Norway	43.1	39.0	17.8
Finland	50.0	36.7	13.3
Sweden	55.6	31.9	12.5
Austria	52.4	31.9	15.7
Switzerland	65.3	25.5	9.2
Mean	**54.1**	**31.6**	**14.3**

Table 9c. Only traditional breeding methods should be used, rather than changing the hereditary characteristics of plants and animals through modern biotechnology

Country	Tend to agree (%)	Tend to disagree (%)	Don't know (%)
Belgium	45.7	32.7	21.6
Denmark	59.0	35.1	5.9
Germany	60.2	25.3	14.5
Greece	68.7	18.6	12.7
Italy	57.8	29.1	13.1
Spain	46.0	30.1	24.0
France	54.0	29.8	16.2
Ireland	58.9	18.3	22.8
Luxembourg	56.3	27.5	16.2
Netherlands	40.7	48.4	10.9
Portugal	48.6	27.7	23.8
UK	60.2	29.6	10.2
Norway	63.0	18.9	18.1
Finland	54.6	36.5	8.9
Sweden	57.9	30.5	11.6
Austria	69.5	16.6	13.9
Switzerland	63.0	25.9	11.1
Mean	**56.7**	**28.3**	**15.0**

Table 9d. The regulation of modern biotechnology should be left mainly to industry

Country	Tend to agree (%)	Tend to disagree (%)	Don't know (%)
Belgium	14.2	67.2	18.6
Denmark	19.2	76.0	4.8
Germany	16.8	67.2	16.0
Greece	14.1	54.8	31.1
Italy	19.3	62.9	17.8
Spain	24.4	42.7	32.9
France	21.7	61.0	17.3
Ireland	26.7	45.8	27.6
Luxembourg	18.3	68.5	13.2
Netherlands	14.7	81.2	4.1
Portugal	20.3	45.1	34.7
UK	27.5	61.9	10.7
Norway	12.6	65.9	21.5
Finland	18.2	72.5	9.3
Sweden	11.8	74.7	13.6
Austria	19.3	58.6	22.1
Switzerland	15.6	74.3	10.1
Mean	**18.5**	**63.4**	**18.1**

Table 9e. Modern biotechnology is so complex that public consultation about it is a waste of time

Country	Tend to agree (%)	Tend to disagree (%)	Don't know (%)
Belgium	24.9	63.4	11.7
Denmark	41.6	53.7	4.7
Germany	20.8	64.8	14.4
Greece	25.8	61.3	12.9
Italy	37.1	52.3	10.6
Spain	27.3	54.1	18.6
France	30.5	62.2	7.3
Ireland	31.9	45.2	23.0
Luxembourg	36.5	55.3	8.2
Netherlands	20.4	76.4	3.2
Portugal	26.8	55.0	18.2
UK	34.6	56.6	8.8
Norway	17.7	65.8	16.5
Finland	19.8	74.1	6.1
Sweden	23.0	69.1	7.8
Austria	29.8	47.6	22.6
Switzerland	19.8	73.0	7.3
Mean	**27.3**	**60.7**	**12.0**

Table 9f. It is not worth putting special labels on genetically modified foods

Country	Tend to agree (%)	Tend to disagree (%)	Don't know (%)
Belgium	17.0	73.7	9.3
Denmark	11.5	85.8	2.7
Germany	21.8	72.0	6.2
Greece	9.5	81.6	8.9
Italy	23.1	66.9	9.9
Spain	13.7	69.7	16.6
France	16.1	78.1	5.8
Ireland	18.4	60.2	21.4
Luxembourg	26.2	66.5	7.3
Netherlands	17.6	79.6	2.8
Portugal	18.2	62.3	19.5
UK	12.1	81.5	6.4
Norway	13.3	76.4	10.3
Finland	13.3	82.0	4.7
Sweden	13.8	81.5	4.7
Austria	18.5	73.4	8.1
Switzerland	19.3	78.5	2.1
Mean	**16.4**	**74.9**	**8.7**

Table 9g. I would buy genetically modified fruits if they tasted better

Country	Tend to agree (%)	Tend to disagree (%)	Don't know (%)
Belgium	29.2	50.9	20.0
Denmark	25.5	66.8	7.7
Germany	25.5	57.0	17.5
Greece	19.6	66.4	14.0
Italy	26.7	59.2	14.1
Spain	28.2	50.3	21.5
France	30.0	53.5	16.5
Ireland	29.3	43.3	27.4
Luxembourg	19.0	68.8	12.2
Netherlands	33.2	56.6	10.2
Portugal	37.4	45.2	17.4
UK	36.8	50.5	12.8
Norway	23.6	57.0	19.3
Finland	29.8	61.4	8.8
Sweden	21.5	67.7	10.8
Austria	16.6	69.7	13.7
Switzerland	22.9	67.9	9.2
Mean	**26.9**	**58.1**	**14.9**

Table 9h. Religious organisations need to have their say in how modern biotechnology is regulated

Country	Tend to agree (%)	Tend to disagree (%)	Don't know (%)
Belgium	22.6	61.5	15.9
Denmark	14.7	77.8	7.5
Germany	35.5	47.5	17.2
Greece	37.2	38.5	24.2
Italy	59.3	30.4	10.3
Spain	33.4	42.1	24.5
France	25.0	62.4	12.6
Ireland	39.3	38.5	22.2
Luxembourg	35.7	51.3	13.0
Netherlands	35.4	59.5	5.1
Portugal	45.7	37.4	16.9
UK	45.0	45.1	10.0
Norway	38.8	40.9	20.3
Finland	23.8	65.9	10.3
Sweden	34.8	49.4	15.8
Austria	50.1	29.7	20.2
Switzerland	32.1	59.3	8.6
Mean	**35.8**	**49.2**	**15.0**

Table 9i. We have to accept some degree of risk from modern biotechnology if it enhances economic competitiveness in Europe

Country	Tend to agree (%)	Tend to disagree (%)	Don't know (%)
Belgium	26.8	53.3	20.0
Denmark	30.4	63.1	6.5
Germany	27.9	59.5	12.6
Greece	32.4	42.9	24.7
Italy	22.7	61.6	15.7
Spain	31.8	36.4	31.8
France	17.8	70.0	12.2
Ireland	41.5	35.5	22.9
Luxembourg	26.0	57.3	16.7
Netherlands	35.5	57.0	7.5
Portugal	25.7	48.6	25.7
UK	39.9	48.4	11.7
Norway	20.9	62.3	16.9
Finland	32.3	58.2	9.5
Sweden	22.8	67.7	9.5
Austria	27.3	56.5	16.2
Switzerland	24.2	67.6	8.3
Mean	**28.6**	**55.6**	**15.8**

Table 9j. Traditional breeding methods can be as effective as modern biotechnology, in changing the hereditary characteristics of plants and animals

Country	Tend to agree (%)	Tend to disagree (%)	Don't know (%)
Belgium	35.0	29.9	35.0
Denmark	49.3	36.9	13.8
Germany	45.9	29.3	24.8
Greece	41.0	32.6	26.4
Italy	32.1	37.1	30.7
Spain	38.3	26.5	35.2
France	37.8	30.4	31.9
Ireland	49.5	18.2	32.3
Luxembourg	32.5	35.2	32.3
Netherlands	41.0	40.2	18.9
Portugal	50.5	18.3	31.2
UK	54.7	26.6	18.8
Norway	45.7	20.7	33.6
Finland	53.6	33.2	13.2
Sweden	45.7	28.4	25.9
Austria	47.7	29.1	23.3
Switzerland	42.1	34.4	23.5
Mean	**43.9**	**29.7**	**26.4**

Table 10. Responses to the question: 'Which one of the following bodies do you think is best placed to regulate modern biotechnology?' (Eurobarometer code Q12)

Country	International organisations e.g. the United Nations (%)	Public bodies in our country (%)	Ethics committee (%)	Our national Parliament (%)	The European Union and public bodies therein (%)	Scientific organi-sations (%)	None of these (spon-taneous) (%)	Don't know (%)
Belgium	35.8	8.5	8.7	1.5	6.4	24.4	4.8	9.8
Denmark	29.7	9.7	25.3	5.9	2.1	19.9	3.1	4.2
Germany	29.5	13.3	10.1	5.4	6.4	18.3	6.8	10.2
Greece	31.7	14.4	5.2	5.8	7.0	25.5	2.5	7.9
Italy	39.3	2.7	5.4	4.4	6.6	30.2	2.6	8.8
Spain	35.0	14.5	2.7	2.9	7.6	23.9	1.6	11.9
France	28.8	11.5	15.6	2.8	7.5	25.5	2.8	5.5
Ireland	24.7	28.2	9.6	3.4	5.9	12.0	1.5	14.7
Luxembourg	38.3	7.0	5.3	4.7	9.0	14.6	7.3	13.8
Netherlands	50.2	5.4	8.1	3.9	10.1	19.0	1.2	2.1
Portugal	30.3	11.1	3.4	6.6	6.4	26.7	1.4	14.1
UK	39.6	9.4	5.2	8.9	9.4	17.2	2.7	7.5
Norway	31.6	28.0	6.0	10.3	2.4	11.4	1.1	9.3
Finland	41.0	4.4	5.2	4.2	10.0	29.7	1.4	4.1
Sweden	34.4	8.4	9.9	5.7	1.9	32.1	1.9	5.7
Austria	29.4	16.9	12.4	10.4	5.9	12.3	4.8	7.9
Switzerland	39.4	8.1	10.2	2.3	3.4	21.6	7.8	7.0
Mean	**34.5**	**12.0**	**8.8**	**5.3**	**6.3**	**21.6**	**3.2**	**8.4**

Table 11. Responses to the question: 'For each of the following developments in modern biotechnology, please tell me whether you think it is likely or unlikely to happen within the next 20 years' (Eurobarometer code Q13)

Table 11a. Substantially reducing environmental pollution

Country	Likely (%)	Unlikely (%)	Don't know (%)
Belgium	53.5	38.4	8.1
Denmark	61.4	36.1	2.5
Germany	37.3	54.5	8.2
Greece	38.7	46.7	14.6
Italy	46.0	45.5	8.4
Spain	55.1	35.7	9.2
France	53.5	41.4	5.1
Ireland	58.8	29.3	11.9
Luxembourg	51.5	39.8	8.7
Netherlands	40.9	55.8	3.3
Portugal	56.7	35.9	7.4
UK	49.2	45.6	5.2
Norway	50.2	39.3	10.5
Finland	38.3	59.6	2.1
Sweden	52.2	43.1	4.7
Austria	32.6	56.4	11.0
Switzerland	29.4	65.2	5.4
Mean	**47.2**	**45.4**	**7.4**

Table 11b. Allowing insurance companies to ask for a genetic test before they set a person's premium

Country	Likely (%)	Unlikely (%)	Don't know (%)
Belgium	36.5	48.9	14.6
Denmark	49.4	46.3	4.3
Germany	41.6	43.6	14.8
Greece	37.0	30.6	32.4
Italy	40.9	40.3	18.8
Spain	37.5	29.8	32.7
France	40.5	52.2	7.3
Ireland	48.2	30.7	21.2
Luxembourg	35.3	51.5	13.2
Netherlands	45.7	51.6	2.7
Portugal	42.1	35.8	22.1
UK	60.3	31.7	8.0
Norway	37.7	44.7	17.6
Finland	25.7	67.3	7.0
Sweden	34.2	59.9	5.9
Austria	28.0	52.4	19.6
Switzerland	50.7	41.5	7.8
Mean	**40.8**	**44.5**	**14.7**

Table 11c. Substantially reducing world hunger

Country	Likely (%)	Unlikely (%)	Don't know (%)
Belgium	36.6	54.2	9.2
Denmark	31.9	64.7	3.4
Germany	27.7	62.9	9.3
Greece	33.8	54.3	11.9
Italy	46.6	44.5	8.8
Spain	42.2	49.7	8.2
France	44.3	50.6	5.1
Ireland	43.7	44.4	11.9
Luxembourg	42.7	49.0	8.3
Netherlands	21.2	76.6	2.2
Portugal	49.4	45.3	5.3
UK	35.4	61.0	3.6
Norway	30.7	57.4	11.8
Finland	16.9	80.5	2.6
Sweden	16.4	78.5	5.1
Austria	30.3	56.0	13.7
Switzerland	34.1	60.3	5.6
Mean	**34.2**	**58.4**	**7.4**

Table 11d. Creating dangerous new diseases

Country	Likely (%)	Unlikely (%)	Don't know (%)
Belgium	60.6	23.9	15.5
Denmark	75.4	19.0	5.6
Germany	60.7	25.4	13.9
Greece	86.9	7.3	5.8
Italy	64.7	24.3	10.9
Spain	74.0	13.1	12.9
France	70.6	18.3	11.1
Ireland	63.3	19.8	16.9
Luxembourg	68.8	19.7	11.5
Netherlands	79.9	15.5	4.6
Portugal	74.6	18.7	6.7
UK	67.7	25.3	7.0
Norway	70.9	13.3	15.7
Finland	67.5	27.5	5.0
Sweden	74.2	17.1	8.6
Austria	60.2	22.9	16.9
Switzerland	66.5	22.6	10.9
Mean	**69.8**	**19.6**	**10.5**

Table 11e. Solving more crimes through genetic fingerprinting

Country	Likely (%)	Unlikely (%)	Don't know (%)
Belgium	59.3	27.3	13.4
Denmark	89.7	7.3	3.0
Germany	71.6	17.1	11.3
Greece	55.2	22.2	22.6
Italy	54.3	32.4	13.3
Spain	73.3	11.6	15.1
France	74.1	18.9	7.0
Ireland	80.5	8.8	10.7
Luxembourg	57.0	31.0	12.0
Netherlands	64.1	32.4	3.5
Portugal	76.4	11.9	11.7
UK	84.7	12.0	3.3
Norway	72.1	11.4	16.4
Finland	79.4	14.5	6.1
Sweden	75.1	15.6	9.3
Austria	50.3	30.4	19.3
Switzerland	65.1	26.3	8.6
Mean	**69.8**	**19.2**	**10.9**

Table 11f. Reducing the range of fruit and vegetables we can get

Country	Likely (%)	Unlikely (%)	Don't know (%)
Belgium	26.5	58.1	15.4
Denmark	24.9	68.1	7.0
Germany	20.6	67.2	12.2
Greece	45.7	39.0	15.3
Italy	26.6	61.7	11.7
Spain	27.9	51.6	20.5
France	40.3	49.1	10.6
Ireland	24.0	56.4	19.6
Luxembourg	31.2	56.8	12.0
Netherlands	18.1	79.2	2.7
Portugal	38.4	46.9	14.7
UK	25.7	65.8	8.6
Norway	22.5	51.4	26.1
Finland	18.3	75.0	6.7
Sweden	29.9	58.3	11.8
Austria	25.3	56.2	18.5
Switzerland	25.6	66.7	7.7
Mean	**27.7**	**59.3**	**13.0**

Table 11g. Curing most genetic diseases

Country	Likely (%)	Unlikely (%)	Don't know (%)
Belgium	54.2	32.5	13.3
Denmark	57.1	37.0	5.9
Germany	42.0	41.9	16.1
Greece	74.6	11.7	13.7
Italy	68.0	21.5	10.5
Spain	69.9	15.5	14.6
France	66.2	24.8	9.0
Ireland	52.1	30.1	17.8
Luxembourg	53.2	34.7	12.2
Netherlands	40.9	53.1	6.0
Portugal	70.3	18.8	10.9
UK	48.0	47.1	4.9
Norway	32.2	41.6	26.3
Finland	68.4	26.8	4.8
Sweden	42.3	48.9	8.8
Austria	37.6	39.3	23.1
Switzerland	46.5	43.2	10.3
Mean	**54.4**	**33.4**	**12.2**

Table 11h. Getting more out of natural resources in Third World countries

Country	Likely (%)	Unlikely (%)	Don't know (%)
Belgium	44.7	39.4	15.9
Denmark	64.3	28.5	7.2
Germany	48.2	35.4	16.4
Greece	51.1	24.2	24.7
Italy	56.5	29.7	13.8
Spain	55.2	26.0	18.8
France	53.8	33.1	13.1
Ireland	55.3	24.8	19.9
Luxembourg	46.5	37.3	16.2
Netherlands	60.1	34.5	5.4
Portugal	50.5	33.5	16.0
UK	60.7	30.4	8.9
Norway	56.3	27.6	16.2
Finland	66.9	24.7	8.4
Sweden	64.7	24.8	10.5
Austria	39.2	37.0	23.8
Switzerland	51.1	38.1	10.8
Mean	**54.6**	**31.0**	**14.4**

Table 11i. Producing designer babies

Country	Likely (%)	Unlikely (%)	Don't know (%)
Belgium	32.7	54.5	12.8
Denmark	31.2	64.6	4.2
Germany	41.9	44.6	13.5
Greece	57.0	21.2	21.8
Italy	37.6	54.1	8.3
Spain	48.0	35.4	16.5
France	41.3	52.8	5.9
Ireland	38.3	43.9	17.8
Luxembourg	33.9	56.1	10.0
Netherlands	17.6	80.5	1.9
Portugal	41.4	45.6	13.0
UK	42.7	50.4	6.9
Norway	43.7	40.0	16.3
Finland	26.9	68.7	4.4
Sweden	16.8	77.9	5.3
Austria	29.0	48.9	22.1
Switzerland	39.9	55.1	5.0
Mean	**36.5**	**52.5**	**10.9**

Table 11j. Replacing most existing food products with new varieties

Country	Likely (%)	Unlikely (%)	Don't know (%)
Belgium	46.0	39.5	14.5
Denmark	37.8	58.4	3.8
Germany	40.5	43.9	15.6
Greece	70.3	14.4	15.3
Italy	50.2	39.8	10.0
Spain	44.8	33.9	21.3
France	50.8	41.1	8.1
Ireland	48.2	34.0	17.8
Luxembourg	35.5	53.5	11.0
Netherlands	31.5	64.9	3.6
Portugal	53.4	34.4	12.2
UK	44.2	48.4	7.3
Norway	31.4	49.6	19.0
Finland	30.7	63.8	5.5
Sweden	30.9	62.3	6.8
Austria	31.9	50.7	17.4
Switzerland	42.8	48.7	8.5
Mean	**42.6**	**45.8**	**11.6**

Table 12. Responses to the question: 'Which of the following sources of information do you have confidence in to tell you the truth about modern biotechnology?' (Eurobarometer code Q14 split ballot A)

Country	Consumer organisations (%)	Environmental organisations (%)	Animal welfare organisations (%)	Political organisations (%)	Trade unions (%)	Religious organisations (%)	Public authorities (%)	Industry (%)	School or university (%)	Don't know (%)
Belgium	34.7	24.8	5.1	1.0	1.0	1.2	3.4	1.4	17.6	9.7
Denmark	36.2	17.5	7.9	0.0	0.6	0.2	6.3	1.0	23.6	6.7
Germany	40.3	25.9	7.5	1.6	1.0	2.6	3.4	0.6	8.1	8.9
Greece	20.5	22.7	2.4	2.2	1.6	5.4	13.1	3.4	20.3	8.4
Italy	25.3	25.5	6.6	0.6	1.6	6.6	4.4	2.4	14.3	12.7
Spain	26.0	24.7	3.8	1.4	0.6	1.8	12.3	1.0	12.1	16.3
France	46.0	19.9	6.0	2.0	2.0	2.6	6.4	1.0	5.8	8.4
Ireland	23.8	23.4	8.1	1.4	1.9	7.0	7.5	0.8	12.0	14.1
Luxembourg	32.4	23.1	5.0	4.0	4.3	2.7	4.7	0.0	13.0	10.7
Netherlands	39.0	21.8	4.4	1.2	1.9	2.7	6.0	1.5	16.2	5.4
Portugal	30.4	20.4	1.4	1.0	1.8	4.4	6.3	0.8	13.1	20.4
UK	27.5	23.1	5.5	1.4	2.6	6.1	5.1	2.0	16.8	9.9
Norway	19.5	17.9	6.0	2.4	2.2	1.4	20.9	1.0	15.1	13.5
Finland	24.0	15.6	2.4	0.4	0.6	1.0	15.4	1.2	33.7	5.6
Sweden	25.6	24.9	5.8	0.4	1.2	2.0	9.7	1.6	23.5	5.2
Austria	37.6	27.0	5.6	2.6	1.1	5.2	5.4	0.6	6.4	8.4
Switzerland	30.2	22.1	6.8	0.7	0.8	2.6	4.3	4.3	18.9	9.2
Mean	**30.5**	**22.4**	**5.4**	**1.3**	**1.5**	**3.3**	**7.8**	**1.6**	**16.1**	**10.2**

Table 13. Responses to the question: 'Which of the following sources of information do you have confidence in to tell you the truth about modern biotechnology, genetically modified food crops grown in fields, and introducing human genes into animals to produce organs for human transplants?' (Eurobarometer code Q14 split ballot B)

Table 13a. Modern biotechnology

Country	Consumer organisations (%)	Environmental organisations (%)	Animal welfare organisations (%)	The medical profession (%)	Farmers' organisations (%)	Religious organisations (%)	National public bodies (%)	Internal public bodies (%)	Industry (%)	Universities (%)	Political parties (%)	Television and newspapers (%)	None of these (%)	DK (%)
Belgium	19.4	17.8	3.8	10.7	1.8	0.6	3.2	2.8	2.0	15.8	0.2	7.5	6.1	8.3
Denmark	28.1	15.1	7.5	9.6	0.4	0.8	5.5	2.4	0.8	16.1	0.6	2.4	6.3	4.3
Germany	24.8	21.6	5.2	7.6	2.4	1.0	6.6	3.8	1.4	6.8	0.2	4.0	7.8	7.0
Greece	15.1	16.7	2.6	12.9	1.4	3.0	4.2	2.8	2.4	12.4	0.4	9.2	6.8	10.2
Italy	14.4	14.0	5.2	21.0	2.4	5.4	1.6	7.6	0.6	7.8	0.6	2.8	6.4	10.0
Spain	13.3	14.9	3.4	17.5	1.2	1.0	3.8	7.6	2.0	7.2	0.6	2.8	4.8	19.7
France	29.0	12.2	3.0	23.2	2.6	1.2	3.0	3.2	3.0	4.4	0.2	2.8	7.4	4.8
Ireland	15.5	16.3	4.3	21.7	1.4	4.1	3.5	4.8	0.6	6.8	0.4	3.3	3.9	13.6
Luxembourg	17.7	16.3	1.7	16.0	3.3	1.3	3.3	3.0	1.7	12.3	0.7	6.3	8.0	8.3
Netherlands	25.7	14.5	4.2	12.0	0.6	0.6	1.5	5.2	0.2	17.2	0.4	6.6	7.7	3.7
Portugal	19.7	13.9	1.8	11.4	2.4	2.4	3.4	1.4	1.6	7.0	0.4	6.6	4.8	23.1
UK	18.2	18.4	4.8	16.6	0.8	1.2	4.8	1.2	0.4	13.5	0.2	7.5	7.2	5.2
Norway	17.1	15.9	4.5	7.7	2.0	2.2	17.5	6.5	1.0	8.9	0.4	2.6	4.7	8.9
Finland	11.8	10.4	1.0	23.3	1.8	0.0	4.0	6.2	1.4	22.3	0.0	7.6	4.6	5.8
Sweden	12.0	25.5	4.8	11.2	1.3	1.3	5.3	5.9	0.8	18.9	0.0	2.9	4.0	6.1
Austria	22.4	19.4	2.2	10.1	1.7	3.7	7.1	6.5	0.4	6.9	1.1	0.6	8.6	9.5
Mean	**19.0**	**16.4**	**3.8**	**14.5**	**1.7**	**1.9**	**4.9**	**4.5**	**1.3**	**11.5**	**0.4**	**4.7**	**6.1**	**9.3**

Table 13b. Genetically modified food crops grown in fields

Country	Consumer organisations (%)	Environmental organisations (%)	Animal welfare organisations (%)	The medical profession (%)	Farmers' organisations (%)	Religious organisations (%)	National public bodies (%)	Internal public bodies (%)	Industry (%)	Universities (%)	Political parties (%)	Television and newspapers (%)	None of these (%)	DK (%)
Belgium	14.0	22.7	2.2	9.9	13.6	1.2	2.6	1.8	0.4	11.7	0.4	5.1	5.5	8.9
Denmark	18.1	17.3	1.4	4.5	22.0	0.6	6.1	3.0	1.2	10.0	0.4	3.0	6.7	5.7
Germany	14.0	34.6	3.2	2.8	15.0	1.4	5.4	2.4	1.0	4.4	0.2	1.8	7.4	6.4
Greece	13.9	21.5	2.8	6.8	16.1	1.2	4.2	3.0	1.0	8.3	1.0	4.6	5.6	10.1
Italy	20.1	24.5	2.2	7.0	17.1	1.0	1.2	3.0	2.0	4.8	0.0	1.6	6.4	8.9
Spain	12.9	19.6	3.0	8.3	15.1	0.0	4.2	4.4	1.6	3.8	0.4	3.0	4.0	19.6
France	24.4	22.8	2.2	9.2	13.0	0.2	2.8	3.0	2.2	3.0	0.2	4.2	6.2	6.4
Ireland	16.8	24.5	3.5	7.1	13.3	2.1	2.9	2.1	1.0	5.4	0.6	2.7	4.4	13.5
Luxembourg	33.2	16.9	1.7	5.0	10.3	0.7	3.0	2.0	1.3	5.3	0.0	5.0	7.3	8.3
Netherlands	26.9	20.5	0.6	1.9	10.3	0.6	4.6	4.3	1.2	12.2	0.2	4.1	5.4	7.4
Portugal	13.3	20.9	2.0	3.8	16.9	1.6	2.4	1.4	1.0	4.6	0.6	5.0	3.6	22.7
UK	16.0	26.6	0.2	3.1	19.5	0.4	2.9	0.6	1.2	6.9	0.6	6.2	8.5	7.1
Norway	15.1	19.2	1.4	2.2	22.5	0.2	13.7	3.3	0.6	4.7	0.4	1.6	4.7	10.2
Finland	11.5	16.5	0.8	11.5	14.7	0.4	4.0	4.0	0.8	20.1	0.0	5.4	4.0	6.4
Sweden	11.8	28.8	2.1	2.1	22.5	0.4	3.4	4.6	0.4	11.8	0.2	3.2	3.4	5.3
Austria	7.3	24.9	1.9	8.6	13.5	2.2	8.0	6.2	1.1	6.7	0.4	0.6	7.5	11.0
Mean	**16.5**	**22.7**	**2.0**	**5.9**	**16.1**	**0.9**	**4.5**	**3.1**	**1.1**	**7.8**	**0.4**	**3.5**	**5.6**	**9.9**

Table 13c. Introducing human genes into animals to produce organs for human transplants

Country	Consumer organisations (%)	Environmental organisations (%)	Animal welfare organisations (%)	The medical profession (%)	Farmers' organisations (%)	Religious organisations (%)	National public bodies (%)	Internal public bodies (%)	Industry (%)	Universities (%)	Political parties (%)	Television and newspapers (%)	None of these (%)	DK (%)
Belgium	3.0	1.6	11.4	43.8	1.6	1.4	1.6	2.2	0.8	13.6	0.0	5.3	5.5	8.1
Denmark	3.9	2.4	10.8	41.1	0.0	0.6	6.3	1.8	0.2	10.4	1.0	2.6	8.3	10.6
Germany	5.6	5.8	13.6	36.6	1.4	1.8	4.6	3.4	1.0	7.4	0.2	1.8	9.6	7.2
Greece	1.6	2.8	9.7	48.3	0.6	2.2	3.6	3.4	0.6	6.8	0.0	5.4	5.8	9.3
Italy	1.0	2.0	12.4	53.2	0.2	1.8	1.0	2.2	0.0	8.8	0.0	1.0	6.4	9.8
Spain	2.0	1.6	5.8	52.7	0.8	0.2	1.8	3.8	0.8	6.6	0.2	3.2	4.2	16.1
France	3.8	2.6	10.2	51.4	1.4	1.0	3.2	1.8	1.2	2.6	0.4	3.6	8.6	8.0
Ireland	3.9	4.8	11.9	42.0	0.2	3.3	2.1	2.5	0.8	5.4	0.4	3.3	4.4	15.0
Luxembourg	1.7	2.0	9.0	47.0	2.3	1.7	0.3	1.0	0.7	11.7	0.3	2.3	9.0	11.0
Netherlands	4.6	2.9	13.9	40.0	0.0	1.0	1.9	2.1	0.2	13.5	0.4	6.0	7.0	6.4
Portugal	1.2	1.0	8.5	51.9	0.0	1.4	2.0	1.6	0.4	3.4	0.4	5.8	2.0	20.3
UK	2.9	3.3	16.8	41.4	0.8	0.8	4.2	0.4	0.4	8.3	0.2	6.0	7.5	6.9
Norway	3.1	3.9	7.3	29.6	0.6	0.6	16.9	2.9	0.6	10.4	1.0	2.4	6.1	14.5
Finland	1.6	0.2	6.4	52.4	0.2	1.0	2.0	4.2	0.4	12.8	0.0	6.6	4.0	8.2
Sweden	1.1	6.1	19.2	37.0	1.3	1.5	3.2	4.2	0.0	11.6	0.0	3.0	6.3	5.5
Austria	2.8	2.8	7.1	36.9	2.4	3.9	5.2	4.5	2.4	6.7	1.5	2.4	9.9	11.8
Mean	**2.8**	**2.9**	**10.9**	**44.1**	**0.8**	**1.5**	**3.8**	**2.7**	**0.6**	**8.7**	**0.4**	**3.8**	**6.5**	**10.6**

Table 14. Responses to the question: 'How important are issues to do with modern biotechnology to you personally?' (10-point scale where 1 = not at all important, 10 = extremely important) (Eurobarometer code Q15)

Country	Not at all important (%)	2 (%)	3 (%)	4 (%)	5 (%)	6 (%)	7 (%)	8 (%)	9 (%)	Extremely important (%)
Belgium	4.6	9.0	4.1	5.3	11.2	17.4	17.3	16.8	7.9	6.3
Denmark	3.1	1.7	5.5	5.9	13.7	9.2	13.6	23.9	8.7	14.6
Germany	4.9	4.2	7.9	8.2	13.6	14.6	15.5	14.9	7.3	8.9
Greece	3.9	4.2	4.1	5.6	8.1	10.4	16.2	17.1	10.3	20.1
Italy	2.2	1.7	3.8	6.0	14.4	14.9	19.6	18.6	6.3	12.5
Spain	4.5	3.8	6.7	5.4	16.3	16.0	18.3	15.2	4.7	9.2
France	2.6	2.1	3.7	5.8	18.2	13.2	19.3	15.9	7.7	11.5
Ireland	8.3	4.0	7.1	7.1	16.2	12.3	13.9	12.8	6.9	11.4
Luxembourg	5.6	4.6	5.4	6.7	17.8	12.1	11.2	13.9	6.0	16.7
Netherlands	1.3	1.2	2.6	3.7	9.0	14.4	25.8	26.4	6.0	9.6
Portugal	2.1	3.4	5.0	6.1	18.9	13.9	16.5	13.5	7.0	13.5
UK	3.6	2.6	5.9	6.7	14.8	11.9	18.4	18.4	5.3	12.3
Norway	5.1	4.0	10.8	7.4	16.1	11.9	15.0	13.7	5.0	11.1
Finland	2.0	2.4	5.2	5.9	12.6	11.9	17.3	24.2	9.7	8.7
Sweden	1.7	1.5	3.3	5.4	12.9	10.8	17.6	22.8	8.7	15.4
Austria	5.6	6.0	6.1	5.1	8.8	11.9	9.0	16.3	10.1	21.1
Switzerland	4.6	3.1	6.8	6.6	16.4	9.5	13.1	17.1	7.6	15.0
Mean	**3.8**	**3.5**	**5.5**	**6.0**	**13.9**	**12.7**	**16.5**	**18.0**	**7.4**	**12.7**

Table 15. Responses to the question: 'Over the last three months, have you heard anything about issues involving modern biotechnology? If yes, was it in newspapers, in magazines, on television, or on the radio?' (Eurobarometer code Q16)

Table 15a. Heard anything about biotechnology issues

Country	No (%)	Yes (%)
Belgium	45.1	54.9
Denmark	61.2	38.8
Germany	60.2	39.8
Greece	30.3	69.7
Italy	51.0	49.0
Spain	40.5	59.5
France	53.8	46.2
Ireland	36.8	63.2
Luxembourg	60.0	40.0
Netherlands	52.5	47.5
Portugal	39.0	61.0
UK	54.7	45.3
Norway	66.3	33.7
Finland	71.9	28.1
Sweden	61.0	39.0
Austria	74.5	25.5
Switzerland	78.8	21.2
Mean	**55.1**	**44.9**

Table 15b. In newspapers

Country	No (%)	Yes (%)
Belgium	85.9	14.1
Denmark	72.3	27.7
Germany	77.2	22.8
Greece	90.4	9.6
Italy	83.6	16.4
Spain	85.9	14.1
France	82.8	17.2
Ireland	82.0	18.0
Luxembourg	71.7	28.3
Netherlands	65.5	34.5
Portugal	88.4	11.6
UK	72.2	27.8
Norway	66.3	33.7
Finland	69.7	30.3
Sweden	73.4	26.6
Austria	65.8	34.2
Switzerland	44.1	55.9
Mean	**75.2**	**24.8**

Table 15c. In magazines

Country	No (%)	Yes (%)
Belgium	88.7	11.3
Denmark	90.3	9.7
Germany	81.2	18.8
Greece	93.9	6.1
Italy	83.7	16.3
Spain	90.1	9.9
France	86.5	13.5
Ireland	94.8	5.2
Luxembourg	79.5	20.5
Netherlands	87.5	12.5
Portugal	91.6	8.4
UK	89.9	10.1
Norway	89.6	10.4
Finland	88.2	11.8
Sweden	93.5	6.5
Austria	82.1	17.9
Switzerland	75.2	24.8
Mean	**87.6**	**12.4**

Table 15d. On television

Country	No (%)	Yes (%)
Belgium	75.9	24.1
Denmark	62.1	37.9
Germany	65.2	34.8
Greece	81.5	18.5
Italy	69.9	30.1
Spain	75.6	24.4
France	64.4	35.6
Ireland	82.3	17.7
Luxembourg	59.3	40.7
Netherlands	62.1	37.9
Portugal	72.9	27.1
UK	61.6	38.4
Norway	65.9	34.1
Finland	46.8	53.2
Sweden	61.7	38.3
Austria	54.6	45.4
Switzerland	41.6	58.4
Mean	**65.0**	**35.0**

Table 15e. On radio

Country	No (%)	Yes (%)
Belgium	95.7	4.3
Denmark	84.5	15.5
Germany	93.3	6.7
Greece	98.2	1.8
Italy	97.1	2.9
Spain	94.6	5.4
France	90.7	9.3
Ireland	94.3	5.7
Luxembourg	86.2	13.8
Netherlands	86.9	13.1
Portugal	97.7	2.3
UK	87.7	12.3
Norway	91.3	8.7
Finland	85.5	14.5
Sweden	86.6	13.4
Austria	83.0	17.0
Switzerland	67.4	32.6
Mean	**89.5**	**10.5**

Table 15f. Do not remember where[a]

Country	No (%)	Yes (%)
Belgium	92.5	7.5
Denmark	91.0	9.0
Germany	92.3	7.7
Greece	95.0	5.0
Italy	92.6	7.4
Spain	95.3	4.7
France	92.7	7.3
Ireland	95.4	4.6
Luxembourg	94.8	5.2
Netherlands	99.0	1.0
Portugal	94.1	5.9
UK	95.2	4.8
Norway	87.2	12.8
Finland	95.0	5.0
Sweden	91.1	8.9
Austria	90.3	9.7
Mean	**93.3**	**6.7**

a This question was not asked in Switzerland

Table 16. Responses to the question: 'Before today, had you ever talked about modern biotechnology with someone?' (*Eurobarometer code Q17*)

Country	No (%)	Yes (%)
Belgium	41.4	58.6
Denmark	15.9	84.1
Germany	24.1	75.9
Greece	60.1	39.9
Italy	38.0	62.0
Spain	51.3	48.7
France	36.6	63.4
Ireland	51.8	48.2
Luxembourg	32.0	68.0
Netherlands	36.8	63.2
Portugal	52.1	47.9
UK	36.0	64.0
Norway	21.2	78.8
Finland	20.9	79.1
Sweden	24.1	75.9
Austria	20.6	79.4
Switzerland	10.0	90.0
Mean	**33.7**	**66.3**

Table 17. Means for responses to the questions: 'Have you heard about or talked about biotechnology with someone'? (Eurobarometer code Q16 and Q17)

Country	No, never (%)	Yes, frequently (%)	Yes, occasionally (%)	Yes, only once or twice (%)	Don't know (%)
Belgium	59.1	3.0	20.3	16.3	1.3
Denmark	24.1	11.7	41.1	21.4	1.7
Germany	33.5	7.4	33.3	20.8	5.1
Greece	72.0	4.3	13.0	10.7	0.0
Italy	57.9	3.6	24.7	12.2	1.6
Spain	66.1	2.6	19.9	10.2	1.2
France	53.3	5.7	24.9	14.5	1.6
Ireland	62.5	3.3	14.1	17.0	3.1
Luxembourg	47.5	6.5	31.3	13.7	1.0
Netherlands	56.2	8.6	25.2	9.9	0.1
Portugal	68.5	4.4	14.8	11.7	0.6
UK	52.1	7.5	27.9	12.1	0.4
Norway	36.8	5.4	30.8	22.7	4.2
Finland	44.9	6.1	38.0	10.3	0.6
Sweden	41.1	4.8	28.6	24.5	0.9
Austria	38.9	14.4	24.7	15.0	7.0
Switzerland	20.6	13.3	47.1	18.9	0.1
Mean	**49.1**	**6.6**	**27.0**	**15.5**	**1.8**

Table 18. Responses to the question: 'Which issue do you think will most influence your vote at the next general election?' (Eurobarometer code Q18)

Table 18a. Protection of the environment and nature, cutting down pollution

Country	No (%)	Yes (%)
Belgium	56.6	43.4
Denmark	34.8	65.2
Germany	56.8	43.2
Greece	58.8	41.2
Italy	68.7	31.3
Spain	68.6	31.4
France	62.3	37.7
Ireland	72.4	27.6
Luxembourg	46.0	54.0
Netherlands	40.5	59.5
Portugal	70.5	29.5
UK	62.7	37.3
Norway	53.2	46.8
Finland	60.5	39.5
Sweden	40.7	59.3
Austria	59.6	40.4
Switzerland	57.9	42.1
Mean	**57.4**	**42.6**

Table 18b. Fight for workers' rights

Country	No (%)	Yes (%)
Belgium	72.9	27.1
Denmark	87.8	12.2
Germany	67.5	32.5
Greece	57.4	42.6
Italy	67.1	32.9
Spain	63.3	36.7
France	69.7	30.3
Ireland	83.5	16.5
Luxembourg	70.7	29.3
Netherlands	81.9	18.1
Portugal	62.4	37.6
UK	83.0	17.0
Norway	78.4	21.6
Finland	80.9	19.1
Sweden	81.0	19.0
Austria	67.6	32.4
Switzerland	79.7	20.3
Mean	**73.9**	**26.1**

Table 18c. Protection of social benefits and health care

Country	No (%)	Yes (%)
Belgium	62.2	37.8
Denmark	47.5	52.5
Germany	40.8	59.2
Greece	50.0	50.0
Italy	77.8	22.2
Spain	70.5	29.5
France	52.7	47.3
Ireland	66.9	33.1
Luxembourg	68.2	31.8
Netherlands	32.5	67.5
Portugal	55.7	44.3
UK	46.7	53.3
Norway	30.4	69.6
Finland	34.9	65.1
Sweden	27.7	72.3
Austria	58.0	42.0
Switzerland	45.9	54.1
Mean	**50.7**	**49.3**

Table 18d. Education

Country	No (%)	Yes (%)
Belgium	82.6	17.4
Denmark	62.0	38.0
Germany	81.6	18.4
Greece	54.5	45.5
Italy	78.3	21.7
Spain	70.5	29.5
France	80.1	19.9
Ireland	77.5	22.5
Luxembourg	64.0	36.0
Netherlands	59.5	40.5
Portugal	64.5	35.5
UK	44.0	56.0
Norway	64.8	35.2
Finland	75.5	24.5
Sweden	70.1	29.9
Austria	75.6	24.4
Switzerland	68.9	31.1
Mean	**69.2**	**30.8**

Table 18e. Fight against racism

Country	No (%)	Yes (%)
Belgium	84.2	15.8
Denmark	86.1	13.9
Germany	88.1	11.9
Greece	94.4	5.6
Italy	88.0	12.0
Spain	87.3	12.7
France	82.6	17.4
Ireland	94.2	5.8
Luxembourg	79.7	20.3
Netherlands	70.0	30.0
Portugal	91.1	8.9
UK	93.1	6.9
Norway	83.6	16.4
Finland	94.9	5.1
Sweden	81.5	18.5
Austria	85.0	15.0
Switzerland	83.5	16.5
Mean	**86.5**	**13.5**

Table 18f. Protection of pensioners' rights

Country	No (%)	Yes (%)
Belgium	78.8	21.2
Denmark	71.5	28.5
Germany	73.5	26.5
Greece	77.3	22.7
Italy	75.5	24.5
Spain	74.3	25.7
France	85.2	14.8
Ireland	85.9	14.1
Luxembourg	83.7	16.3
Netherlands	81.8	18.2
Portugal	74.2	25.8
UK	77.0	23.0
Norway	67.9	32.1
Finland	75.7	24.3
Sweden	76.2	23.8
Austria	64.4	35.6
Switzerland	82.3	17.7
Mean	**76.6**	**23.4**

Table 18g. Fight against unemployment

Country	No (%)	Yes (%)
Belgium	51.4	48.6
Denmark	64.2	35.8
Germany	33.2	66.8
Greece	35.7	64.3
Italy	33.4	66.6
Spain	34.5	65.5
France	32.7	67.3
Ireland	46.0	54.0
Luxembourg	51.7	48.3
Netherlands	58.3	41.7
Portugal	39.1	60.9
UK	61.7	38.3
Norway	57.6	42.4
Finland	32.5	67.5
Sweden	22.2	77.8
Austria	45.1	54.9
Switzerland	34.3	65.7
Mean	**42.9**	**57.1**

Table 18h. Fight against homelessness and poverty

Country	No (%)	Yes (%)
Belgium	80.9	19.1
Denmark	76.8	23.2
Germany	83.9	16.1
Greece	71.8	28.2
Italy	87.6	12.4
Spain	87.2	12.8
France	71.2	28.8
Ireland	72.3	27.7
Luxembourg	82.7	17.3
Netherlands	81.0	19.0
Portugal	76.8	23.2
UK	67.1	32.9
Norway	74.0	26.0
Finland	76.9	23.1
Sweden	82.7	17.3
Austria	83.0	17.0
Switzerland	80.5	19.5
Mean	**78.5**	**21.5**

Table 18i. Fight against crime and delinquency

Country	No (%)	Yes (%)
Belgium	53.1	46.9
Denmark	56.7	43.3
Germany	67.3	32.7
Greece	83.9	16.1
Italy	68.5	31.5
Spain	73.8	26.2
France	73.7	26.3
Ireland	49.2	50.8
Luxembourg	67.8	32.2
Netherlands	59.2	40.8
Portugal	61.5	38.5
UK	55.1	44.9
Norway	52.0	48.0
Finland	74.4	25.6
Sweden	62.4	37.6
Austria	70.0	30.0
Switzerland	76.4	23.6
Mean	**65.0**	**35.0**

Table 18j. Taxation

Country	No (%)	Yes (%)
Belgium	56.7	43.3
Denmark	77.6	22.4
Germany	61.6	38.4
Greece	75.6	24.4
Italy	52.9	47.1
Spain	75.7	24.3
France	72.7	27.3
Ireland	74.0	26.0
Luxembourg	85.8	14.2
Netherlands	78.6	21.4
Portugal	86.5	13.5
UK	82.1	17.9
Norway	86.6	13.4
Finland	61.7	38.3
Sweden	82.1	17.9
Austria	67.2	32.8
Switzerland	80.3	19.7
Mean	**73.7**	**26.3**

Table 18k. Don't know

Country	No (%)	Yes (%)
Belgium	95.6	4.4
Denmark	95.9	4.1
Germany	96.4	3.6
Greece	98.8	1.2
Italy	96.0	4.0
Spain	94.0	6.0
France	95.8	4.2
Ireland	91.9	8.1
Luxembourg	90.2	9.8
Netherlands	96.2	3.8
Portugal	97.1	2.9
UK	95.5	4.5
Norway	96.1	3.9
Finland	95.3	4.7
Sweden	99.5	0.5
Austria	93.8	6.2
Switzerland	95.3	4.7
Mean	**95.6**	**4.4**

Table 20. Age of respondents

Table 19. Gender of respondents

Country	Female (%)	Male (%)
Belgium	51.6	48.4
Denmark	51.1	48.9
Germany	52.0	48.0
Greece	51.1	48.9
Italy	51.9	48.1
Spain	51.5	48.5
France	51.4	48.6
Ireland	50.7	49.3
Luxembourg	52.0	48.0
Netherlands	50.9	49.1
Portugal	52.5	47.5
UK	51.8	48.2
Norway	52.6	47.4
Finland	51.5	48.5
Sweden	51.0	49.0
Austria	52.5	47.5
Switzerland	49.4	50.6
Mean	**51.5**	**48.5**

Country	15–24 years (%)	25–39 years (%)	40–54 years (%)	55+ years (%)
Belgium	16.2	28.3	22.8	32.7
Denmark	18.7	27.3	23.3	30.8
Germany	14.7	29.2	24.1	32.0
Greece	18.5	26.0	22.5	33.0
Italy	17.9	26.7	22.9	32.5
Spain	20.2	27.8	21.1	30.9
France	18.3	29.3	22.9	29.5
Ireland	23.3	28.1	22.3	26.3
Luxembourg	18.2	28.8	23.8	29.2
Netherlands	17.6	30.3	24.6	27.5
Portugal	20.5	26.1	22.5	30.9
UK	16.7	28.2	23.0	32.1
Norway	18.4	27.7	21.9	32.0
Finland	15.3	28.4	27.1	29.2
Sweden	15.3	26.1	25.1	33.5
Austria	16.7	30.0	23.1	30.2
Switzerland	13.9	32.7	29.8	23.6
Mean	**17.7**	**28.3**	**23.7**	**30.4**

Appendix 3. Guidelines for the time-series press analysis, 1973–96

Aim

The aim of the time-series study was to establish a media indicator of public debates on biotechnology, allowing for international comparison of: **1** the initiation and onset of public debates about biotechnology, **2** the timing of these debates in different countries, and **3** the changing argumentation (frames) of these debates in a basic content analysis.

The selection of medium: newspapers or magazines

It was not realistic to devise a comprehensive analysis of the press coverage over the entire period from 1973 to 1996. Thus a prominent press medium was selected, either newspaper(s) or weekly magazine(s) that had an *opinion-leader function*. This allowed us to make a functional equivalent comparison, given the variety of the press across countries. The test of the opinion leader-function was: which papers or magazines are a country's elites (politicians, business, military, etc.) most likely to read? In some countries this opinion-leader function was assumed by a single medium (e.g. *Le Monde* in France) and, in others several newspapers were considered (e.g. *The Times* and the *Independent* in the UK).

We constructed time-series indicators of biotechnology in the press from 1973 to 1996, producing a graph with 23 or more time points. The start year '1973' was selected because of the Asilomar I conference held in January 1973, rDNA controversy at the Gordon Conference on Nucleic Acids in June 1973, and Berg's letter to *Science* in July 1974: Asilomar II was held in February 1975. Each country calculates an indicator (peak year = 100) based on the local sample.

The sampling of articles

A *random cluster sample* was constructed for the chosen medium. Individual countries used combinations of three different procedures to obtain samples: hands-on sampling, index sampling and on-line sampling.

Hands-on sampling

As well as the question of how many random dates should be generated (a single newspaper issue is a sampling cluster), we faced the complication of sampling rare events in the 1970s and early 1980s. Sampling error depends on the expected proportion of newspapers that contain relevant material per year. Assuming that about 5 of the newspaper issues contain relevant material, an acceptable estimate for the year would have to be based on about 100 scanned newspapers, or about a third of the total number of dates. Generally speaking, the smaller the proportion of relevant material, the larger the sample needs to be. For the sampling of weekly magazines an analogous sampling strategy is applied. The following recommendations were made:

Sampling 1970s, scanning minimum 100 random days (30% of possible dates)
Sampling early 1980s, scanning minimum 50 random days (15% of possible dates)
Sampling 1986 onwards, scanning minimum 12 random days (5% of possible dates)

An 'aim and shoot' approach was also recommended. According to this, the 1973 and 1975 (Asilomar conferences) were sampled first, then 1980. Based on these results, 1977 was sampled, then 1983. This was continued until a clear picture for the time-series emerged. If several media were used, dates were taken from a list of 100 yearly random dates supplied.

Newspaper index sampling

Some newspapers such as *The Times* and *Le Monde* have an index for the articles they publish. This index was used for purposes of 'targeting and shooting' relevant articles. It was necessary to identify the relevant keywords used to index articles that were relevant to the purposes of this project. For example, in *The Times* index during the 1970s relevant material needed to be found under the entry for 'genetics'. The keywords 'biotech*' or 'genetic engineering' are not used before 1980.

On-line sampling

On-line data were available for some media, from sources such as FT-Profile (UK) or NEXIS (USA) or CD-ROM archives of newspapers. These allowed for keyword searches and the creation of a *simple random sample* of the total amount of material on biotechnology retrieved.

Calibrating the time-series

Newspaper index and on-line sampling allowed a total enumeration of relevant articles with a low error rate while hands-on sampling produces a larger sampling error. The overall sampling strategy was a combination of manual sampling, index-based enumeration and on-line enumeration for recent years. Those who had access to on-line data files were able to compare the different procedures in overlapping years to validate the time-series.

Five sampling windows on biotechnology

The intensity index may identify up to several thousand relevant articles over the years. We could not analyse all of them, therefore a sample ($n = 500$) was selected for further content analysis. About 100 articles were selected for each five-year window: 1973–77, 1978–82, 1983–87, 1988–92 and 1993–96.

Selection of relevant articles

Definition of biotechnology

As agreed in London (September 1995) biotechnology was understood to be 'interventions at the level of the gene' (new genetics; new biotechnology).

Our minimal comparative focus was on *new genetics and biotechnology*. Researchers may have widened their definition for the purpose of national studies. The inclusion of 'human reproductive techniques' was warranted by the fact that, in national politics, these issues are regulated together, e.g. Swiss 'Gene Referendum' of 17 May 1992. Under the heuristic of 'observing observers' we avoided making distinctions where the media or policy process does not make them, even if there were good reasons based on science to distinguish terms.

Search terms for manual or on-line retrieval

National public debates and different languages influence the kinds of keywords that were relevant. Experimentation was recommended to test out whether a set of keywords appeared to be comprehensive enough to cover all relevant material.

Basic: 'biotech*', 'genetic*', 'genome', 'DNA'.

Additional: 'clon*ing'. IVF [*in-vitro* fertilisation]. 'test tube babies' ,'genetic tests', 'genetic engineering', 'genetic manipulation', 'DNA', 'gene', 'DNA fingerprinting'. 'Genome analysis', 'Human Genome Project', 'genetic disorder', 'gene therapy', 'Cancer gene'. 'biotech industry', 'loan mother', 'oncomouse', 'tissue engineering', 'Flavr Savr tomato', AIDS.

Content analysis

A basic coding frame was compiled for manual and computer-assisted coding of press articles over all five-year 'windows'. For comparative purposes a *minimal coding frame*, comparable across all studies, was agreed (shown in Appendix 5). National teams may have developed their own codings in line with their particular analytical focus.

Address for correspondence

Martin W. Bauer, Methodology Institute, London School of Economics and Political Science, London WC2A 2AE. Further details and a complete codebook are available.

Appendix 4. Sampling procedures

Table 1. Sampling procedure for longitudinal analysis of opinion-leading press

Country	Population coverage (N) (weight)	Sampling procedure for coding (n)	Newspapers (opinion-leader)	Size of corpus[1]	Comments on corpus condition
Austria	variable ratio; 1973–79 *Presse* 100/yr (×3.6) *Profil* complete (×1); 1980–85 *Presse* 65/yr (×5.1) *Profil* complete (×1); 1986–96 *Presse* 35/yr (×10) *Profil* 26/yr (×2); 1973–93: manual	data ≠ index sampling, i.e. yearly frequencies are not absolute intensity measures	*Presse* (d)[2], right *Profil* (w), left	$N = 1396$ (est.) $n = 302$ $N_{Presse} = 1454$ $N_{Profil} = 191$	hard copies available
Denmark	1973–89: total coverage Arhus data base manual index; supplemented backwards from on-line database 1990–96: total enumeration of large articles only from on-line service; 1996: all articles	data ≠ index yearly frequencies are not absolute intensity measures	*Information* (d) left, small *Politiken* (d), mainstream	$N = 150$ (est.) $n = 300$ $N_{Information} = 168$ $N_{Politiken} = 132$	hard copies available
Finland	1973–96 total coverage < 93 coded from fiche > 94 hard copy	data = index total enumeration	*Savon Sanomat* (d)	$N = 375$ $n = 375$	>94 only on hard copy
France	1970 & 1975: all issues 1982–86: enumeration 1987–96: every fourth day (×4) 1970–86: index manual 1987–1996: from CD-ROM	data ≠ index yearly frequencies are not absolute intensity measures	*Le Monde* (d), centre-left	$N = 1483$ (est.) $n = 622$	hard copies available

Table 1. Continued

Country	Population coverage (N) (weight)	Sampling procedure for coding (n)	Newspapers (opinion-leader)	Size of corpus[1]	Comments on corpus condition
Germany	1973–96: fixed ratio *Spiegel* manual, 50%, 2nd issue (×2) *FAZ* archive (×3), *FAZ* artificial weeks (×3)	data ≠ index yearly frequencies are not absolute intensity measures	*FAZ* (d), conservative *Spiegel* (w), left-liberal	$N = 1254$ (est.) $n = 588$ $n_{Faz} = 418$ $n_{Spiegel} = 170$ $N_{Faz} = 1254$ $N_{Spiegel} = 340$	hard copies available
Greece	1973–85: 100 random days, every 2nd year (×3.6) 1986–96 index manual (×1)	data ≠ index yearly frequencies are not absolute intensity measures	*Kathimerini* (d), right, quality *Eleftherotypia* (d), left	$N = 88$ (est.) $n = 65$	hard copies available
Italy	1973–1996: total coverage 1973–84: manual 1984–91: on-line, headline + abstract 1992–96: full text on-line	data ≠ index yearly frequencies are not absolute intensity measures 1973–86: whatever found, some missing	*Corriere della Sera* (d)	$N = 936$ (est.) $n = 340$	hard copies available
Netherlands	1973–85: every 2nd year; 50% of issues (×2) 1986–96: 100% 1986–92 manual, 1993–96 on-line)	data ≠ index yearly frequencies are not absolute intensity measures	*De Volkskrant* (d), 3rd largest paper (overall), largest quality paper	$N = 1185$ (est.) $n = 1119$	hard copies available large number of non–focus from on-line
Poland	1973–82: *Polityka* all issues; *Trybuna Ludu* one issue per week, random 1983–96: *Polityka* all issues; *Rzeczpospolita* one issue a week, 1996 all issues	data ≠ index yearly frequencies arc not absolute intensity measures	*Polityka* (w), critical opinion *Trybuna Ludu* (d), official line *Rzeczpospolita* (d), official line	$N = 113$ (est.) $n = 208$ includes some local newspapers	
Sweden	1973–96: total coverage 1973–92: index, manual 1992–96: on-line search	sample = index total enumeration	*Dagens Nyheter* (d)	$N = 734$ $n = 734$	hard copies available
Switzerland	1973–96 175 artificial weeks (×7) (1026 issues sampled from *c.* 7800)	data ≠ index prop. sampling, i.e. yearly frequencies are not absolute intensity measures	*NZZ* (d), liberal-right	$N = 1537$ (est.) $n = 211$	hard copies available

Table 1. Continued

Country	Population coverage (N) (weight)	Sampling procedure for coding (n)	Newspapers (opinion-leader)	Size of corpus[1]	Comments on corpus condition
UK	1973–96: total coverage *The Times* 1973–80 manual, index; 81–87 on-line *Independent* 88–96 on-line	data ≠ index variable sampling ratios: yearly frequencies are not absolute intensity measures *The Times* 1973–87: 20/yr *Independent* 1988–95: 15/yr 1996: 20/yr	*The Times* (d), centre-right (to 1987) *Independent* (d), centre-left (from 1988)	$N = 5471$ $n = 539$ $N_{Times} = 256$ $N_{Independent} = 283$	hard copies available *The Times* and *Independent* form opinion leader *Independent* replaced *The Times* as 'paper of record' at British Library

1 N is the absolute number of articles over the whole period, calculated to the comparable base of one newspaper. In some countries this figure is an estimate.
   n is the actual size of the corpus as it is used in the content analysis.
2 (d), daily; (w), weekly

Table 2. Summary of sampling process

Country	Population coverage	Sampling procedure for corpus	Newspapers (opinion-leader)	Estimated population size (N)[1]	Size of corpus (n)	Availabilty of hard copies
Austria	Variable ratio; total coverage	Data ≠ index, fixed sampling ratio	*Presse* (d)[2], right *Profil* (w), left	1396 (est.)	302	yes
Denmark	Total coverage enumeration + sampling	Data ≠ index,	*Information* (d), left *Politiken* (d)	150 (est.)	300	yes
Finland	Total coverage enumeration	Data = index,	*Savon Sanomat* (d)	375	375	no
France	1975, 1982–86 total coverage; >87 sampling	Data = index, enumeration + sampling	*Le Monde* (d)	1483 (est.)	622	yes
Germany	Fixed ratio	Data ≠ index, fixed sampling ratio	*FAZ* (d) *Der Spiegel* (w)	1254 (est.)	588	yes
Greece	Fixed ratio; total coverage	Data ≠ index, fixed sampling ratio	*Kathimerini* (d), right, quality	88 (est.)	65	yes
Italy	Total coverage	Data ≠ index, variable sampling ratio	*Corriere della Sera* (d)	936 (est.)	340	yes
Netherlands	Fixed ratio; total coverage	Data ≠ index, enumeration + sampling	*De Volkskrant* (d)	1185 (est.)	1119	yes
Poland	Variable ratio	Data ≠ index, variable ratio sampling	*Polityka* (w) *Trybuna Ludu* (d) *Rzezpolita* (d)	113 (est.)	208	no
Sweden	Total coverage	Data = index enumeration	*Dagens Nyheder* (d)	734	734	yes
Switzerland	175 artificial weeks	Data ≠ index, artificial weeks	*Neue Zurcher Zeitung* (NZZ) (d)	1537 (est.)	211	yes
UK	Total coverage	Data ≠ index, variable sampling ratio	*The Times* (d) *The Independent*	5471	539	yes

1   *N* is the estimate based on a single opinion leader newspaper. The two sampling problems were handled differently in different countries: **a** estimating the total number of biotechnology coverage (population), and **b** sampling the corpus for the purpose of content analysis. Only Sweden and Finland picked up all articles and analysed all articles; all other countries applied a mixture of sampling procedures suitable to cope with the different local archival facilities and the quantity of relevant materials.

2   (d), daily; (w), weekly

Table 3. Description of newspaper circulation and readership

Country	Population in 1992 in 000 (percentage adult newspaper-reading public, if available)	National papers circulation in 000 (total circulation per 100 people)	Newspaper sample (opinion-leader)	Circulation freq. (000)	Readership[a] of sample in 000 (%;15+)
Austria	7,800	2,459 (35)	Die Presse Profil	78	210; 3.3%
Denmark	5200	922 (35)	Information Politiken	 152	
Finland	500 (70)	2,276 (56)	Savon Sanomat	90.5	
France	56,600	2,111 (21)	Le Monde	316	
Germany	80,200 (60)	26,500 (39)	FAZ Der Spiegel	496	2.4%
Greece	10,200	839 (14)	Kathemerini Eleftherotypia	36 114	
Italy	57,800 (32)	6,764 (11)	Corriere della Sera		
Netherlands	15,200 (64)	2,047 (31)	Volkskrant	334	
Poland	38,500	4,875 (13)	Polityka Zycie Warszawy & Rzeczpospolita Trybuna Ludu & Trybuna		
Sweden	8,600 (76)	2,358 (53)	Dagens Nyheter	415	
Switzerland	6,800	3,217 (46)	Neue Zürcher Zeitung	150	441; 11.9%
UK	57,700 (58)	16,498 (40)	The Times Independent	697 255	1904; 4% 867; 2%

Sources: consistent and strictly comparable indicators are hard to come by. Population figures and newspapers per 100 people: 'Human Development Report, 1994'; circulation and readership figures from Euromedia Research Group, *The Media in Western Europe. The Euromedia Handbook* (London: Sage, 1992); Eurobarometer 46.1.

a   Certified readership in percentage of population aged 15+ years; no comparative documentation available.

Appendix 5. Coding frame for media analysis

Summary of variables

Basic information

V1	country
V2	coder
V3	item number
V4	newspaper name
V5	month
V6	day of month
V7	year
V8	weekday

Attention structure

V9	page type, exposition
V10	headline
V11	size of article

Content

V12	newspaper section
V13	news format
V14	author
V15abc	main themes
V16ab	main actors
V17	controversy
V18	location of event
V19	likelihood of benefit
V20ab	type of benefit
V21	likelihood of costs/risks
V22ab	type of costs/risks

Judgements and ratings

V23	valuation tone
V24	metaphors
V25	focus
V26	frames

Detail of variables

Basic information

V1. Country

Britain	01
France	02
Germany	03
Netherlands	04
Austria	05
Sweden	06
Spain	07
Greece	08
Denmark	09
Italy	10
Finland	11
Portugal	13
Ireland	14
Luxembourg	15
Switzerland	16
Poland	17
Canada	18
USA	19
Norway	20
Japan	21

V2. Coder

[country code + coder]

Commentary. In some countries several coders may be at work. Each coder is identified. Use [country code + a running number], e.g. UK coders are '1' for country plus one digit for coders hence 11, 12, 13, etc.

V3. Item number

[coder number + 3 digits]

Commentary. Each article is assigned an identification number: the country number + four digits, e.g. UK uses numbers 10000–19999, France 20000–29999.

V4. Newspaper name

[country code + 1 digit]

Commentary. The first two digits indicate the country, e.g. 01 for the UK, the third digit indicates the newspaper.

283

V5. Month

[2 digits]

Commentary. number the months from January (1) to December (12).

V6. Day of month

[2 digits]

Commentary. Number the days of the month, 1

V7. Year
[2 digits]

Commentary. Use a 2-digit number, e.g. [73], [85]

V8. Weekday

not applicable	0
Monday	1
Tuesday	2
Wednesday	3
Thursday	4
Friday	5
Saturday	6
Sunday	7

Commentary. For a daily newspapers this will vary; for a weekly magazine the day of the week will be a constant.

Attention structure

These variables measure the editorial importance given to articles and the means by which the reader's attention is attracted.

V9. Page type/exposition

info not available	0
front page (absolute)	1
second or third page	2
font page of folder	3
in the middle	4
back page (absolute)	5
back page of folder	6

Commentary. Some newspapers have several thematic folders with front and back page. The front page can be either the first page of the whole paper, or the front of any folder of the paper. The same is true for the back page.

V10. Headline content

Commentary. Record verbatim as a string variable. In case of several headlines, use the largest.

V11. Size of the article

Commentary. Size of the article is obviously relative to the size of the newspaper. Define according to local practice.

small	1
medium	2
large	3

Content

Content has two aspects: a classification from the point of view of journalistic work; and a classification from the point of view of the event and the issues involved.

Journalistic features

V12. Newspaper section

not applicable	0
front page/general news	1
international news	2
national news	3
local news	4
other media quotations	5
debate	6
'light' page	7
consumer	8
editorial page	9
culture; feuilleton	10
business pages	11
science and technology, environment, medicine	12
letters to the editor	13
entertainment: TV, radio, film, theatre	14
sport	15
lifestyle, health	16
other	17

Commentary. Some newspapers provide markers to identify the newspaper sections; for other papers the categories may be approximate.

V13. News format

article with latest news	1
investigation, reportage, background	2
interview, mainly	3
column, commentary (regular)	4
editorial (paper's editor)	5
commentary from external sources	6
reviews of books, films, etc.	7
other	8

Commentary. Here we are attempting a distinction between facts, opinion and interview. This may not apply equally well in all countries as traditions of journalistic classifications may differ.

Authorship

V14. Author

wire service text, e.g. Reuters	1
in-house journalist	
political journalist	2
science journalist	3
other journalist	4
contributed	
other authors: scientists	5
other authors: party politician	6
other authors: special interests	
consumer	7
religious	8
industry, business	9
patient groups	10
environment group	11
agricultural, farming	12
civil service	13
labour union	14
regulatory, ethics committee	15
military	16
judicial, legal voice	17
other special interest	18
general public voice	19
no signature, anonymous, unkown	20

Commentary. the list of authors is more or less the same as the actors (see V16) in biotechnology; each biotechnology actor is a potential author of media material.

Biotechnology events

V15abc. Main theme of story

[three codings]

transgenic	
microorganisms	1
plants breeding	2
animal breeding	3
humans (general)	4
human genome research	5
gene therapy	6
xenotransplants	7
GMO release, e.g. field tests of plants	8
safety/risks	
laboratory, workers	9
environment	10
public, local community	11
food	12
identification	
genetic 'fingerprinting' for crime	13
genetic 'fingerprinting' for other purposes	14
diagnosis, testing, predictive medicine	15
screening of large populations	16
insurance issues	17
privacy, protection of genetic information	18
other issues	
patenting, property rights	19
economic prospects, opportunities	20
pharmaceuticals, vaccines	21
reproduction, child bearing,	
e.g. *in-vitro* fertilisation	22
DNA research (unspecific)	23
public opinion of genetics, biotechnology	24
biodiversity	25
legal regulation	26
voluntary regulations	27
science policy for biotechnology, genetics	28
education, genetic literacy	29
human inheritance	30
eugenics	31
military, defence issues	32
ethical issues	33
other	34
labelling	35

Commentary. Most articles will cover several themes. Code a *maximum of three themes* (V15abc).

V16ab. Main reference to actors

[two codings per article]

not applicable, unkown	0
public sector actors	
parliament	1
government	2
government agencies (other)	
environment	3
health	4
industry	5
agriculture	6
government research institutions	7
universities (scientists)	8
ethics committees	9
hospitals	10
national patent office	11
police	12
military	13
judicial, legal voice	14
technology assessment agency	15
public, public opinion	16
media, published opinion	17
private sector	
political parties	18
religious organisations	19
consumer groups	20
environmental organisations	21
labour unions	22
agriculture, farming	23
professional organisations: medical, legal, etc.	24
scientific organisations	25
patient groups, lobbies	26
industry, producers	27
distributors	28
scientists in private laboratories	29
other	30
stock exchange	37
institutions	
developing countries	31
European Patent Office	32
European Union, European Commission, European Parliament	33
OECD, EFTA/EEU	34
UN organisations	35
other international organisations	36

Commentary. what type of actor, who is mentioned? *Two actors* are coded: first the main actor, i.e. the one mentioned first or mentioned in the lead of the article, in V16a; then code one other actor in V16b.

Contexts

V17. Controversy

none	0
if a controversy, is the report balanced?	1
imbalanced, advocating?	2

V18ab. Location of event (two codings per article)

not mentioned	0
national regions	1–20
international	
EU	
Austria	21
Belgium	22
Denmark	23
Finland	24
France	25
Germany	26
Greece	27
Italy	28
Ireland	29
Netherlands	30
Luxembourg	31
Portugal	32
United Kingdom	33
Spain	34
Sweden	35
Other Europe	36
USA	37
Canada	38
Latin America	39
USSR, Russian Federation area	40
Japan	41
Other East Asia, inclusive of China	42
Other Asia	43
North Africa	44
Sub-Saharan Africa	45
South Africa	46
Australia	47
New Zealand	48
Developing world, Third World	49
not identified	50
Switzerland	51
Europe	52
World	53

Commentary. National regions obviously vary; reserved codes are 1–19. We code a maximum of two locations per article.

Impacts/outcomes

V19. Benefits, usefulness

not mentioned	0
possible benefit	1
mentioned, and quantified	
unlikely	2
likely	3
very likely	4

V20ab. Type of benefits   [two codings]

not mentioned	0
economic growth, development	1
'Third World development'	2
health	3
legal	4
social (in)equality	5
moral, ethical	6
environmental, ecological	7
war and peace, military	8
research	9
consumer	10
other	20

Commentary. Various benefits may be mentioned; only two, the main benefits, are coded – i.e. the ones mentioned first or in the lead of the story.

V21. Risks, costs

not mentioned	0
possible risks	1
mentioned, and quantified	
unlikely	2
likely	3
very likely	4

V22ab. Type of costs and risks

[two codings]

not mentioned	0
economic growth, development	1
'Third World development'	2
health	3
legal	4
social (in)equality	5
moral, ethical	6
environmental, ecological	7
war and peace, military	8
research	9
consumer	10
other	20

Commentary. Various risks/costs may be mentioned; only two, the main risks/costs are coded – i.e. the ones mentioned first or in the lead of the story.

Ratings, judgements

In this section we obtain more judgemental data. Valuation ratings may be linked with survey data on public opinion.

V23a. Negative valuation of biotechnology, genetic developments

not applicable	0
[–] critical, some discourse of concern	1
[– – – – –] very critical: discourse of	
great concern, doom	5

V23b. Positive valuation of biotechnology, genetic developments

not applicable	0
[+] affirmative; discourse of promise	1
[+ + + + +] very enthusiastic, discourse	
of great promise, progress	5

Commentary. Code in any case: tone as set by the author of the article; use rating scale from 1 to 5 for 'negativity' and 'positivity'.

V24. Metaphors used

no metaphors	0
metaphors used	1

V25. Focus

main biotechnology, genetics focus	1
other story with biotechnology/genetic reference	2

Commentary. Two distinct types of articles occur: those with a focus on biotechnology or a genetic issue, and articles with a different focus, where the reference to biotechnology/genetics is mentioned in passing.

V26. Frames

1 'Progress': celebration of new development, breakthrough; direction of history; conflict between progressive/conservative–reactionary.

2 'Economic prospect': economic potential; prospects for investment and profits; R&D arguments.

3 'Ethical': call for ethical principles; thresholds; boundaries; distinctions between acceptable/ unacceptable risks in discussions on known risks; dilemmas. Professional ethics.

4 'Pandora's box': call for restraint in the face of the unknown risk; the 'opening of flood gates' warning; unknown risks as anticipated threats; catastrophe warning.

5 'Runaway': fatalism after the innovation; having adopted the new technology/products a price may well have to be paid in the future; no control any more after the event.

6 'Nature/nurture': environmental vs genetic determination; inheritance issues.

7 'Public accountability': call for public control, participation, public involvement; regulatory mechanisms; private versus public interests.

8 'Globalisation': call for global perspective; national competitiveness within a global economy; opposite: splendid isolation.

Commentary. We suggest to use the term 'frame' with the following preliminary definition: a frame is a structure that **1** organises central ideas on an issue, **2** deploys particular symbolic devices and metaphors, and **3** defines a particular controversy within the frame (i.e. an agreement about how to disagree). A frame's function is to construct meaning, incorporating new events into its interpretative envelope.

We suggest distinguishing 'theme' and 'frame' as figure–ground ambiguity: the frame is ground, the theme is figure. The same theme, e.g. genetic testing, could be presented in different frames, and the same frame can accommodate different themes. There may be however statistical associations between themes and frames.

Common coding sheet

Headline:

Basic information	V1	V2	V3	V4
	country	coder	item no.	paper
	V5	V6	V7	V8
	month	day of month	year [2-digit]	weekday
Attention structure	V9	V10 (string)	11	
	pagetype	headline	size	
Content	V12	V13	V14	V15a
	section	format	author	theme 1
	V15b	V15c	V16a	V16b
	theme 2	theme 3	actor 1	actor 2
	V17	V18	V19	V20a
	controversy	location	benefit	type of benefit
	V20b	V21	V22a	V22b
	type of benefit	risk	type of risk	type of risk
Judgement and ratings	V23a	V23b	V24	V25
	negative	positive	metaphors	focus
	V26			
	frames			

Appendix 6. Basic frequencies of corpus variables of biotechnology in the European opinion-leading press, 1973–96

Table 1. Frequency of sampled biotechnology articles in corpus by newspaper name

Country	Newspaper	Frequency	Total
UK	*The Times*	256	**539**
	Independent	283	
France	*Le Monde*	623	**623**
Germany	*Frankfurter Allg. Zeitung*	418	**588**
	Der Spiegel	170	
Netherlands	*Volkskrant*	1119	**1119**
Austria	*Die Presse*	191	**302**
	Profil	111	
Sweden	*Dagens Nyheter*	734	**734**

Country	Newspaper	Frequency	Total
Greece	*Kathemerini*	16	**65**
	Eleftherotypia	49	
Denmark	*Information*	168	**300**
	Politiken	132	
Italy	*Corriere della sera*	340	**340**
Finland	*Savon sanomat*	375	**375**
Switzerland	*NZZ*	211	**211**
Poland	*Polityka*	76	**208**
	Rzeczpospolita	118	
	Trybuna Ludu & Trybuna	14	
Total			**5404**

Table 2. Estimated absolute frequencies by year (– = not sampled)

Year	UK	France	Germany	Netherlands	Austria	Sweden	Greece	Denmark	Italy	Finland	Switzerland	Poland	Total
1973	12	(–)	3	0	7	0	0	1	3	6	0	2	34
1974	29	(–)	21	0	11	3	0	0	7	7	7	4	89
1975	31	17	15	0	11	3	0	2	3	1	14	3	100
1976	36	(–)	3	2	7	3	2	1	9	3	21	3	90
1977	85	(–)	12	4	11	7	2	1	10	1	7	1	141
1978	99	(–)	39	5	22	20	6	2	11	4	28	6	242
1979	27	(–)	21	6	30	36	8	1	7	1	21	1	159
1980	72	(–)	12	10	45	27	12	1	19	3	21	3	225
1981	46	(–)	21	14	45	16	6	5	22	4	14	1	194
1982	109	27	32	29	25	32	8	5	26	5	21	1	320
1983	164	32	51	44	45	53	10	3	16	15	14	6	453
1984	159	17	39	37	45	47	8	9	19	11	42	6	439
1985	169	19	81	30	80	40	6	16	21	14	49	8	533
1986	272	17	66	35	70	15	1	13	34	29	54	7	613
1987	330	176	60	39	60	25	1	10	73	6	84	2	866
1988	325	240	54	66	50	26	1	5	71	13	147	6	1004
1989	351	172	117	64	60	63	1	14	52	18	54	2	968
1990	409	204	60	68	60	45	2	3	67	42	112	4	1076
1991	350	216	33	39	50	28	3	9	42	8	63	5	846
1992	463	124	60	48	100	31	3	4	110	10	112	7	1072
1993	554	176	64	143	70	49	13	10	105	66	147	12	1409
1994	666	128	108	190	110	35	9	14	96	26	84	3	1469
1995	698	244	111	174	210	80	8	14	77	43	147	12	1818
1996	878	297	174	204	230	50	2	14	36	39	210	22	2156
Total	5471	1483	1254	1185	1396	734	88	150	936	375	1537	113	16,316

Table 3. Frequencies in corpus by year (– = not sampled)

Year	UK	France	Germany	Netherlands	Austria	Sweden	Greece	Denmark	Italy	Finland	Switzerland	Poland	Total
1973	2	(–)	1	0	2	0	0	1	3	6	0	3	18
1974	12	(–)	8	0	3	3	0	0	8	7	1	8	50
1975	9	17	5	0	5	3	0	3	3	1	2	7	55
1976	13	(–)	1	0	2	3	0	1	7	3	3	6	39
1977	19	(–)	8	2	6	7	1	2	6	1	1	2	55
1978	20	(–)	14	0	11	20	0	3	11	4	4	11	98
1979	12	(–)	11	3	10	36	4	1	3	1	3	2	86
1980	20	(–)	7	0	14	27	6	2	10	3	3	6	98
1981	18	(–)	13	7	10	16	3	9	22	4	2	2	106
1982	20	27	12	0	6	32	0	9	21	5	3	1	136
1983	20	32	20	22	11	53	5	6	16	15	2	11	213
1984	19	17	14	0	13	47	0	17	10	11	6	11	165
1985	19	19	33	15	22	40	3	32	9	14	7	15	228
1986	20	17	25	35	10	15	1	26	10	29	8	14	210
1987	20	44	31	39	10	25	0	19	20	6	12	4	230
1988	20	60	28	66	15	26	1	9	20	13	21	11	290
1989	22	43	43	64	12	63	1	28	20	18	8	5	327
1990	26	51	24	68	13	45	2	5	20	42	16	7	319
1991	28	54	21	39	8	28	3	18	20	8	9	10	246
1992	24	31	30	48	14	31	3	7	19	10	16	13	246
1993	41	44	39	143	12	49	13	20	21	66	21	23	492
1994	44	32	57	190	26	35	9	27	20	26	12	7	485
1995	46	61	63	174	27	80	8	28	20	43	21	23	594
1996	45	74	80	204	40	50	2	27	21	39	30	6	618
Total	539	623	588	1119	302	734	65	300	340	375	211	208	5404

Table 4. Percentage of biotechnology articles within countries by year in corpus

Year	UK	France	Germany	Netherlands	Austria	Sweden	Greece	Denmark	Italy	Finland	Switzerland	Poland	Total
1973	0.4	(–)	0.2	0	0.7	0	0	0.3	0.9	1.6	0	1.4	0.3
1974	2.2	(–)	1.4	0	1.0	0.4	0	0	2.4	1.9	0.5	3.8	0.9
1975	1.7	2.7	0.9	0	1.7	0.4	0	1.0	0.9	0.3	0.9	3.4	1.0
1976	2.4	(–)	0.2	0	0.7	0.4	0	0.3	2.1	0.8	1.4	2.9	0.7
1977	3.5	(–)	1.4	0.2	2.0	1.0	1.5	0.7	1.8	0.3	0.5	1.0	1.0
1978	3.7	(–)	2.4	0	3.6	2.7	0	1.0	3.2	1.1	1.9	5.3	1.8
1979	2.2	(–)	1.9	0.3	3.3	4.9	6.2	0.3	0.9	0.3	1.4	1.0	1.6
1980	3.7	(–)	1.2	0	4.6	3.7	9.2	0.7	2.9	0.8	1.4	2.9	1.8
1981	3.3	(–)	2.2	0.6	3.3	2.2	4.6	3.0	6.5	1.1	0.9	1.0	2.0
1982	3.7	4.3	2.0	0	2.0	4.4	0	3.0	6.2	1.3	1.4	0.5	2.5
1983	3.7	5.1	3.4	2.0	3.6	7.2	7.7	2.0	4.7	4.0	0.9	5.3	3.9
1984	3.5	2.7	2.4	0	4.3	6.4	0	5.7	2.9	2.9	2.8	5.3	3.1
1985	3.5	3.0	5.6	1.3	7.3	5.4	4.6	10.7	2.6	3.7	3.3	7.2	4.2
1986	3.7	2.7	4.3	3.1	3.3	2.0	1.5	8.7	2.9	7.7	3.8	6.7	3.9
1987	3.7	7.1	5.3	3.5	3.3	3.4	0	6.3	5.9	1.6	5.7	1.9	4.3
1988	3.7	9.6	4.8	5.9	5.0	3.5	1.5	3.0	5.9	3.5	10.0	5.3	5.4
1989	4.1	6.9	7.3	5.7	4.0	8.6	1.5	9.3	5.9	4.8	3.8	2.4	6.1
1990	4.8	8.2	4.1	6.1	4.3	6.1	3.1	1.7	5.9	11.2	7.6	3.4	5.9
1991	5.2	8.7	3.6	3.5	2.6	3.8	4.6	6.0	5.9	2.1	4.3	4.8	4.6
1992	4.5	5.0	5.1	4.3	4.6	4.2	4.6	2.3	5.6	2.7	7.6	6.3	4.6
1993	7.6	7.1	6.6	12.8	4.0	6.7	20.0	6.7	6.2	17.6	10.0	11.1	9.1
1994	8.2	5.1	9.7	17.0	8.6	4.8	13.8	9.0	5.9	6.9	5.7	3.4	9.0
1994	8.5	9.8	10.7	15.5	8.9	10.9	12.3	9.3	5.9	11.5	10.0	11.1	11.0
1996	8.3	11.9	13.6	18.2	13.2	6.8	3.1	9.0	6.2	10.4	14.2	2.9	11.4

100%

Table 5. Percentage of biotechnology themes by country (V15)[a]

No.	Main theme	UK	France	Germany	Netherlands	Austria	Sweden	Greece	Denmark	Italy	Finland	Switzerland	Poland	Mean[b]
1	Microorganisms	5.0	4.3	2.8	4.5	3.1	7.0		11.0	3.0	2.6	4.8	8.4	4.9
2	Plants breeding	3.4	4.7	5.1	6.8	3.3	3.9	1.9	18.5	4.4	3.6	4.6	7.5	5.9
3	Animal breeding	2.0	2.8	1.8	8.6	1.8	2.7	1.9	8.3	7.3	8.5	3.7	3.6	4.9
4	Humans/general	1.2	5.2	0.6	3.4	0.9	4.9	4.9	1.4	7.6	3.2	0.4	5.1	3.2
5	Human genome	1.1	3.1	2.0	1.2		5.1	3.9	2.4	3.9	0.2	0.6	5.5	2.3
6	Gene therapy	0.8	3.3	3.1	2.1	1.3	3.9	6.8	3.7	2.7	2.1	2.3	5.1	2.7
7	Xenotransplantation	0.2	0.3	0.5	0.4	0.4	0.2		0.3	0.6	0.2	0.8	3.9	0.5
8	GMO release	0.7	1.3	2.7	1.0	1.5	1.4	1.9	2.3	1.7		1.7	3.6	1.5
9	Laboratory safety	0.3	1.5	1.0	0.2	0.2	2.4	2.9	0.6	0.3	0.2	0.4	0.2	0.8
10	Environmental safety	0.6	2.0	3.2	1.5	1.3	0.9		1.0	2.3	1.7	2.1	1.0	1.6
11	Public saftey	1.5	0.2	1.0	1.6	1.7	3.3			1.4	0.2	0.2	0.2	1.2
12	Food safety	1.1	0.3	1.7	0.6	0.2	0.8	3.9	0.7	1.7	0.8	1.0	0.2	0.9
13	Genetic fingerprinting/crime	2.4	1.1	0.8	4.6	2.2	2.3	1.0		0.5	0.4	0.4	1.2	1.8
14	Genetic fingerprinting/other	8.3	2.1	1.1	3.3	0.6	0.5	1.0	0.1	1.0	0.5	1.2	1.2	2.0
15	Diagnosis	5.3	5.4	5.6	4.7	2.2	8.2	6.8	1.6	6.9	8.9	3.3	7.0	5.6
16	Screening	1.4	0.4	0.5	1.4	0.6	0.4	1.0	0.2	0.6	1.0		4.1	0.9
17	Insurance	0.6	0.2	0.1	0.6	0.4	1.1		0.1	0.1			0.2	0.4
18	Privacy of genetic information	0.9	0.3	0.6	0.6		0.4			0.1	0.2	0.2	0.2	0.4
19	Patenting	2.2	1.4	1.9	1.4	1.3	2.2		2.1	1.9	1.2	2.3	1.2	1.7
20	Economic prospect	11.5	13.7	8.2	8.4	7.9	3.4		3.8	6.2	10.9	6.4	3.4	7.8
21	Pharmaceuticals	8.6	4.7	5.9	5.1	8.6	5.1	6.8	1.5	5.6	8.0	7.2	8.0	5.9
22	Reproduction	1.3	2.2	1.3	3.0	2.4	1.7	7.8	7.2	6.7	5.9	6.4	4.1	3.5
23	DNA research	6.2	11.1	12.0	8.9	11.2	6.4	24.3	8.8	6.8	2.0	18.2	8.9	8.9
24	Public opinion	0.5	1.2	4.7	1.2	2.6	0.2		2.2	1.5	1.7	1.9	0.7	1.7
25	Biodiversity	3.0	0.6	0.5	1.2	0.9	0.5	1.0	0.3	0.8	0.9	1.2	0.7	1.0
26	Legal regulations	3.9	3.1	7.6	4.6	5.7	6.0	2.9	4.1	3.2	3.0	10.6	2.2	4.9
27	Voluntary regulations	0.4	0.5	0.8	0.3	0.4	0.1		0.5	0.9	0.2	0.2		0.4
28	Science policy	3.5	5.2	4.8	0.8	4.8	1.1		4.2	1.7	3.5	2.7	1.9	2.9
29	Education	1.4	0.1	0.9	1.5	0.4	0.7		0.3	1.3	2.4	1.0	0.7	1.0
30	Human inheritance	9.9	2.4	0.7	5.9	4.8	1.5	3.9		6.0	4.8	6.6	2.2	4.0
31	Eugenics	0.9	1.9	0.5	0.5	3.5	1.7	1.9	1.4	1.8	1.7	0.8	1.7	1.3
32	Military	0.3		0.4	0.2	0.6	0.4	1.0	0.3	0.1	0.1		0.5	0.3
33	Ethical issues	5.2	7.1	3.7	4.8	6.8	9.6	5.8	5.4	8.6	4.1	4.3	4.1	5.8
34	Other	3.9	5.9	11.0	5.1	15.3	9.1	6.8	5.6	0.6	14.9	0.4	1.4	7.1
35	Labelling	0.5	0.4	0.9	0.1	1.5	1.0		0.1	0.1	0.4	1.9		0.5

a Combined multiple codings.   100%
b Weighted mean.

Table 6. Percentage of biotechnology actors by country (V16)[a]

No.	Main actor	UK	France	Germany	Netherlands	Austria	Sweden	Greece	Denmark	Italy	Finland	Switzerland	Poland	Mean[b]
0	Not applicable	1.2	7.3	0.3	2.8	6.5	5.0	1.1	0.2	1.1	0.6	0.6	2.4	2.7
1	Parliament	1.1	1.4	1.9	1.5	0.7	0.8		12.2	1.7	0.3	4.2	0.3	2.2
2	Government	5.2	7.3	9.2	6.5	9.3	4.6	4.3	20.9	1.5	5.4	3.9	0.3	7.0
3	Environment	0.1	0.1	0.5	0.5		0.6		5.7	0.4		0.6		0.7
4	Health	1.2	1.4	3.2	2.5	1.4	1.5		2.1	2.0	3.8	1.8	2.0	2.1
5	Industry	0.8	0.3	0.2	0.1	2.4	0.6		0.5	0.2			0.7	0.4
6	Agriculture	1.1	0.1	0.7	0.4	0.7	0.4	1.1	0.5	0.2		0.3	1.0	0.6
7	Government research institutes	6.0	17.0	4.1	3.2	2.4	1.4	9.8	4.2	5.2	5.2		20.1	5.6
8	Universities	37.3	15.6	40.2	31.1	37.3	42.7	42.4	20.9	38.1	42.6	43.3	45.1	34.6
9	Ethics committees	2.3	2.3	0.4	1.2	2.2	5.0	2.2	6.1	3.0	0.5	0.3	1.4	2.2
10	Hospitals	2.5	5.7	0.2	1.8	0.5	1.5	7.6	1.6	2.0	10.1	0.6	8.2	2.9
11	National patent office		0.2	0.7	0.2	0.2	0.3		0.3	0.6	0.2		0.7	0.3
12	Police	1.5	0.2		0.7	0.2	1.5	1.1			0.3	0.6	0.3	0.6
13	Military		0.1	0.3	0.2	0.2	0.4		0.2	0.2	0.2		0.3	0.2
14	Judicial	1.6	2.1	0.6	4.2	2.9	1.6			0.6	1.7	0.6	1.0	1.9
15	Technology assessment agencies		0.5	1.0	0.5	0.5			0.7					0.3
16	Public opinion	2.9	1.0	2.6	0.4	1.4	0.7		1.2	3.7	2.1	1.5	1.4	1.5
17	Media	1.1	0.9	2.5	5.9	1.4	3.0	1.1	6.6	5.9	1.6	3.6	0.7	3.3
18	Political parties		0.5	2.4	0.7	5.3	4.9	1.1	0.7	0.9	0.8	6.1		1.8
19	Religious organisations	0.4	0.9	0.5	0.3	1.0	0.3			1.5	0.2	0.3	0.3	0.5
20	Consumer groups	0.4		1.1	0.6	0.2	0.4		0.9		0.8	0.6		0.5
21	Environmental organisations	1.8	0.5	2.4	3.3	3.3	1.0		0.3	1.1	0.3	7.9	0.3	1.9
22	Labour unions			0.1	0.1	0.2	0.1			0.2	0.6			0.1
23	Agriculture	0.4	2.4	0.1	1.1		0.6		0.9	1.3	4.0	0.6	0.3	1.1
24	Professional organisations	1.1	0.6	0.2	2.6		0.6		0.5	0.7	1.1	1.2		1.0
25	Scientific organisations	0.5	0.9	1.4	2.5	3.3	0.2	2.2	0.2	1.7	3.6		1.4	1.5
26	Patient groups	2.2	1.6	0.2	0.5		0.1		0.5	2.8	0.3	0.6		0.8
27	Industry	16.3	15.2	13.1	17.2	8.6	10.6	2.2	8.0	7.8	7.8	15.5	2.7	12.4
28	Distributors	0.1	0.1		0.2		0.2			0.2	0.2		0.3	0.1
29	Scientists in private laboratories	4.3	2.6	1.7	0.7	2.4	0.8	18.5	0.9	8.9	0.3		6.5	2.3
30	Other private institutions	2.6	5.9	3.2	4.3		5.6	2.2	0.7	3.3	2.8	1.2		3.4
31	Developing countries	0.7	0.1	0.1	0.3					0.6				0.2
32	European Patent Office	0.7	0.2	0.1	0.2		0.5	1.1	0.2			0.3	0.3	0.2
33	EU, EC, EP	1.8	3.1	2.7	1.2	3.1	1.8	1.1	1.6	1.9	1.3	1.5	0.3	1.9
34	OECD, EFTA		0.2	0.2			0.1	1.1				0.3		0.1
35	UN	0.4	0.6	0.6	0.3		0.2		0.2	0.4	0.3	0.9		0.4
36	Other international	0.1	0.8	1.4	0.2		0.5		0.5	0.2	0.9		1.4	0.6

a Combined multiple codings.
b Weighted mean.

100%

Table 7. Percentage of biotechnology locations by country (V18)[a]

Location of event	UK	France	Germany	Netherlands	Austria	Sweden	Greece	Denmark	Italy	Finland	Switzerland	Poland	Mean[b]
Not mentioned	1.5	12.6	2.3	9.9	7.0	5.2	1.2	1.4	6.8	1.8	13.7	2.4	6.0
Austria		0.3	0.1	0.2	39.6	0.1	1.2				0.4		2.6
Belgium	0.8	0.1	0.5	2.5	0.5	1.5		4.1			1.3		1.2
Denmark	0.3	0.3	0.5	0.2	0.2	0.6	1.2	50.9					3.9
Finland						0.3	2.3		0.3	68.3		0.5	6.2
France	1.7	49.8	1.5	1.1	1.4	0.6	2.3	1.2	1.6	0.5	1.3	6.3	6.6
Germany	2.3	0.7	37.6	1.6	4.7	1.3	3.5	2.1	0.5	0.8	7.7	0.5	6.4
Greece			0.1	0.2	0.2		12.8	0.2	0.3				0.2
Italy	0.2	0.5	0.6	1.8	0.5	0.2	2.3	0.4	31.8	0.7	1.3	1.9	2.5
Ireland	0.2			0.1									0.0
Netherlands	0.9	0.3	0.2	46.5	0.2	0.7		0.6	1.1	1.2	0.4		9.3
Luxembourg				0.2		0.2							0.1
Portugal	0.2								0.3				0.0
UK	51.1	3.9	6.5	4.9	3.7	3.4	15.1	3.3	7.9	3.3	4.3	9.2	9.3
Spain		0.5	0.1	0.1					0.3		0.4		0.1
Sweden	0.8	0.8	1.0	0.7	0.9	59.0	4.7	1.2	0.8	2.5	0.4		8.5
Other Europe	0.9	0.7	0.9	1.8	0.5	1.0	2.3	0.8	0.5	0.8		24.2	1.7
USA	27.8	15.9	29.7	20.7	22.5	16.5	33.7	23.5	30.4	11.4	12.8	28.0	21.6
Canada	0.9	0.5	0.3	0.8	0.7	0.9	1.2	0.2				1.0	0.6
Latin America	0.6	0.7	0.2	0.8	0.2	0.3	1.2		0.5	0.3	0.4	0.5	0.5
USSR	0.9	0.4	0.8	0.6	0.5	0.5	2.3	0.6	0.3	1.8	0.4	3.9	0.8
Japan	1.1	1.1	1.3	1.0	0.9	0.3	3.5	1.2	0.3	0.8	1.7	1.0	1.0
Other East Asia/China	0.3	0.7	0.1	0.5	0.7	0.8	2.3	0.4	0.3	0.3	0.9		0.5
Other Asia	0.8	0.5	0.5	0.3	1.2	0.7					0.9	0.5	0.5
North Africa	0.2			0.2		0.1		0.2	0.5				0.1
Sub Saharan Africa	0.5	0.1		0.8	0.2			0.2	0.3		1.7		0.3
South Africa			0.1	0.1	0.1								0.0
Australia	0.8	0.4	0.7	0.5	0.2	0.3	2.3	0.2	0.8	0.5	0.4	0.5	0.5
New Zealand				0.2						0.2			0.0
Third World	0.3	0.1	0.3	0.7	0.5			2.3			0.9		0.4
Not identified		0.8	0.1	0.2		1.3		3.1				0.5	0.5
Switzerland		0.8	2.2		2.1	0.2	1.2	0.2	1.4		48.3	1.0	2.3
Europe	2.7	3.9	6.6		5.2	1.8	1.2	0.2	3.3	1.7		1.4	2.5
World	2.6	3.7	5.2		5.6	2.0	2.3	1.2	9.8	3.2	0.4	16.9	3.4

a Combined multiple codings.
b Weighted mean.

100%

Table 8. Percentage risks and/or benefits in biotechnology articles by country (V19, V21 combined)

Risks and benefits	UK	France	Germany	Netherlands	Austria	Sweden	Greece	Denmark	Italy	Finland	Switzerland	Poland	Mean[a]
Risks only	6.3	11.6	8.2	9.0	13.2	16.1	6.2	18.5	9.7	5.3	19.0	1.9	10.5
Benefits only	56.2	42.4	51.5	28.6	51.0	35.6	63.1	21.2	41.8	53.2	27.0	63.0	41.5
Both	22.1	24.4	30.3	15.8	16.6	23.7	24.6	57.9	42.1	34.0	12.8	31.3	25.9
Neither	15.4	21.5	10.0	46.6	19.2	24.6	6.2	2.4	6.5	7.5	41.2	3.8	22.1

a Weighted mean.

100%

Table 9. Percentage articles with negative valuation of biotechnology in press by country (V23a)

Negative valuation	UK	France	Germany	Netherlands	Austria	Sweden	Greece	Denmark	Italy	Finland	Switzerland	Poland	Mean[a]
Not applicable	83.7	39.0	65.5	76.6	59.3	69.3	72.3	29.7	55.3	80.5	80.6	79.8	**66.3**
Critical (−)	5.4	23.0	13.4	8.8	6.6	3.0	3.1	10.0	13.2	11.5	19.4	13.0	**10.7**
(− −)	7.6	14.3	6.8	8.1	9.6	5.5	10.8	3.7	7.4	2.1		2.9	**7.2**
(− − −)	2.2	15.2	7.3	3.9	14.9	6.8	6.2	6.7	9.1	2.7		2.9	**6.7**
(− − − −)	0.9	6.7	5.1	2.1	5.3	7.3	6.2	13.0	6.8	2.1		1.4	**4.6**
Very critical (− − − − −)	0.2	1.8	1.9	0.4	4.3	8.1	1.5	37.0	8.2	1.1			**4.5**

a Weighted mean.   100%

Table 10. Percentage articles with positive valuation of biotechnology in presss by country (V23b)

Positive valuation	UK	France	Germany	Netherlands	Austria	Sweden	Greece	Denmark	Italy	Finland	Switzerland	Poland	Mean[a]
Not applicable	32.3	19.4	25.5	62.2	28.8	69.1	24.6	33.0	16.8	21.7	79.1	13.9	**40.4**
Affirmative (+)	9.8	15.2	26.7	12.0	7.9	4.5	9.2	11.0	7.6	23.8	20.9	7.7	**13.2**
(+ +)	33.8	25.8	14.3	13.5	22.2	5.6	18.5	2.0	8.2	29.4		12.5	**16.1**
(+ + +)	19.1	31.6	19.7	10.3	25.5	12.1	33.8	6.3	14.7	17.4		16.8	**16.4**
(+ + + +)	3.9	6.7	10.4	1.8	12.6	8.4	7.7	7.3	21.2	5.9		29.3	**7.9**
Very enthusiastic (+ + + + +)	1.1	1.1	3.4	0.3	3.0	0.3	6.2	40.3	31.5	1.9		19.7	**6.1**

a Weighted mean.   100%

Table 11. Use of metaphors in press coverage of biotechnology by country (V24)

Metaphors	UK	France	Germany	Netherlands	Austria	Sweden	Greece	Denmark	Italy	Finland	Switzerland	Poland	Mean[a]
No metaphors	84.8	88.9	84.5	93.2	74.2	93.7	87.7	89.0	56.5	97.9	99.1	96.6	**88.0**
Metaphors used	15.2	11.1	15.5	6.8	25.8	6.3	12.3	11.0	43.5	2.1	0.9	3.4	**12.0**

a Weighted mean.   100%

Table 12. Focus of press coverage of biotechnology by country (V25)

Focus	UK	France	Germany	Netherlands	Austria	Sweden	Greece	Denmark	Italy	Finland	Switzerland	Poland	Mean[a]
Main biotechnology focus	84.8	46.1	86.1	50.3	36.0	89.8	27.7	94.3	79.7	59.6	58.6	70.7	67.5
Other story with biotechnology	15.2	53.9	13.9	49.7	64.0	10.2	72.3	5.7	20.3	40.4	41.4	29.3	32.5

a Weighted mean.   100%

Table 13. Frames of press coverage of biotechnology by country (V26)

Frame	UK	France	Germany	Netherlands	Austria	Sweden	Greece	Denmark	Italy	Finland	Switzerland	Poland	Mean[a]
Not applicable	10.4	20.8	7.8	37.3	0.0	14.3	0.0	1.0	0.9	7.8	22.7	0.0	15.5
Progress	51.2	38.6	49.3	22.3	45.7	42.0	72.3	24.5	64.1	51.1	29.4	73.6	41.6
Economic prospect	15.4	19.6	7.8	10.4	12.3	6.7	3.1	19.5	5.6	16.8	8.5	9.6	11.7
Ethical	9.3	12.2	7.5	9.9	23.2	15.5	7.7	21.8	10.3	5.9	9.5	6.7	11.6
Pandora's box	2.0	4.5	4.6	0.8	4.6	7.0	7.7	8.1	5.3	2.7	1.9	1.9	3.8
Runaway	0.0	1.6	0.3	1.5	3.3	1.1	1.5	3.4	1.2	1.6	0.0	3.4	1.4
Nature/nurture	5.6	1.1	0.7	1.3	5.0	0.5	6.2	4.7	6.5	5.3	1.9	2.9	2.7
Public accountability	5.8	1.4	15.1	15.9	4.3	12.6	1.5	14.1	4.7	7.5	20.9	0.5	10.1
Globalisation	0.4	0.0	6.8	0.6	1.7	0.1	0.0	3.0	1.5	1.3	5.2	1.4	1.6

a Weighted mean.   100%

Appendix 7. Reliability of content analysis

Table 1. Intercoder reliability within countries (perentage agreement between two coders)

Variable	Name	UK (%)	Netherlands (%)	Austria (%)	Sweden (%)	Average (%)
V5	Month		98	100		99
V6	Day		100	100		100
V7	Year		100	100		100
V8	Weekday		100	96		98
V9	Page type		100	98	70	89
V11	Size	65	89	100	85	85
V12	Section	43	96	86	62	72
V13	News format	68	94	82	69	78
V14	Author	73	96	92	31	73
V15a	Theme	43	87	66		65
V15b	Theme	8	85	30		41
V15c	Theme	31	77	62		57
V15abc	Theme-collective				54	54
V16a	Actor	49	87	78		71
V16b	Actor	24	85	70		60
V16ab	Actor-collective				60	60
V17	Controversy	80	92	70	62	76
V18a	Location	65	94	96		85
V18b	Location	43	94			69
V18ab	Location-collective				62	62
V19	Likelihood of benefit	76	85	70	39	68
V20a	Type of benefit	59	79	78		72
V20b	Type of benefit	51	89	82		74
V20ab	Type of benefit-collective				71	71
V21	Likelihood of risk	86	89	68	65	77
V22a	Type of risk	75	83	82		80
V22b	Type of risk	73	96	86		85
V22ab	Type of risk-collective				77	77
V23a	Negative valuation	80	83	84	20	67
V23b	Positive valuation	76	79	92	8	64
V24	Metaphors	65	98	90	89	86
V25	Focus	94	96	92	85	92
V26	Frame	75	94	64	85	80
Median		65	92	84	64	74
Mean		61	87	78	61	72

Table 2. Intercoder reliability between countries[a]

| Variable | 'Biological heritage' | | 'Watson' | | Average |
	I[b]	II[c]	I	II	
V11	0.35	0.35	0.715	0.715	0.553
V12	0.755	0.43	1.0	0.674	0.715
V13	0.714	0.523	0.824	0.209	0.568
V14	1.0	1.0	0.757	0.353	0.778
V15	0.727	0.388	0.798	0.336	0.562
V16	0.737	0.209	0.807	0.495	0.562
V17	0.307	0.307	0.654	0.654	0.481
V18	0.612	0.612	0.612	0.612	0.612
V19	1.0	0.423	0.904	0.519	0.712
V20	0.88	0.478	0.759	0.438	0.639
V21	1.0	0.423	0.904	0.423	0.688
V22	0.715	0.383	0.916	0.373	0.597
V23a	0.571	0.314	0.815	0.354	0.514
V23b	0.722★	0.445	0.63★	0.354	0.538
V24	1.0	0.384	1.0	0.538	0.731
V25	1.0	1.0	1.0	0.846	0.962
V26	1.0	1.0	0.74	0.74	0.870
Median	**0.76**	**0.42**	**0.81**	**0.50**	**0.63**
Mean	**0.77**	**0.51**	**0.814**	**0.508**	**0.65**

a The formula used:
$r = (k / (k-1)) (p - 1/k)$
k = number of categories
p = average agreement
b High frequency categor**ies** vs other (category set agreement)
c Highest frequency categor**y** vs other (single-category set agreement)

Index

Index

A

abortion 90, 91
acceptance *see* public perceptions
agriculture 3
  see also transgenic animals; transgenic plants
  Austria 15, 17, 20, 21
  Denmark 29
  Finland 46
  France 51, 58
  Germany 64
  Greece 77, 78, 79, 83, 84
  Italy 92
  Netherlands 103
  organic 17, 18
  Poland 118, 122
  Sweden 130
Aktuelt (Denmark) 33
animal welfare 46, 104
Asilomar Conference (1975) 17, 30, 82, 103, 138,
  166, 180
attitudes 194–5, 195–200, 207, 209, 247–61
  Austria 19, 22, 26, 27
  Denmark 38, 40
  Finland 46–7, 49
  France 57, 61
  Germany 69, 71, 74
  Greece 81–4, 86, 87
  Italy 95, 96, 199, 102
  Netherlands 107, 108–9, 112, 113, 114, 115
  Poland 124–5, 126
  Sweden 136, 139, 142
  Switzerland 149, 151, 152, 157, 159, 160
  United Kingdom 168, 169, 170, 171, 175
'Australian Group' 120
Austria 15–28
  Academy of Sciences, Commission for
    Recombinant DNA 16
  agriculture 15, 17, 18, 20, 21
  attitudes 19, 22, 26, 27
  crossnational comparison 19, 21, 224
  economy 5
  Eurobarometer 46.1 survey results 239–75
    see also public perceptions
  Genetic Engineering Act (1994) 19, 24
  industry 16
  knowledge 19–20
  media coverage 18–19, 25, 26, 221, 278, 281,
    290–6
  Parliamentary Inquiry Commission 16, 17, 19
  people's initiative 17, 21
  political system 15–16
  public perceptions 19–21, 222, 224, 225
  public policy 16–17, 220
  regulatory profile 24
  research and development 15, 24
  'social partnership' 16, 18
  trust 20, 21, 22

B

barley 47
Belgian Blue cattle 136, 139
Belgium, Eurobarometer 46.1 survey results 239–75
'Beltsville pig' 45
'Berg Letter' 163
Berlingske Tidende (Denmark) 33, 36
Bild (Germany) 66
biodiversity 3
bioethics
  see also moral acceptability
  France 53, 54–5
  Greece 80–1
  Italy 90–1, 93, 95
  Netherlands 104
  Sweden 134
  United Kingdom 164
Børsen (Denmark) 33
BST (bovine somatotropin) 104, 108, 116–17
BSE (bovine spongiform encephalopathy) 17, 18,
  23, 46, 164, 167, 170

C

Catholic Church
  Italy 90, 91, 96
  Poland 119
chymosin 104, 108, 116–17, 164
Ciba-Geigy 147, 148, 153
cloning 32, 34, 80, 97, 125, 152
coding frame 8, 9
confidence *see* trust
consensus conferences 34, 35, 54, 104, 165
content analysis (media) 8–9, 277
  reliability of 297–8
corpus (media coverage) 8, 9
Corriere della Sera (Italy) 92, 279, 282, 290

D

Dagens Nyheter (Sweden) 133–4, 138, 279, 282, 290
Dansk Landbrug (Denmark) 32

data quality   9, 11
Denmark   29–42
   agriculture   29
   Act on Gene Technology and the Environment
      (1986)   29, 30, 31, 34, 37, 38–9
   attitudes   38, 40
   crossnational comparison   224
   Danish Board of Technology   31, 32, 34, 35
   Eurobarometer 46.1 survey results   35–8, 239–75
      see also public perceptions
   industry   29, 37–8
   knowledge   35
   media coverage   32–4, 40, 41, 221, 278, 281,
      290–6
   NOAH   30–1, 34, 37
   political system   29–30
   public debate   37
   public perceptions   34–6, 222, 224, 225
   public policy   30–32, 219
   Registration Committee concerning Genetic
      Engineering (RUGE)   30, 39
   regulatory profile   38–9
Dolly the sheep   3, 11, 32, 34, 36, 38, 80, 97, 120,
   123, 125, 135, 150, 151, 152, 164, 191

E

EC *see* European Community
Eleftherotypia (Greece)   29, 79, 87, 279, 282
engagement with biotechnology   10, 191, 204, 222,
   223–5
   Austria   19–21
   Finland   46
   France   58
   Germany   69
   Greece   81, 84
   Italy   96–7
   Switzerland   151
ethics frame   220
EU *see* European Union
eugenics   3, 169
Eurobarometer 35.1 (1991)   4, 35, 168, 182
Eurobarometer 39.1 (1993)   4, 35, 168, 182
Eurobarometer on Biotechnology 46.1 (1996)   5,
   182, 189–214
   attitudes   189, 194–5, 196–200
   Austria   19, 20
   conceptual framework   4
   crop plants   247, 250, 253, 255, 259
   data quality   11
   Denmark   35–6, 37
   engagement with biotechnology   191, 204
   France   56–8
   general findings   189–195
   genetic testing   195, 207, 249, 252, 255, 258, 260
   Germany   68, 71
   GM food   195, 207, 213, 247, 250, 253, 256,
      259, 260

Greece   81
   impact of technologies   189–91, 202–3
   Italy   94–5
   knowledge   192–3, 199–200, 204, 208
   logics   196–8, 200–1, 210, 211–12
   media coverage   7–9, 221–2, 223–6
   medical applications   248, 251, 254, 257
   medicines and vaccines   195, 207, 248, 251,
      254, 257, 259
   public perceptions   9–11, 207, 209, 210,
      219–20, 223–6
   public policy   5–7
   questionnaire   10–11, 189, 208–9, 232–6
   regulation   193–4, 201, 261–4
   sampling procedure   231
   socio-demographics   198–9
   survey results   189–213, 239–75
   Sweden   136, 139
   Switzerland   151–2
   technical specifications   231–8
   transgenic animals   195, 207, 212
   transgenic plants   195, 207
   trust   194, 201, 205, 206, 208, 267–8
   United Kingdom   168–9, 170
   xenotransplantation   195, 207, 213, 249, 252,
      255, 258, 259, 260
Europe   177–85
   Biotechnology Coordination Committee 181
   EUROPABIO   182
   European Bioethics Committee   182
   European Molecular Biology Organisation
      (EMBO) 177, 179
   European Patent Organisation (EPO)   179–80
   European Science Foundation (ESF)   179
   Group of Advisers on Ethical Implications of
      Biotechnology (GAIEB)   180
   industry   177–8, 182
   media coverage   220–2
   non-governmental organisations (NGOs)   182
   policy debate   219–20
   political parties   182
   public perceptions   223–6
   public policy   178–9, 180–1
   research and development   177, 183
   regulatory profile   183–5
   scientists   182
European Community (EC)
   'Biotechnology and the European Public'
      Concerted Action   4
   directives on GMO use (90/219) and release
      (90/220)   3, 22, 30, 53, 91, 120–1, 149, 164,
      178, 181, 183, 220
European Union (EU)   177, 220
   Commission   177, 178, 179, 189, 220
   Committee of the Regions   178
   Council of Europe   6, 55, 178, 179, 180, 220
   Council of Ministers   177, 178
   Court of Justice   178

Directorate Generals (DGs)   178, 180–2
'Euroscepticism'   30, 37, 131, 139, 171
Parliament   6, 177, 178, 179, 182, 220
public policy   163, 177, 178–9, 220
Single European Act (1989)   178, 180
Social and Economic Committee   178, 180
Switzerland and the   149

F

Le Figaro (France)   55
Finland   43–50
   Advisory Board for Technology   45
   agriculture   46
   attitudes   46–7, 49
   Board for Gene Technology   454
   Eurobarometer 46.1 survey results   239–75
     see also public perceptions
   Gene Technology Act (1995)   44–5, 47
   industry   43
   knowledge   46
   media coverage   45–46, 47, 49, 50, 221, 278, 281, 290–6
   political system   43–4
   public perceptions   45, 46–47, 222, 224, 225
   public policy   44–45, 219
   regulatory profile   48
   research and development   43
   trust   47
Flavr Savr tomato   3, 33, 168
focus groups   169
food biotechnology *see* GM food
frame laws (Finland)   44
frames   6, 9, 288, 296
   Austria   19, 25
   ethics   220
   France   56
   Germany   67–8
   Greece   80
   Italy   93
   Netherlands   106, 108
   Poland   122
   progress   220
   Sweden   135
   Switzerland   150
   United Kingdom   166–7
France   51–62
   agriculture   51, 58
   attitudes   57, 61
   bioethics   53, 54–5
   crossnational comparison   57, 224
   economy   51
   Eurobarometer 46.1 survey results   56–8, 239–75
     see also public perceptions
   industry   51, 53
   Interministerial Group for Chemical
     Products   52
   knowledge   57, 58
   media coverage   55–6, 61, 62, 221, 278, 281, 290–6
   political system   51–2
   public perceptions   56–8, 62, 222, 225
   public policy   51–5, 219
   regulatory profile   59–61
   research and development   51
   trust   57
France Soir (France)   55
Frankfurter Allgemeine Zeitung (Germany)   66, 279, 282, 290

G

Gazeta Wyborcza (Poland)   121
gene laws   31, 37, 44–5, 47, 66–7, 133, 219
gene technology   217
   Denmark   31, 35
   Finland   44, 46
   Sweden   133, 138, 139
   Switzerland   147, 151–2, 155
gene therapy   3, 7
   Austria   17, 24
   Denmark   32, 38
   Europe   183
   Finland   45, 46, 48
   France   59
   Germany   73
   Greece   85
   Italy   97
   Netherlands   104, 110, 116–17
   Poland   127
   Sweden   133, 134, 140
   Switzerland   147, 155
   United Kingdom   175
genetic fingerprinting   3, 134, 162, 166
genetic screening *see* genetic testing
genetic testing   3, 7, 195, 197, 249, 252, 255, 258, 259
   Austria   17, 24
   Denmark   32, 38
   Europe   183
   Finland   48
   France   59
   Germany   74
   Greece   82, 85
   Italy   97
   Netherlands   104, 107, 110
   Poland   127
   Sweden   133, 140
   Switzerland   152, 155
   United Kingdom   172
genetically modified organisms *see* GMO
Germany   63–76
   agriculture   64
   attitudes   69, 71, 74
   Association for the Protection of the Environment
     and Nature (BUND)   64
   Benda Commission   65, 72
   crossnational comparison   68–9, 224

Enquete Commission   65, 73–4
Eurobarometer survey results   68, 73, 239–75
   see also public perceptions
Gene Law (1990)   65–6
industry   63–4
knowledge   69
media coverage   66–8, 71, 74, 75, 221, 279,
   281, 290–6
political system   63
public perceptions   68–70, 71, 222, 224
public policy   64–6, 70–1, 219
regulatory profile   73–4
research and development   63
trust   70
globalisation   22, 67–8
Głos Wielkopolski (Poland)   121, 122
GM *see* genetic modification
GM food   3, 7, 11, 195, 198, 247, 250, 253, 256
Austria   17–18, 24
Denmark   32, 36, 38
Europe   183
Finland   46, 48
France   57, 59
Germany   64, 73
Greece   78, 79, 81, 82, 85
Italy   95, 97
Netherlands   104, 108, 110
Poland   120, 127
Sweden   135, 139
Switzerland   146, 151, 152, 155
United Kingdom   164, 167, 169
GMO release   7
Austria   16–19, 23
Denmark   31, 38
Europe   183
France   59
Germany   73
Greece   85
Italy   97
Netherlands   110, 116
Poland   120, 127
Sweden   132, 140
Switzerland   146, 148, 153
United Kingdom   165, 172
GMO use   7
Austria   17
Denmark   29, 30, 31, 33–4, 38
Europe   183
Finland   48
France   52, 59
Germany   73
Greece   85
Italy   97
Netherlands   108, 110
NIH (National Institutes of Health) guidelines   146
Poland   120, 124, 127
Sweden   140
Switzerland   146, 155

United Kingdom   163, 172
Greece   77–88
agriculture   77, 78, 79, 83, 84
attitudes   81–4, 86, 87
Biohellas SA   78
crossnational comparison   224
economy   77
Eurobarometer 46.1 survey results   81, 239–75
   see also public perceptions
Greek Association for Biotechnology   79
knowledge   81, 82, 83
industry   78
Institute for Technical Applications   79
media coverage   79–81, 83, 86, 87, 88, 221,
   279, 281, 290–6
political system   77
public perceptions   81–2, 222, 224, 225
public policy   78–9, 83, 220
regulatory profile   85
research and development   77, 78, 83
trust 82,   83, 84
Green party/movement
Austria   16, 17, 20, 22–3
in European Parliament   182
Finland   44, 46
France   53, 54
Germany   63, 64, 65
Italy   97
Poland   119, 120
United Kingdom   165
Greenpeace   38, 44, 53, 54, 64, 79, 151, 165

H

hands-on sampling   276
Human Genome Project   91, 93, 191
Herman the bull   104, 108, 116–17
Hoechst   65–6

I

imagination, role of   10
images of biotechnology   21, 22, 56, 68, 95, 96,
   107, 191, 243
in-vitro fertilisation   30, 33, 36, 47, 123, 125
Independent (United Kingdom) 15, 165–6, 280,
   282, 290
index sampling   8, 55, 92, 276
Information (Denmark)   33, 278, 282, 290
Ingeniøren (Denmark)   32
insulin   30, 33, 103
intellectual property   91, 96
Ireland, Eurobarometer 46.1 survey results   239–75
Italy   89–102
agriculture   92
attitudes   95, 96, 100, 102
bioethics   90–1, 93
crossnational comparison   224

Eurobarometer 46.1 survey results   94–5, 239–75
    see also public perceptions
Federation of Technical and Scientific
    Associations (FAST)   90
industry   89, 90
knowledge   94
media coverage   92–4, 97, 100, 102, 221, 279,
    281, 290–6
National Research Council (CNR)   89
political system   89
public perception   94–7, 222, 224, 225
public policy   22, 89, 90–2
regulatory profile   98–9
research and development   89
trust   94–5

J

Jyllandsposten (Denmark)   33

K

KabiGen   131, 310
Kathimerini (Greece)   79, 87, 279, 282, 290
keywords   8, 55, 77, 277
knowledge   10, 192–3, 199–200, 204, 208, 223
    Austria   20–2
    Denmark   35
    Finland   46
    France   57, 58
    Germany   69
    Greece   81, 82, 83
    Italy   94
    Netherlands   107, 109, 114, 115
    Poland   123–6
    Sweden   137
    United Kingdom   168
Kristeligt Dagblad (Denmark)   33
Kronenzeitung (Austria)   23

L

labelling   6, 17, 32, 36, 54, 81, 104, 105, 108, 120,
    135, 139
Libération (France)   55
logics of support   196–8, 200–1, 210, 211–12
    see also moral acceptability; risk
Luxembourg, Eurobarometer 46.1 survey results
    239–75

M

Maastricht Treaty   30, 89, 178
maize   3, 17, 18, 19, 22, 23, 32, 36, 38, 47, 54, 97,
    105
media coverage   4, 7–9, 220–3
    actors   293
    basic frequencies in opinion-leading press   290–6

benefit only   220
CD-ROMs   8, 106
coding frame   8, 9, 283–9
evaluations   9, 295
frames   9, 221, 296
framing   220, 288, 296
intensity   7, 221
manual sampling   8, 121
metaphors   295
microfiche   106
newspaper circulation and readership   282
opinion-leading press   8, 276
reliability   297–8
risk-only   220
sampling guidelines   276–7
sampling procedures   9, 278–82
themes   9, 292
medical applications   248, 251, 254, 257
    Austria   21
    Denmark   35
    Finland   46
    France   56, 57, 58
    Greece   80, 82, 83
    Italy   93, 95
    Netherlands   106
    Poland   122
    Switzerland   150
    United Kingdom   169
media profiles   9
medicines and vaccines   20, 103–4, 108, 118, 195,
    197, 248, 251, 254, 257
methodology   5, 9, 10
Le Monde (France)   55, 56, 61, 278, 282, 290
Monsanto   34, 36, 38, 153, 164
moral acceptability   10, 196, 200, 260
    Austria   20
    Denmark   36
    Eurobarometer 46.1 survey   200
    Germany   70
    Italy   95
    Netherlands   107, 109
    Sweden   139
    Switzerland   152

N

nature/nurture debate   10, 166, 167, 192, 244–6
Nestlé   144, 145
Netherlands   103–17
    agriculture   103
    attitudes   107, 108–9, 112, 113, 114, 115
    Brede Committee   103
    crossnational comparison   224
    Eurobarometer 46.1 survey results   239–75
        see also public perceptions
    Foundation for Public Information on Science,
        Technology and the Humanities   104
    knowledge   107, 109, 114, 115

industry   103
media coverage   105–6, 112, 113, 108, 116–17, 221, 279, 281, 290–6
political system   103
public perceptions   106–8, 116–17, 222, 224, 225
public policy   103–5, 116–17, 219
regulatory profile   110–11
research and development   103
Schroten Committee   104
VCOGEM   104, 110–11
Neue Zürcher Zeitung (Switzerland)   150–1, 279, 282, 290
newspaper index sampling   276
NOAH-bladet (Denmark)   32
non-governmental organisations (NGOs)   6
Austria   17, 22–3
Denmark   30
Finland   44
Germany   64, 67, 70
Netherlands   104
Poland   119
Sweden   131, 133, 138
Switzerland   147, 150, 153
United Kingdom   165, 167
Nordisk Gentofte   30, 34, 35, 37
Norway, Eurobarometer 46.1 survey results   239–75
Novartis   144, 147
novel foods   32, 36, 45, 54, 66, 77, 78, 108
Novo   30, 33, 34, 35, 37
nuclear power   3
Austria   16, 22, 23
Denmark   30, 36–7
Finland   47
Germany   72
Sweden   130, 138
Switzerland   147

O

on-line sampling   8, 276
opinion *see* public perceptions
optimism   10, 189–91, 203, 204, 239–43
Austria   19, 20, 21
Denmark   33
Finland   46, 47
Greece   81, 82, 83
Italy   93
United Kingdom   168
Ouest France (France)   55

P

parallel content analysis   8
Le Parisien (France)   55
patents   32, 34, 36, 38, 53, 120, 132, 135, 148, 153, 179–80, 177
pessimism   10, 83, 189–91, 239–43
Austria   19, 20, 21

Greece   81, 82, 83
Poland   118–29
agriculture   118, 122
attitudes   124–5, 126
Biotechnology Committee   118–19, 120
crossnational comparison   224
knowledge   123–6
industry   126
media coverage   121–3, 126, 128, 221, 279, 281, 290–6
political system   118–19
public perceptions   123–6, 222, 224, 225
public policy   119–21, 220
research and development   118–19, 120
regulatory profile   127
policy cultures   6, 7
Politiken (Denmark)   33, 278, 282, 290
Polityka (Poland)   121, 279, 282, 290
Portugal, Eurobarometer 46.1 survey results   239–75
potatoes   19, 104, 148
Die Presse (Austria)   18, 278, 282, 290
profil (Austria)   18, 278, 290
progress frame   220
Przegl√d Techniczny (Poland)   121
public discourse   4–5
public perceptions   4, 9–11, 200, 222–6
public policy   5–7, 223–6

R

random cluster sample   276
rDNA *see* recombinant DNA
recombinant DNA   3, 15, 33, 64, 103, 108, 131–2, 138, 153, 154, 163, 170, 217
referendum   130, 138, 145, 146, 154
regulation   10, 193–4, 201, 261–4
see also gene laws
Austria   20, 22, 29
Denmark   36–7
Europe   180, 205
France   53, 54, 57
Germany   64–6, 70
Greece   78–9, 81
Italy   90, 91, 92, 94–5
Netherlands   108–9, 115
Poland   120, 121
Sweden   131, 132, 133, 134, 137
Switzerland   146, 147, 148, 154, 159
United Kingdom   163, 164, 169
regulatory profiles   7
Austria   24
Denmark   38–9
Europe   183–5
Finland   48
France   59–61
Germany   73–4
Greece   85
Italy   90–1

Netherlands 110–11
Poland 127
Sweden 140–1
Switzerland 155–6
United Kingdom 172–4
reproductive technologies 7, 277
  Austria 22, 24
  Denmark 31, 22, 38
  Europe 183
  Finland 45, 48
  France 59
  Germany 73
  Greece 85
  Italy 98
  Netherlands 110
  Poland 127
  Sweden 134, 140
  Switzerland 147, 155
  United Kingdom 165, 170, 172
La Repubblica (Italy) 92
research and development 7
  Austria 15, 24
  Denmark 31, 34, 38
  Europe 177, 183
  Finland 43
  France 51
  Germany 63
  Greece 77
  Italy 89, 95, 96
  Poland 118–19, 120
  Switzerland 144, 155
  United Kingdom 163, 172
risk 10, 196, 200
  attitude to 222, 259
  Austria 19, 20
  Denmark 35–6
  Eurobarometer 46.1 survey 200
  Finland 46, 47
  France 56
  Germany 67, 69, 75
  Greece 80, 81, 82
  Italy 94–5, 97, 101
  Poland 123
  laboratory workers 163
  Netherlands 109
  Sweden 132, 134, 137, 139
  Switzerland 150, 158, 159, 160
  United Kingdom 167, 169
risk laboratories 134, 135
risk-only media coverage 220
Roche 144, 147
Roundup herbicide 31, 34
'Roundup-Ready' soyabeans 3, 34, 164
Rzeczpospolita (Poland) 121, 279, 282, 290

S

sampling 276, 278–82

Samvirke (Denmark) 32
Sandoz 147
Savon Sanomat (Finland) 278, 282, 290
Der Schweizerische Beobachter (Switzerland) 147
semantics 9, 11, 55, 58, 81, 168, 217
social horizon 123–6
socio-demographics 10, 169, 198, 198–9
Il Sole (Italy) 92
soya(beans) 3, 11, 17, 19, 32, 34, 46, 47, 79, 80,
  104, 105, 108, 136, 139, 148, 151, 153, 164
Spain, Eurobarometer 46.1 survey results 239–75
Der Spiegel (Germany) 66, 67, 279, 282, 290
sugar beets 31, 34
support 10, 210–13, 222–3
  Austria 21–2
  Denmark 35
  Finland 46
  Germany 68
  Sweden 139
  United Kingdom 169
Sweden 130–43
  Agency for Technical Development 131
  agriculture 130
  attitudes 136, 139, 142
  Commission on Gene Ethics 132
  crossnational comparison 136, 224
  'environmental election' 138
  Eurobarometer 46.1 survey results 136, 139,
    239–75
    see also public perceptions
  Gene Technology Law (1995) 133
  knowledge 137
  industry 130–1
  media coverage 133–5, 138–9, 142, 143, 221,
    279, 281, 290–6
  political systems 130
  public perceptions 135–7, 222
  public policy 131–3, 138–9, 219
  Recombinant DNA Advisory Committee 131,
    133, 135
  regulatory profile 140–1
  trust 137
Switzerland 144–61
  Action Group on Gene Technology 147, 148
  attitudes 149, 151, 152, 157, 159, 160
  Basle Appeal against Gene Technology 147
  Beobachter Initiative 147, 148, 149, 150, 153,
    154
  crossnational comparison 224
  economy 144
  Eurobarometer survey results 151–2, 239–75
    see also public perceptions
  Federal Ordinance on Incidents in Industrial
  Plants and Installations (1991) 146
  Gen-Schutz Initiative 146, 148, 149, 150, 153–4
  industry 144–5, 153
  media coverage 149–51, 152, 153, 157, 158,
    221, 279, 281, 290–6

political system   145
popular initiative   145
Priority Programme for Biotechnology   144
public perceptions   151–2, 157, 158, 159, 160,
   222
public policy   146–9, 219
research and development   144, 155
regulatory profile   155–6
Swiss Action Group on Gene Technology   148

T

technological optimism   222
technology
   awareness   22
   impact of   189–90, 202, 203
technology culture   190, 207
The Times (United Kingdom)   165–6, 280, 282, 290
tomatoes   19, 36, 47
   see also Flavr Savr
transgenic animals   7, 195
   Austria   24
   Denmark   34, 38
   Europe   183
   Finland   43, 46, 48
   France   59
   Germany   73
   Greece   85
   Italy   95, 98
   Netherlands   104, 110, 116–17
   Poland   118, 120, 127
   Sweden   132, 133, 140
   Switzerland   148, 152, 153, 155
   United Kingdom   172
transgenic plants   43, 195, 197
   France   57
   Italy   95
   Poland   118
trust   10, 194, 201, 205, 206, 208, 267–8
   Austria   20, 21, 22
   Denmark   35
   Eurobarometer 46.1 survey   201, 205, 206
   Finland   47
   France   57
   Germany   70
   Greece   82, 83, 84
   information sources   194
   Italy   94–5
   Sweden   137

United Kingdom   169
Trybuna Ludu (Poland)   121, 279, 282, 290

U

United Kingdom   162–75
   Advisory Committee on Novel Foods and
      Processes (ACNFP)   164
   Advisory Committee on Releases to the
      Environment (ACRE)   164
   attitudes   168, 169, 170, 171, 175
   Biotechnology Directorate   163
   crossnational comparison   166, 223
   Eurobarometer survey results   168–9, 170,
      239–75
   Genetic Manipulation Advisory Group (GMAG)
      163, 164
   industry   162
   knowledge   168
   media coverage   165–7, 170, 171, 174, 175,
      221, 280, 281, 290–6
   National Consensus Conference on Plant
      Biotechnology (1994)   165, 167
   Nuffield Council for Bioethics   164
   political system   162
   public perceptions   167–70, 174, 222, 224
   public policy   162–5, 170, 219
   regulatory profile   172
   research and development   162, 163, 172
   trust   169
USA   18, 166, 171, 177, 180

V

'vegetarian' cheese   3, 164
De Volkskrant (Netherlands)   105–6, 279, 282, 290

W

Weekendavised (Denmark)   33
women's rights   20
Wprost (Poland)   121

X

xenotransplantation 3, 36, 57, 82, 95, 107, 152, 169,
   171, 195, 197
xenotransplants 249, 252, 255, 258, 268